SHAKESPEAREAN TERRITORIES

SHAKESPEAREAN TERRITORIES

STUART ELDEN

THE UNIVERSITY OF CHICAGO PRESS
CHICAGO AND LONDON

The University of Chicago Press, Chicago 60637
The University of Chicago Press, Ltd., London

Published 2018
Printed in the United States of America

27 26 25 24 23 22 21 20 19 18 1 2 3 4 5

ISBN-13: 978-0-226-55905-6 (cloth)
ISBN-13: 978-0-226-55919-3 (paper)
ISBN-13: 978-0-226-55922-3 (e-book)
DOI: https://doi.org/10.7208/chicago/9780226559223.001.0001

Library of Congress Cataloging-in-Publication Data

Names: Elden, Stuart, 1971– author.
Title: Shakespearean territories / Stuart Elden.
Description: Chicago : The University of Chicago Press, 2018. | Includes bibliographical references and index.
Identifiers: LCCN 2018004669 | ISBN 9780226559056 (cloth : alk. paper) | ISBN 9780226559193 (pbk. : alk. paper) | ISBN 9780226559223 (e-book)
Subjects: LCSH: Shakespeare, William, 1564–1616—Criticism and interpretation. | Shakespeare, William, 1564–1616—Themes, motives. | English drama—Early modern and Elizabethan, 1500–1600—History and criticism. | English drama—17th century—History and criticism. | Geography in literature. | Geopolitics in literature.
Classification: LCC PR3014 .E48 2018 | DDC 822.3/3—dc23
LC record available at https://lccn.loc.gov/2018004669

♾ This paper meets the requirements of ANSI/NISO Z39.48–1992 (Permanence of Paper).

CONTENTS

Acknowledgments *vii*

Introduction: Shakespearean Territories *1*

1. Divided Territory: The Geo-politics of *King Lear* *11*

2. Vulnerable Territories: Regional Geopolitics in *Hamlet* and *Macbeth* *29*

3. The Territories: Majesty and Possession in *King John* *63*

4. Economic Territories: Laws, Economies, Agriculture, and Banishment in *Richard II* *88*

5. Legal Territories: Conquest and Contest in *Henry V* and *Edward III* *113*

6. Colonial Territories: From *The Tempest* to the Eastern Mediterranean *136*

7. Measuring Territories: The Techniques of Rule *165*

8. Corporeal Territories: The Political Bodies of *Coriolanus* *197*

9. Outside Territory: The Forest in *Titus Andronicus* and *As You Like It* *218*

Coda: Beyond Pale Territories *240*

References to Shakespeare's Plays *253*

Notes *255*

Index *325*

ACKNOWLEDGMENTS

This book has been in development for several years. I have spoken about Shakespeare at New York University (February 2012), the Exterritory conference, Paris (May 2012), University of Nottingham (May 2012), Edinburgh University (July 2012), University of Warwick (October 2012, October 2013, November 2015), Durham University (November 2012), Oxford University (November 2012), University of York (January 2013), University of Aberystwyth (February 2013), Memorial University of Newfoundland (March 2013), the Shakespeare Birthplace Trust (October 2013), University of New South Wales (March 2015), Purchase College, SUNY (April 2015), King's College London (July 2015), Cambridge University (November 2015, November 2016), University College London (November 2015), University of Memphis (September 2016), Huntington Library (October 2016), and Kingston University/Rose Theatre (March 2017). Several people discussed or responded to the work, and I am grateful to them all, especially Neil Brenner, Klaus Dodds, Paul Ennis, Tamar Garb, Helen Hackett, Anselm Haverkamp, Daniel Hoffman-Schwartz, Morris Kaplan, James Kneale, Steve Mentz, Adam David Morton, Fionnuala O'Neill, Julie Sanders, Garrett Sullivan, and Richard Wilson. I would also like to thank the Institute of Advanced Studies at University College London, where much of this book was drafted.

An earlier version of chapter 1 appeared as "The Geopolitics of *King Lear*: Territory, Land, Earth," *Law and Literature* 25, no. 2 (2013): 147–65; and an earlier version of chapter 8 was published as "The Political Bodies of Coriolanus," in *International Politics and Performance: Critical Aesthetics and Creative Practice*, ed. Jenny Edkins and Adrian Kear (London: Routledge, 2013), 182–203. I am grateful to the University of California Press and Taylor & Francis for permission to reuse material.

I would also like to thank Mary Laur, Rachel Kelly, and their colleagues at the University of Chicago Press for their support for and advice on this project. The reports commissioned on the draft manuscript were generous, thoughtful, critical, and helpful—an all-too-rare, but much appreciated, combination. Marian Rogers did an excellent job with the copyediting, Lisa Scholey compiled the index, and Susan Karani served as production editor.

Although the book rarely discusses performances, I appreciate theater companies, especially the Royal Shakespeare Company and Shakespeare's Globe, for continual sources of inspiration and, at times, frustration.

I am grateful to Susan, as ever, for her love; and this book is dedicated to my mother, Rosemary, who—among many other things—taught me to read.

INTRODUCTION

Shakespearean Territories

Reading Shakespeare politically is a long-established practice in the literature. Several of his plays can be analyzed in this way, from those tracing a long period of English history to the tragedies of *Othello*, *Macbeth*, *Hamlet*, and *King Lear*.[1] Some of the tragedies based on historical events, such as the Roman plays *Coriolanus*, *Julius Caesar*, and *Antony and Cleopatra*, are also open to political analysis. *Measure for Measure* is a play about authority and its misuse; *The Tempest* and some other plays raise questions concerning colonialism. Reading Shakespeare geographically is a newer move in the scholarship. As Andrew Gurr has noted, while Shakespeare's grasp of specific geographies could be shaky, his plays frequently stress their location very early in the text.[2] Yet the geographical aspects of Shakespeare go some way beyond mere situation. *Shakespearean Territories* makes the case that his plays, and some of his poetry, exhibit a profound political geographical imagination, and that read in this way they can shed considerable light on the word, concept, and practice of territory.

PURPOSE OF THIS STUDY

Discussions of Shakespeare and geography often begin with a passage from *The Comedy of Errors* where Dromio of Syracuse likens Nell the kitchen maid to a globe, and the different countries to parts of her body.[3] The passage is offensive, certainly, but its sexism comes from equating the corporeal with the geophysical, a "geographical anatomization."[4] The "globe" may refer back to Gerardus Mercator's versions from the 1530s, but is perhaps a more topical allusion to Emory Molyneux's 1592 version.[5] The passage certainly trades on the current vogue for maps of the world. In its figuration of geography in a woman's body, the explicit reference may well be to the famous image in Sebastian Munster's *Cosmographica* of Europa Regina from 1588, though there were other similar images at that time, including one of Elizabeth I as Europe held by the Ashmolean Museum in Oxford.[6] The passage also has a wider geopolitical sense, referring to the civil war in France between 1589 and 1594, and is the only time the word America appears in Shakespeare's writings.[7]

Fortunately, most of Shakespeare's geographical allusions are not so gendered and unpleasant. Throughout his plays there are a number of discussions of geographical ideas, specific places, and topics. A wide range of literature has developed discussing these themes. Two fundamental studies that set much of the agenda for future work were John Gillies, *Shakespeare and the Geography of Difference* and Garrett A. Sullivan Jr., *The Drama of Landscape*; two recent studies with a more specific focus are Steve Mentz, *At the Bottom of Shakespeare's Ocean* and Kristen Poole, *Supernatural Environments in Shakespeare's England*.[8] Other studies with a focus on spatial concerns have looked at themes such as exile, borders, national geographies, exploration, and environment.[9] There has also been a large literature on spatial practices in Shakespeare's time, on themes such as cartography, surveying, geometry, and navigation.[10] The spatiality of the stage is also an important theme, though not one discussed in detail here.[11]

I draw extensively on the work of Gillies, Sullivan, Mentz, and Poole and many of the other studies in what follows, but my focus is both more specific and potentially more wide-ranging. The focus is the question of territory. It is striking that while there has been a burgeoning interest in Shakespeare's geographies, and many of these works touch on territory or some of its aspects, this has not before become a major theme. In part this is because, as I have argued before, territory is assumed to be a simple concept, which is complicated only when it comes to actually drawing lines and establishing rule.

Much of my work has been concerned with challenging this assumption. In *The Birth of Territory* I provided a genealogy of the emergence of territory in Western political thought, and in *Terror and Territory* discussed the contemporary contours and tensions of the post–Cold War world, especially in the "war on terror."[12] In those works I tried to show how territory is a complex concept as well as always contested and convoluted in practice. Territory, in my argument, is something that cannot be simply understood as a bounded space, as it is generally is defined, but encompasses a variety of different, multiple, and contested processes.[13] Territory is not a product, but a process, formed by a range of practices and techniques, including bordering, dividing, conquering, excluding, enclosing, controlling, surveying, and mapping. In order to understand territory, we need to move beyond the simplistic definition, and to examine multiple registers—political and geographical issues certainly, but also economic, strategic, legal, and technical concerns. This is a means of grasping the complexities of how territory has been understood and practiced in different ways in diverse times and places. My argument is that it is helpful to understand territory as a political, calculative space, as a political technology. Political technology is a notion developed by Michel Foucault to analyze bodies and population, which I adopt to make sense of the concept and practice of territory. It emerges in his book *Surveiller et punir*, which we know in English as *Discipline and Punish*, to examine "the metamorphosis of punitive methods" through a study of power relations on the body.[14] It is then used in his later lectures as a way of making sense of governmental strategies, especially as conducted toward population.[15] The same kinds of calculative techniques that Foucault sees as crucial to population are central to understanding territory. These include techniques such as population statistics, censuses, cartography, land surveying, and aspects of political rule.[16] In my terms, political technology is a way of making sense of "the processual, multiple, and conflictual nature of the bundle of political techniques . . . that make up and transform the contested and diverse notion of territory."[17] This understanding helps us to comprehend boundaries, which in their modern sense are a calculated set of points surveyed and mapped with technical precision, rather than boundaries defining territory.

Shakespeare, I believe, strikingly illustrates these different aspects of territory in his plays. We can find political and geographical themes in multiple works, but different plays put an emphasis on themes such as the strategic, the economic, the legal, and technical. Yet Shakespeare is not read here instrumentally, where his plays simply provide examples of themes I had previously identified. Shakespeare's plays open up new ways of thinking about these questions, providing depth and illustration of these themes at a significant historical juncture. Even more significantly, Shakespeare's plays highlight aspects that my own previous work insufficiently acknowledged—the colonial, the geophysical, and the corporeal. These crucial themes have been highlighted in some critical engagement with my work, and I use Shakespeare to push me further in developing this account of the contested and complicated concept and practice of territory.

This book is therefore both a book about Shakespeare for a geography and territory audience, and a book about territory for a Shakespeare audience. I have tried to keep a balance between writing for these two, largely discrete, audiences throughout. But the essential claims are ones that work for both audiences. Territory is a multifaceted notion, which can only be grasped in its complexity if we recognize that the standard definition is restrictive at best and misleading at worst. The different aspects of territory help us to understand the political-geographical relations between people, power, and place. In this book, Shakespeare is a guide to the complexity of territory, while at the same time demonstrating how deeply a political-geographical sensibility runs through his dramatic work.

To accomplish this, I provide close readings of a number of Shakespeare's plays to explore the complex imbrications of these different aspects of territory. While this book is a study of literary texts, it is not just a study of the word "territory."[18] Indeed, as I will go on to explore in much more detail in subsequent chapters, Shakespeare only uses the word "territory" twice—in *As You Like It* and in *King Lear*, the latter only in the Folio text of that play. The plural "territories" is a little more common, but only appears in a handful of plays. And yet, many of Shakespeare's plays, especially but not exclusively his histories and tragedies, are concerned with issues that relate to the question of territory. The word may be rare, but the concept and practice of territory are not at all marginal to his work. A number of his plays are structured around questions of exile, banishment, land politics, spatial division, contestation, conquest, and succession. The work conducted in *The Birth of Territory*, which looked at "territory" as word, concept, and practice, is continued here.

Shakespeare was born in 1564, in the reign of Queen Elizabeth. At that time, almost no political theory used the word "territory," and even Latin texts that used *territorium* generally meant something quite different from the way we use the term today. By the time he died in 1616, the term was much more common. It can be found in Jean Bodin's *Six Books of the Commonwealth*, written in French in 1576 and revised in Latin in 1586; has an important role in Richard Hooker's *On the Laws of Ecclesiastical Polity*, first published in 1594; and is prominent in Germanic debates about the status of constituent parts of the Holy Roman Empire, often in Latin texts such as Andreas Knichen's *De sublimi et regio territorii iure* (1600) and Johannes Althusius's *Politica Methodice Digesta* (1603). Shakespeare died just two years before the Thirty Years' War broke out in continental Europe, in part over the issues the German thinkers had been debating. The Peace of Westphalia, which ended that war in 1648, is often seen as the beginning of the modern territorial state system. This emphasis is misleading and its influence overdone, of course, but it was a significant moment, and led to further debates in political theory—Thomas Hobbes, Theodor Reinking, Bogislaw Philipp von Chemnitz, Samuel Pufendorf, and, especially, Gottfried Leibniz are all working in its wake. Asked by his employer, the Duke of Hanover, to qualify what rights he had as a ruler following Westphalia, Leibniz came up with a strikingly modern formulation: "Sovereign or potentate is that Lord or State who is master of a territory."[19]

Shakespeare's period of writing, at the end of the sixteenth and beginning of the seventeenth century, was a time when the word used to designate the concept that describes a practice was both becoming more common and shifting its meaning. In Robert Cawdrey's *Table Alphabeticall*, a 1604 text that has been described as the first English dictionary, the term "teritorie" is defined as a "region, or the countrie lying about the citie."[20] The first part of this is very general, but the second, indicating the surrounding lands around a city, is a definition that would have made sense to an ancient Roman, being close to the classical use of the term *territorium*. The more specific sense of how we would understand the term today is absent. Leibniz's formulation from 1678 may come sixty-two years after Shakespeare's death, but its argument is anticipated in some of the writers contemporary with his dramatic work, such as Knichen and Althusius.

Shakespeare, this book contends, helps us understand territory's variant aspects, tensions, ambiguities, and limits. He reads some of those concerns back into English history, especially in the second tetralogy of *Richard II*, the two parts of *Henry IV*, and *Henry V*. Such rivalries continued in the European continent in which he set so many of his tragedies and comedies. Jean Howard notes that of Shakespeare's tragedies, only *King Lear* is set in England, temporally rather than spatially distant.[21] Yet this discounts seeing the English histories as tragedies, the way some of them were described in their original titles. Equally, while most of the comedies are set abroad, with local flavor adding to the dramatic action, most fairly transparently transplant English concerns into superficially foreign locations. In this England the dominant form of political power was conducted in a space that was, by his time, relatively ordered and bordered, even if who should rule, and their relation to wider geopolitical forces such as the papacy and the empire was far from settled. Late in his career the key question became the relation of England to the recreated Britain, being forged through the union of the crowns with King James. We see that explored, in different ways, in Jacobean plays such as *King Lear* and *Cymbeline*, though there are elements of a nascent national identity in the Elizabethan history plays, including *King John*, *Richard II*, and *Henry V*. England's recent past—explored in the history plays—was certainly not stable. We can see such disputes, for example, in *Richard II*, which reveals a great deal about the political economies of land and the politics of banishment.

Banishment is a major theme. As Jane Kingsley-Smith has noted, "Fourteen of Shakespeare's 38 plays represent the banishment of one or more central characters."[22] While most of those banished are elite figures, in Shakespeare's likely contribution to the collaborative *Sir Thomas More*, the eponymous character urges the crowd to have sympathy for the displaced:

> Imagine that you see the wretched strangers,
> Their babies at their backs, with their poor luggage,
> Plodding to th' ports and coasts for transportation.[23]

The question of territory figures in all the other history plays, including *King John* and *Edward III*, a play of disputed authorship. In *The Tempest* Shakespeare explores what this might mean when Europe came into contact with its outside, as he does in different ways in *Pericles*, *Othello*, and *Antony and Cleopatra*. Technologies of territory are explored in a number of plays, looking at cartography, surveying, calculating, and measuring. The theme of the outside of territory is explored in plays such as *Titus Andronicus* and *As You Like It*. While the early *Romeo and Juliet* lacks a wider frame beyond the feuding families, even some of the comedies have a sense of a world outside. The returning soldiers at the beginning of *Much Ado About Nothing* are asked, "how many gentlemen have you lost in this action?" and reply, "but few of any sort and none of name."[24] In *All's Well That Ends Well*, Bertram travels from France to Italy to escape from marriage, joining the "Tuscan wars."[25] More significant is the wider situation of *Troilus and Cressida*, in which the tragic love story is set in the context of the siege of Troy.[26]

The earlier setting of *King Lear* shows a place that is historically distant and spatially disrupted. In that it is more similar to the Europe in which Shakespeare set most of his tragedies, and his comedies. This was a space that was contested and fractured, both politically and spatially. We see that in *Coriolanus*, but it also functions in important ways in *Hamlet* and *Macbeth*. In *Coriolanus* territory as the body of the state is only one aspect of its corporeal nature. It is a play about the political body of the polity itself, its inside and outside, the aggressive wars to keep it safe externally, and its internal health and well-being. The play also raises a range of questions about what it is to contribute to a political community, and who should rule. But literally and figuratively these are often the physical bodies of its characters, with language invoking wounds, contagion, animals, and a variety of body parts.

ORGANIZATION OF THIS STUDY

This book begins with two chapters that treat three of the major tragedies. The first is a sustained reading of *King Lear* around the themes of territory, land, and earth, what I call the play's *geo*-politics. The second chapter reads *Hamlet* and *Macbeth* with a focus on regional relations, decentering the title characters in an analysis that privileges the wider worlds outside. In *Hamlet* this is the relations between Denmark, Norway, and Poland, and, more peripherally with England and France. In *Macbeth* it is the war with Norway, and the relations with England and Ireland. Showing the vulnerability of territories, and how they are embedded in a set of strategic relations, these plays demonstrate how Shakespeare frequently situates his plays within a wider network of connections. To focus on the court politics alone, as is frequently done especially with *Hamlet*, is to miss a crucial part of the action.

Having established the way that territory explicitly, and territorial politics generally, figure in the tragedies, the book shifts to examine some of Shakespeare's histories. While *King Lear*, *Hamlet*, and *Macbeth* indicate multiple important aspects of the question of territory, the sense of the term that emerges from these texts is still quite a traditional, restrictive one. Shakespeare's plays, however, provide plenty of resources to show how territory is more complicated than the standard, straightforward, definition. *King John* is a play with an intriguing use of the term "the territories," which has caused its editors some confusion. Carefully tracking the different ways that Shakespeare uses the words "territory" and "territories," the chapter shows how the vast majority of uses are possessive, lands held by people or kingdoms. But in this play there is a hint of a more modern sense, of the territory itself. What might that territory be? The next chapters examine different plays to focus on specific aspects. These demonstrate that Shakespeare's contribution to a theory of territory goes far beyond his rare use of the word itself. The economic sense of territory and its relation to land is explored through a reading of *Richard II*, the legal aspects through a discussion of *Henry V*, especially its opening scenes, and *Edward III*. The colonial aspects of territory are discussed in *The Tempest*, of course, though this is balanced here by a reading of other plays that have imperial themes, notably in the eastern Mediterranean—*Pericles*, *Othello*, and *Antony and Cleopatra*.

One of the key aspects of the birth of territory within modern political thought and practice was the development of political technologies—cartography, surveying, measuring, and counting. These territorial practices are examined through a reading of a range of plays that treat concerns around calculation, navigation, cartography, and surveying, especially focusing on the map division in *Henry IV, Part 1*. That play demonstrates clearly how the geophysical runs alongside the geopolitical in thinking about territory. In a related vein, corporeal aspects of territory—the body of the state and the body of its rulers—are examined through the Roman play *Coriolanus*. Finally, the book examines what it might mean to be outside of territory, through a reading of *Titus Andronicus* and *As You Like It*. In these and other plays, Shakespeare helps us to understand what it means to be part of territory or outside territory, conceptually, historically, and politically. The coda begins with a reading of national identity, especially through *Cymbeline*, and then turns to the question of the relation of territory to a pale, a bounded area. A pale is a marked-out area of forest, fenced off and thus an enclosed space. The term came to be used of political jurisdictions. While a pale can therefore be a territory, not all territories are pales, nor is this meaning sufficient to understand the complexity of territory.

THE AUTHORSHIP QUESTION

There is, of course, a conspiracy theory disputing that William Shakespeare was actually the author of the plays that bear his name. Other candidates are put forward, often on the basis that Shakespeare did not have the background or ability to write such remarkable texts. I am largely uninterested in these debates, which are often para-academic, and if readers want to silently put scare quotes around every instance of "Shakespeare" that is fine: it would not alter my substantive argument.[27] I am interested in the plays ascribed to this specific author, rather than his biography.[28] But this does raise two authorship questions that I *am* interested in, that of the Apocrypha, and the question of collaborative writing.[29] There seems to be general consent that Shakespeare collaborated on some of his earliest plays, and again at the end of his career. This would fit the model of an apprenticeship, followed by his mature years, and then his own role as mentor and tutor to younger playwrights. Some of the plays I discuss in *Shakespearean Territories*, notably *Titus Andronicus*, *Macbeth*, *Measure for Measure*, *Timon of Athens*, *Pericles*, and *Henry VIII*, bear marks of a second author. These are all plays printed in the First Folio. In addition, there are anonymously printed plays or ones where attribution is disputed that are of interest.

As Richard Proudfoot notes, "Five plays omitted by the editors of the First Folio have been strongly backed as wholly or in part his work. Three of them now generally find a place in collected editions." The three are *Pericles*, probably a collaboration with George Peele; *The Two Noble Kinsmen*, with John Fletcher; and *Sir Thomas More*, with several others. *Double Falsehood* is the fourth, supposedly a version of Shakespeare and Fletcher's lost play *Cardenio*, though Proudfoot notes that the version of this text that is preserved "may even be a double palimpsest—to wit, Theobald's reworking of Thomas Betterton's adaptation of the Jacobean original."[30] The fifth is *Edward III*, a play that is only slowly being incorporated into the canon, and not without dispute. This play, whether or not Shakespeare did have a hand in its composition, anticipates many of the themes of *Henry V*, and is discussed in chapter 5. In addition, *Arden of Faversham* looks likely to have had at least some material by Shakespeare, and it is possible that the 1602 revisions to Thomas Kyd's *The Spanish Tragedy* were also his work.[31] I am less convinced that *Edmund Ironside*, an anonymous play of the 1580s, is an early Shakespeare work, though given its likely date Eric Sams rightly suggests it has a good claim to be "the very first chronicle history and hence in its own right a work of seminal significance in the history of English and world drama."[32] Sams has advanced the most sustained argument on behalf of its Shakespearean attribution, and sees it as an apprentice work, "plainly the precursor of *Henry VI* as of *Titus Andronicus*."[33]

There is also the question of anonymous texts such as *The Troublesome Reign of King John*, and its relation to the canonical play *The Life and Death of King John*; or *Thomas of Woodstock*, sometimes called *King Richard II, Part 1*. These are discussed in detail in chapters 3 and 4, in relation to Shakespeare's *King John* and *Richard II*. Other texts discussed here exist in markedly different forms—the Quarto and Folio editions of *King Lear* and *Henry V* or the variant texts of *Hamlet*—or where our only source is believed to be a corrupted and edited version of some lost original, such as is the case, for different reasons and to varying degrees, with *Macbeth* or *Pericles*. Such textual questions are discussed in detail in the respective chapters.[34] These concerns matter not just because of their intrinsic interest and importance, but because they bear directly on the subject matter. To pick just a few examples, only the Folio edition of *King Lear* includes the important phrase "interest of territory, cares of state," and the identity of the invading army seems to change between the Quarto and the Folio. There are important differences between the Quarto and Folio of *Henry V*, with the Quarto losing most of the passages that hint at Henry's immoral conduct in war, as well as the opening scenes that are my focus in chapter 5. The complicated textual history of *Hamlet* also has important consequences for my argument in chapter 2.

Of Shakespeare's contemporaries, Christopher Marlowe is perhaps the one with the most pronounced geographical sensibility.[35] The two parts of *Tamburlaine the Great*, notably, stretch across almost the whole world of antiquity, almost a narrative extending across Abraham Ortelius's famous 1570 work, *Theatrum Orbis Terrarum*. For Jonathan Bate, "Tamburlaine is a land play. It name-checks the whole of known Asia and Africa, but ignores the islands of the Mediterranean. Marlowe fills that gap in *The Jew of Malta*."[36] Marlowe's plays use the word "territories," like Shakespeare, only rarely, but they are clearly suffused with a wide geopolitical potential.[37] As Tamburlaine says,

> I will confute those blind geographers
> That make a triple region in the world,
> Excluding regions that I mean to trace,
> And with this pen reduce them to a map.
> Calling the provinces, cities and towns
> After my name and thine, Zenocrate.
> Here at Damascus will I make the point
> That shall begin the perpendicular.[38]

The pen, of course, is his sword; his self-aggrandizing gesture, like that of colonizers for generations, is a mainstay of toponyms, while the "triple region" is the old style "T-in-O" maps that showed Europe, Asia, and Africa. This map is important in Shakespeare's *Antony and Cleopatra*, as discussed in chapter 6. The "perpendicular" will be a new division, either reinscribing the "T" in the map or drawing a new meridian.[39] There are other examples, of course, including Michael Drayton's topographical poem, *Poly-Olbion*, published in two parts in 1612 and 1622.[40] Shakespeare's contemporaries will, occasionally, be discussed in what follows, but the focus is on the multifaceted understanding of territory that can be taken from the plays that bear this single author's name.

CHAPTER ONE

Divided Territory: The Geo-politics of *King Lear*

"DIVIDED IN THEIR DIRE DIVISION"

As subsequent chapters will show in more detail, many of Shakespeare's fairly rare uses of the word "territories" have a sense close to "lands." People are banished from territories, welcomed into them, or ownership is asserted or disputed. Some of the instances come in the early *Henry VI* plays. Together with *Richard III* these plays are now known as Shakespeare's first tetralogy, though they are unlikely to have been conceived as that while being written. Division of territories is a key theme through these four texts. The plays are episodic, covering a large part of the reign and overthrow of the king, but some key themes can be identified. In particular, the division of territories becomes ever more tightly focused throughout their dramatic action. Part 1—possibly written after the other two parts—concerns the wars with France, and part 2 the wars internal to England, while part 3 focuses even more tightly on the splits between the rival families of York and Lancaster. In *Richard III* the eponymous character murders many of his own family members to claim and maintain the throne, before being overthrown by Henry Tudor, who becomes King Henry VII and unites the families of York and Lancaster through marriage.

In *Henry VI, Part 1*, the gains made by Henry V are shown to be very fragile, and the Earl of Suffolk promises the territories of Anjou and Maine to France in order to secure Margaret of Anjou's wedding to the young king. He is greeted by the Reignier, the Duke of Anjou, with "Welcome, brave earl, into our territories,"[1] and concedes, "That is her ransom. I deliver her, / And those two counties I will undertake / Your grace shall well and quietly enjoy."[2] Richard, Duke of York, bemoans the loss of these lands, gained by bloodshed and now given away in "effeminate peace," and anticipates "the utter loss of all the realm of France."[3] Charles the Dauphin anticipates this weakness, arguing that he is already "possessed / With more than half the Gallian territories," and intends to become "viceroy of the whole."[4] In other words, he already holds more than half as "lawful king," and is unwilling to exchange the reverence he gets for this status with the control of the whole in a less senior role.[5]

This story continues into *Henry VI, Part 2*, where the "articles of contracted peace" are read in the opening scene. Among them is the clause that says, "the duchy of Anjou and the county of Maine shall be released and delivered to the King her father."[6] The Duke of Gloucester admits he cannot bear to see the territories gained by his brother's wars given away so easily. He fears the loss of all France, something that comes to pass later in the play when the Duke of Somerset says, "That all your interest in those territories / Is utterly bereft you; all is lost."[7] Gloucester is accused of treason, and blamed for the loss, and later murdered. But the King eventually agrees that the Duke of Suffolk is "banished fair England's territories."[8] Salisbury says that this needs to be done to appease the people.

In *Henry VI, Part 3*, the struggle is over who should be king of England. Henry declares:

> I am the son of Henry the Fifth,
> Who made the Dauphin and the French to stoop
> And seized upon their towns and provinces.[9]

Warwick replies that he should "Talk not of France, sith thou has lost it all,"[10] which Henry blames on others in his childhood: "The Lord Protector lost it and not I. / When I was crowned I was but nine months old."[11] While most of the focus in these plays is on elite struggles, *Henry VI, Part 2* also has a class struggle, led by Jack Cade. Cade leads a revolt of working men to London, though he is defeated by the aristocracy. Cade flees to the country, and through hunger seeks food in a private garden. There he is confronted by the owner and killed. As Stephen Greenblatt has noted, in this scene, "status relations . . . are being transformed before our eyes into property relations."[12]

At the very end of *Richard III*, Henry Tudor surveys the battlefield, and seeks to create a new state of affairs in England. He remarks:

> We will unite the white rose and the red. . . .
>
> England hath long been mad, and scarred herself:
> The brother blindly shed the brother's blood;
> The father rashly slaughtered his own son;

The son, compelled, been butcher to the sire.
All this divided York and Lancaster,
Divided in their dire division. . . .

Abate the edge of traitors, gracious Lord,
That would reduce these bloody days again
And make poor England weep in streams of blood.
Let them not live to taste this land's increase
That would with treason wound this fair land's peace
Now civil wounds are stopped; peace lives again.
That she may long live here, God say amen.[13]

This is an interesting and powerful passage, seeking to mend the broken land of which he has become king. The two roses of York and Lancaster symbolized the division of the Plantagenet family. One couplet has caused editors and commentators some pause, because it seems like the Folio's "Divided, in their dire division" may be corrupted. York and Lancaster are "united" in their division, the one thing that they share; and indeed some editions of the text emend in this way (Norton and Oxford). The Norton edition glosses the emended line as "joined by hatred, having nothing in common but mutual antagonism."[14] Jacques Lezra has provided a particularly interesting reading of these lines.[15] This speech is, for Lezra, "perhaps the least equivocal assertion of the so-called Tudor myth of history to be found in Shakespeare's work and surely his most obscure treatment of political 'division.' "[16]

It is important to recognize that the earlier Quarto text is not quite the same, reading "Deuided in their dire deuision,"[17] which Lezra suggests provides "a nice play on *device*.[18] How important this technology might be is unclear. The division is not just *between* York and Lancaster, but *within* them, as *Richard III* has especially shown. For Lezra, "surely this would suggest, not that 'dire division' divided the two camps, but rather that York and Lancaster *shared* at least this: that both camps were divided, each within each, each against the other."[19] In joining together the two roses, the two families, through his marriage to Elizabeth, Richmond is clearly trying to unite something that has been divided. Lezra argues that "the 'dire' political division between York and Lancaster and within each camp has now been replaced by 'union' between two distinct orders of submission: the 'submission' of the subjects to the sovereign, and the 'submission' of his (or her) speech to the 'fair ordinance' and the 'will' of God."[20] In addition, Richard's brief reign was a division between legitimate kings, "the division that Richard provoked in the fabric of British history, which Richmond now heals,"[21] a lineage that Richmond seeks to rejoin in his accession to the throne. "We who listen to Richmond are in consequence divided from Richard's division and from the divisions shared by York and Lancaster, apart from them and united as a result of this division from 'division.' "[22] But all this is division within the kingdom, not division *of* the kingdom.

In contrast, *King Lear* is divided within, and divided between. It is a remarkable play about space and in particular the politics of space. The focus of this chapter is what might be called the "geo-politics" of *King Lear*, its politics of earth, of land, of the *geo*. This focus has three parts: the opening scene with its division of territory, the wider politics of land in the play, and the more figurative use of the term "earth."

"INTEREST OF TERRITORY, CARES OF STATE"

The most intriguing of the passages from the *Henry VI* plays concerning "territories" is the report from Lord Somerset: "That all your interest in those territories / Is utterly bereft you; all is lost."[23] "Interest" here clearly shows that there is a political issue beyond a more economic one, control over and legal title to, rather than simple possession.[24] It also has a sense of the terrain over which, and for which, a struggle of competing forces may take place. This passage is strikingly similar to the way the word "territory" is used in *King Lear*. In the play's opening scene, Lear is conducting a ceremonial division of his kingdom between his three daughters and their husbands. Along with *As You Like It*, this is one of only two plays in which the singular "territory" is used by Shakespeare, though here notably only in the Folio text. The first thing that Lear says is that Gloucester should "attend the Lords of France and Burgundy,"[25] suitors for his youngest daughter's hand. He then begins:

> Meantime, we shall express our darker purpose.
> Give me the map there. Know that we have divided
> In three our kingdom; and 'tis our fast intent
> To shake all cares and business from our age,
> Conferring them on younger strengths, while we
> Unburdened crawl toward death. Our son of Cornwall,
> And you, our no less loving son of Albany,
> We have this hour a constant will to publish
> Our daughters' several dowers, that future strife
> May be prevented now.
> The two great princes, France and Burgundy,

Great rivals in our youngest daughter's love,
Long in our court have made their amorous sojourn,
And here are to be answered. Tell me, my daughters—
Since now we shall divest us both of Rule,
Interest of territory, cares of state—
Which of you shall we say doth love us most,
That we our largest bounty may extend
Where nature doth with merit challenge.[26]

At least, that is how it is in the composite text that is usually used of *King Lear*, which builds on the later Folio *The Tragedy of King Lear* (1623), but usually incorporates material that was originally in the earlier Quarto *The History of King Lear* (1608) that does not appear in the Folio.[27] There were also several, though fewer, lines found in the Folio but not in the Quarto, including in this passage the remark concerning territory. The differences between the texts have given rise to the contention that there are actually two *King Lears*, with the 1623 text representing Shakespeare's revisions.[28] The couplet specifically invoking "territory" only appears in the Folio edition. Look at the text of each of these editions for this speech.

QUARTO

Meantime, we will express our darker purposes.
The map there. Know we have divided
In three our kingdom; and 'tis our first intent
To shake all cares and business of our state,
Confirming them on younger years,

FOLIO

Meantime, we shall express our darker purpose.
Give me the map there. Know that we have divided
In three our Kingdom; and 'tis our fast intent
To shake all cares and business from our age,
Conferring them on younger strengths, while we
Unburthened crawl toward death. Our son of Cornwall,
And you, our no less loving son of Albany,
We have this hour a constant will to publish
Our daughters' several dowers, that future strife

The two great Princes, *France* and *Burgundy*,	May be prevented now. The Princes, France and Burgundy,
Great rivals in our youngest daughter's love,	Great rivals in our youngest daughter's love,
Long in our Court have made their amorous sojourn,	Long in our court have made their amorous sojourn,
And here are to be answered. Tell me, my daughters,	And here are to be answered. Tell me, my daughters,
	(Since now we shall divest us both of Rule,
	Interest of Territory, Cares of State)
Which of you shall we say doth love us most?	Which of you shall we say doth love us most?
That we our largest bounty may extend	That we our largest bounty may extend
Where merit doth most challenge it.	Where Nature doth with merit challenge.[29]

The Folio certainly appears a more polished text, with several differences in the first few lines: the diffuse "darker purposes" becomes the particular and comparative "darker purpose"; there is the telling replacement of "of our state" with "from our age," presumably because of amplifications to come; and that of "years," already implied by "younger," with "strengths," indicating force and political power. It is difficult to see how the Quarto could simply be a corrupted version of some lost original. In both texts Lear proclaims that the division has already been made, and that the map he asks for presumably shows this.[30] The Folio has lines that include the indication of Lear's wish to live out his remaining time free of concerns; ones that more clearly indicate the relations to Cornwall and Albany, the establishment of the plural dowers, and the attempt to head off future conflict. He wants to hand over the burdens—"cares and business"—to a younger generation, to his sons-in-law and the daughters married to them, and his third as-yet unmarried daughter. Goneril is married to the Duke of Albany, and Regan to the Duke of Cornwall, and in the Folio at least it appears in large part that Lear is aiming this redistribution at them more directly than the daughters themselves. A few lines later there is the explicit, parenthetical noting of what exactly is at stake: rule, territory, state. Finally, in the last line of this passage, the meaning is subtly different: instead of this being simply to do with merit, there is now a reference to the conflict between merit and nature, worth and familial ties.

For Goneril and Regan, the request is straightforward. They are both able to praise their father and state their love for him in hyperbolic terms. Goneril says that her love for him is "Dearer than eyesight, space, and liberty / Beyond what can be valued, rich or rare."[31] Regan begins by saying, "I am made of the self-same mettle that my sister is, / And prize me at her worth." She then quickly adds, "Only she comes too short . . . the most precious square of sense possesses."[32] The language of both sisters is striking—not just professions of love, but spatial, calculative, and economic. These, of course, are also the characteristics of the lands they seek and are quickly rewarded with.[33] It would appear that the two elder daughters are, only now, receiving their dowries. Lear tells Goneril of her portion of the kingdom:

> Of all these bounds, even from this line to this,
> With shadowy forests and with champains riched,
> With plenteous rivers and wide-skirted meads,
> We make thee lady. To thine and Albany's issues
> Be this perpetual.[34]

While the characteristics of the lands are outlined, the most important aspect is their boundary, the division being marked on the map. Lear also, continuing the importance of his lineage, seeks to cement a hereditary principle for Goneril and Albany's descendants. He rewards Regan with "To thee and thine hereditary ever / Remain this ample third of our fair kingdom / No less in space, validity, and pleasure, / Than that conferred on Goneril."[35] The division is clearly predetermined, a division into thirds, even before Cordelia has spoken.

The hereditary aspect is clearly in his mind when he turns to Cordelia too, noting that it is to her "young love / The vines of France and milk of Burgundy / Strive to be interessed."[36] He asks her what she can say "to draw / A third more opulent than your sisters."[37] Given that the previous two parts have already been distributed, this may appear to be a false question. Alternatively Lear may be stressing the malleability of these territorial divisions, with the parts given to the first two sisters provisional rather than fixed, subject to amendment by his sovereign decision. Yet Cordelia refuses to act like her sisters. Instead she responds by saying, "Nothing, my lord." Lear believes this to be a mistake, replying, "nothing will come of nothing," and urging her to "speak again."[38] "Nothing" is a rich term in the play, being both a slang term for the vagina (no thing) and the spatial category of the void, both important in understanding the play's situation in the early seventeenth century.[39] Cordelia clarifies: she loves her father "according to my bond; no more nor less."[40] Cordelia was put in an impossible position: she was required to love her father entirely while at the same time taking a husband. She highlights this concern: "Why have my sisters husbands, if they say / They love you all?"[41] Her future husband, she suggests, "shall carry / Half my love with him, half my care and duty," and she will "never marry like my sisters."[42] Of course, Lear entirely misses that Cordelia probably loves him more than either of her sisters, and simply is unwilling to join their obsequious display. Yet this honesty, and her true love, yield no reward: Lear tells her, "Thy truth, then, be thy dower."[43] The remainder of the play develops from this point onward, leading to the madness of Lear and his late reconciliation with Cordelia, before their deaths. *King Lear* is a play that is fundamentally structured by this division of land.

Lear's initial plans were already known before his entry onto the stage. This is clear because in the opening lines of the play the Dukes of Kent and Gloucester are discussing this. Kent tells Gloucester that he had thought that the King had preferred or "affected the Duke of Albany than Cornwall."[44] Gloucester agrees this was the case in the past:

> It did always seem so to us; but now, in the division of the kingdom, it appears not which of the Dukes he values most; for equalities are so weighed that curiosity in neither can make choice of either's moiety.[45]

"Moiety" means a share or portion, and Kent and Gloucester seem to think there could be no complaint from either for their allotment. Lear's entry is a ceremony, with each daughter expected to play a particular role. Kent and Gloucester's awareness further seems to indicate that the future relations between Cornwall and Albany is the key issue, with lands given to them, rather than to the daughters they have married. Albany was the old name for Scotland, probably including parts of modern-day northern England; Cornwall was much larger than the modern county of that name, and included large parts of the southwest and some of Wales. It too was formerly a separate kingdom. (James Shapiro suggests that this was a topical reference to King James and his two sons.[46]) When Shakespeare wrote the play, the eldest son of King James, Henry Frederick, was Duke of Cornwall and later Prince of Wales.[47] Given their status as dukes in the northern and southwestern extremities of the British Isles, Lear is either giving Albany and Cornwall additional lands contiguous to their existing claims, or stressing their status as first-order rulers rather than just landowners. The plan, clearly, is to reserve the central part of the island for Cordelia, with a view to his own retirement in this region. As such, given this is already so predetermined, the key motive behind his love test may be to decide whom Cordelia marries, presumably following her father's wishes if her love for him is true and absolute.

King Lear certainly acts in a foolish way, mistaking his youngest daughter's honesty for a lack of duty, and her sisters' fawning behaviour for love. But was he foolish in proposing the division in the first place? Lear had worked out a plan, deciding ahead of time on the division of lands, which are given to each daughter in turn before he has heard the next daughter's answer. Lear is therefore not making a comparison at all. It is a strategic decision, used to deal with the potential rivalry between the two dukes. As he explicitly says in the Folio text, his intention is "that future strife / May be prevented now."[48] Harry Jaffa has made a forceful argument that in this Lear was therefore acting in a good, strategic way, trying to deal with the question of succession: "It was an action predestined by the very means required to bring unity to the kingdom. Lear, it appears, delayed the division as long as possible, but he could not put it off indefinitely, any more than he could put off indefinitely his own demise."[49] For Jaffa, Lear's plan was of "living on as king with Cordelia, with Albany and Cornwall acting as his deputies in regions which he could not control without their loyalty anyway." Thus he questions: "Does it seem that Lear was giving up anything that he could in any case have kept to himself much longer?"[50]

But Cordelia upsets this carefully calibrated plan through her refusal to act in the way her father expects. Cordelia will receive nothing, losing her share of the kingdom entirely. It is striking that Lear gives her share to the two dukes, not their wives: "Cornwall and Albany, / With my two daughters' dowers digest the third."[51] The Duke of Kent attempts to intervene, but his attempts at mediation are immediately rebuffed. His injunction to Lear is to "Reserve thy state,"[52] which William Dodd interprets as the only place where he "draws attention to the division of the kingdom or abdication as a *political* error" rather than a personal one. This admonition is, unfortunately, unheeded, and Kent goes into exile, returning disguised to serve Lear.[53] For the Duke of Burgundy, the absence of a dowry is a serious issue. He pleads with the King to "Give but that portion which yourself proposed, / And here I take Cordelia by the hand, / Duchess of Burgundy."[54] This is yet another indication that the proposed division had already been agreed and was known. Because Lear refuses to change his amended decision Burgundy tells Cordelia, "I am sorry, then, you have so lost a father / That you must lose a husband."[55] The King of France appears not to be so concerned. He had earlier told Burgundy, "She is herself a dowry," and now declares, "Thy dowerless daughter, King, thrown to my chance, / Is queen of us, of ours, and our fair France."[56] Lear seems uninterested, allowing France to take her, dismissing both of them, and yet still welcoming Burgundy:

> Thou hast her France; let her be thine, for we
> Have no such daughter, nor shall ever see
> That face or hers again. Therefore, be gone,
> Without our grace, our love, our benison.
> Come, noble Burgundy.[57]

Cordelia leaves the stage and only returns much later in the play when she leads forces in an invasion of Britain.

LAND

The use of the word "land" in the play is worth more attention. As Francis Barker notes, "In one sense, *Lear* is about nothing but land. . . . Yet at best it comes into focus and disappears again."[58] Other writers have pushed the case further: there is a literature on the relation between Shakespeare, land politics, and apartheid South Africa, for example.[59] Akira Kurosawa's film *Ran*, which transplants the play to medieval Japan, and gives the king sons instead of daughters, is also about land.[60] Nicholas Visser has underscored that all the principal male characters in the play, with one major exception, are named after places: Gloucester, Kent, Cornwall, Albany, Burgundy, and France.[61] The major exception, of course, is Lear, who is never called Britain, but of course surrenders his relation to land at the very beginning of the play. That said, he does have an association with a place because Leicester, in Anglo-Saxon Ligoraceastre, is named after him: literally, "Lear's city or fortress." This link between people and place returns in many other instances, such as in *Hamlet*, and more generally, as well as the obvious link between dukes and earls and their lands, "the territories of counts are *counties*, of marquesses are *marches*, of barons are *baronies*." Even the disinherited Hamlet is the name of a person and a small village.[62]

There are three groupings of the use of the word "land" in the play. One concerns the subplot regarding Gloucester and the inheritance of his bastard son Edmund or his legitimate child Edgar.[63] Edgar and Edmund are of course the other exceptions to the point made about names and places, but they are yet to inherit, and the subplot is precisely about this point. As Edmund says, "Legitimate Edgar, I must have your land"; and "Let me, if not by birth, have lands by wit."[64] As foolish as Lear in falling for Edmund's trickery, Gloucester declares that Edgar will not get away—"Not in this land shall he remain uncaught."[65] He listens as Edmund falsely claims that Edgar had described him as "thou unpossessing bastard,"[66] and then declares:

> Hark, the Duke's trumpets; I know not why he comes.
> All ports I'll bar, the villain shall not 'scape;
> The Duke must grant me that. Besides, his picture
> I will send far and near, that all the kingdom
> May have due note of him; and of my land,
> Loyal and natural boy, I'll work the means
> To make thee capable.[67]

The land politics is evident here even when the word is not used: "unpossessing" means unable to inherit or own property. Equally, when Gloucester describes Edmund as his "natural boy," he is stressing both the genetic link and the legal disconnect, with "natural" having a sense at the time of being illegitimate. Nonetheless, Gloucester promises to find a way to make him "capable," legally able to inherit.[68] The "ports" are town gates as well as seaports.[69]

The second use of the word "land" concerns the Fool's mocking of the King for giving away his possessions. Recall the first scene where Cordelia's response when asked to profess her love is "Nothing, my lord," to which Lear replies, "nothing will come of nothing." Now Lear suggests "nothing can be made out of nothing," recalling this reply.[70] The Fool responds by asking Kent, "Prithee, tell him, so much the rent of his / land comes to."[71] As Paul Kahn notes, "Lear had hoped to give away his land to his daughters and to live off, or with, them. But what he receives from them is nothing."[72] The Fool then asks Lear who was the "lord that counselled thee / To give away thy land," and describes him as "the sweet and bitter fool."[73] Lear asks, "thou call me fool, boy?" to which the Fool replies, "All thy other titles thou hast given away; that / thou wast born with."[74] (This last exchange does not appear in the Folio.) The Fool talks of an egg with the middle removed to show the two crowns that remain—the middle that Lear planned for Cordelia and then swiftly gave away to her sisters.[75]

The third instance largely pertains only to the Quarto. In this version of the text the King of France and Cordelia invade Britain, and Shakespeare exploits two senses of the word "land." On the one hand, it applies to Britain itself: "France spreads his banners in our noiseless land,"[76] and "France invades our land";[77] and on the other, to the means of invasion itself: "the army of France is landed" and "the army that was landed."[78] The invasion lands on the land it seeks to control. The invasion is a crucial point in both texts, but for different reasons. The Folio is much more ambiguous than the Quarto, because it does not explicitly state who is making the invasion, though it is clearly led by Cordelia. The invasion appears to be a response to the ill use of Lear by his elder daughters, but the time span means the invasion must have been planned before any news would have reached France. Was it then an invasion to take Cordelia's share by force?[79] Cordelia herself claims that the invasion was only in her father's interest, that France had supported her, and that it was not concerned with any failure to profit herself:

> O dear father,
> It is thy business that I go about;
> Therefore great France
> My mourning and important tears hath pitied.
> No blown ambition doth our arms incite,
> But love, dear love, and our aged father's right:
> Soon may I hear and see him![80]

Kenneth Muir has convincingly argued that Shakespeare keeps this ambiguous as "the result of cunning rather than carelessness," because resolving the ambiguity could have been avoided only by "slowing up the action."[81] He notes that only two or three days have elapsed from the initial division of the kingdom in any case, far too few for an invasion to have been planned for either reason.[82] In this regard, W. W. Greg's interpretation, which suggests that there are effectively two parallel timescales running in the play, is more compelling:[83] there is the day-to-day of dramatic action, and the background events that shape the overall contours of the plot. Where Greg is less convincing is in saying that because of the time frame the invasion preparations must have begun immediately after France and Cordelia left Lear at the end of the opening scene.[84] Even though the action of the invasion does not take place until the fourth act, the Dukes of Gloucester and Cornwall receive letters alerting them to the news in the third act.[85] Greg concludes that "there can be no doubt, therefore, that the French army had actually landed before Lear had any quarrel with his [elder] daughters."[86] In order to account for this, and Cordelia's speech on her motives, Greg concocts a complicated narrative where France fell out with Lear in an unscripted meeting, planned the invasion to avenge this wrong and the loss of Cordelia's portion, and then, learning of the ill-treatment of Lear by her sisters, Cordelia is able to effect the invasion with "a change in its purpose,"[87] which is then used to explain why France returns home, leaving Cordelia as the head of the invasion force.[88] Yet if we follow Greg's own indication of the parallel timescales, and Muir's suggestion of a deliberate ambiguity, no such problems arise. Cordelia, in truth, remains a much more sympathetic character if the more literal interpretation is followed. Indeed, in the reconciliation scene Lear asks, "Am I in France?" to which Kent replies, "In your own kingdom, Sir," showing both respect and that the kingdom is preserved, rather than incorporated.[89]

The Folio seems more carefully considered, as the text largely strips out the France and Britain conflict in order to emphasize the civil dissent. There may have been important reasons for this—at the time of writing and presentation the idea of a French invasion, even if defeated, was politically very suspect. Internal strife would have resonated much more effectively with his audience, and indeed was the core of the history plays. There are some quite substantive changes. In Act IV, an entire scene from the Quarto is missing from the Folio version.[90] In Act V, scene ii, the Quarto's stage direction is for the "powers of France" to enter, which becomes "drum and colours . . . and soldiers" in the Folio; and most of the references to France noted above are lost. In one key scene Kent informs on what is happening, but gives radically different speeches in the two versions. In the Quarto he suggests:

> . . . There is division,
> Although as yet the face of it is covered
> With mutual cunning, 'twixt Albany and Cornwall;
> But, true it is, from France there comes a power
> Into this scattered kingdom; who already,
> Wise in our negligence, have secret feet
> In some of our best ports and are to point
> To show their open banner.

Yet in the Folio, he declares

> . . . There is division,
> Although as yet the face of it is covered
> With mutual cunning, 'twixt Albany and Cornwall;
> Who have—as who have not, that their great stars
> Throned and set high, servants, who seem no less,
> Which are to France the Spies and Speculations
> Intelligent of our state.[91]

These give quite opposite stories. The first is a report of an invasion from France; the second is a report of civil dissension, of which France is aware.[92] Muir speaks for the older textual position when he says in general terms that "the modern editor will, of course, restore these omitted lines, whether his text is based mainly on the Quarto or on the Folio."[93] Indeed, he does just that here in his edition, gluing the two different speeches together as if they are part of a whole. Although Michael Taylor is overstating the case in claiming that "the Quarto and Folio treat the nationality of Cordelia's army in consistently different ways," he is on stronger ground when he suggests that "any conflation of the two [texts] produces incoherence."[94]

The story of Lear also appears in Geoffrey of Monmouth's *The History of the Kings of Britain*,[95] but there the King simply says his intent is to divide the kingdom.[96] There are a range of other sources Shakespeare used in the composition of his own version.[97] In 1565 two politicians in the early years of the reign of Queen Elizabeth, Thomas Norton and Thomas Sackville, published a play entitled *Gorboduc*, or, *Ferrex and Porrex*.[98] It had first been performed three years earlier, perhaps to an audience including the queen. Norton apparently wrote the first three acts; Sackville the last two. It has been described as "the work that initiates English tragedy" by Franco Moretti,[99] who suggests that "half a century later . . . Shakespeare rewrites *Gorboduc* and calls it *King Lear*."[100] The story certainly has some similarity. The King abdicates and divides his kingdom between two sons, Ferrex and Porrex. Porrex kills Ferrex and takes the kingdom, but is killed by his mother, Queen Videna, in revenge. Then the people murder the King and Queen in rebellion. The nobles put down the rebellion, but there is no legitimate heir, and the Duke of Albany seeks to control things himself:

> Fergus, the mighty duke of Albany,
> Is now in arms and lodgeth in the field
> With twenty thousand men. Hither he bends
> His speedy march and minds to invade the crown.
> Daily he gathereth strength and spreads abroad
> That to his realm no certain heir remains,
> That Britain land is left without a guide,
> That he the sceptre seeks for nothing else
> But to preserve the people and the land,
> Which now remain as ship without a stern.[101]

The play ends with a sense of the disorder to come, with the plot synopsis declaring "the land for a long time almost desolate and miserably wasted."[102] This too is a play about division of land, yet it is important to note that this was not an uncommon issue, and that *Gorboduc* is based on the story of Gorbodugo in Geoffrey of Monmouth's *Histories of the Kings of Britain*, rather than his account of Leir.[103] The influence is more in its importance as a literary text, rather than its content.

In his evaluation of the competing virtues of *King Lear* and *Hamlet* as Shakespeare's finest moment, R. A. Foakes suggests that "for the immediate future *King Lear* will continue to be regarded as the central achievement of Shakespeare, if only because it speaks more largely than the other tragedies to the anxieties and problems of the modern world."[104] Yet, like *Gorboduc*, it is also a text that comes from a particular context. The political resonances from Shakespeare's own time are important: Queen Elizabeth had died in March 1603, succeeded by James VI of Scotland as England's James I. *King Lear* has been dated to between March 1603 and 1606.[105] Elizabeth's reign had been characterized by familial relations, among them her role as mother to the people.[106] Dodd has convincingly argued that the opening line of the play—"I thought that the King had more affected the Duke of Albany than Cornwall"—shows the "personalized politics of Elizabethan absolutism," as does the "love test" moments later.[107] James saw himself as father of the kingdom but was equally concerned with the unification of England and Scotland: Lear functions both as his parallel and as a contrast.[108] Martin Orkin has even suggested that some of the passages cut from the Folio version were a result of censorship for too close a parallel with issues at the court of James: particularly the exchange between the Fool and Lear on the giving away of land and titles, especially given the reference a moment later to "monopoly."[109] Orkin equally suggests that the challenge against enclosures and the dispossession from the land this occasioned also frame the context, with the disinherited Edgar, in the guise of Poor Tom, as the key figure representing the landless in the play.[110] The first Enclosure Act had been in 1604, for land at Radipole in Dorset, though the practice preceded this legal expression.[111] It is a theme that recurs in several of Shakespeare's plays.

EARTH

The final theme to be discussed in *King Lear* concerns the word "earth." Apart from Gloucester's throwaway exclamation "Heaven and earth!" when deceived by Edmund concerning Edgar (which only appears in the Quarto),[112] the word first appears in the play when Lear threatens Goneril and Regan early in their betrayal:

> . . . No, you unnatural hags,
> I will have such revenge on you both
> That all the world shall—I will do such things—
> What they are yet I know not, but they shall be
> The terrors of the earth![113]

The word then appears four more times at important moments. First, the Gentleman describes Lear's descent into madness in elemental terms:

> Contending with the fretful elements;
> Bids the winds blow the earth into the sea,
> Or swell the curled water 'bove the main,
> That things might change or cease.[114]

The second use of the word also relates to Lear's madness, when Edgar as Poor Tom describes how "the foul fiend Flibbertigibbet . . . hurts the poor creature of earth,"[115] thus tying himself to this element in opposition to the demons taunting him. But it is the third and fourth instances that are perhaps the most interesting. They both concern Cordelia. She first declares her own good intentions, speaking of her search for Lear and his redemption, and calling for "All blest secrets, / All you unpublished virtues of the earth, / Spring with my tears!"[116] Her antidote to Lear's madness, provoked by the "terrors of the earth" he had wished on her sisters is her love, proven and redeemed at the end of the play. But it has a tragic ending: Cordelia dies, with Lear declaring that "she's dead as earth."[117] Earth, then, is a more figurative term in *King Lear* than the concreteness of land politics. Perhaps an even more powerful use of "earth" comes in *Richard III*, where Lady Anne threatens Richard: "O earth! Which this blood drink'st, revenge his death. . . . Or earth gape open wide and eat him quick."[118] This sense returns later in the play: "Earth gapes, hell burns, fiends roar, saints pray / To have him suddenly conveyed from hence."[119]

King Lear is therefore one of Shakespeare's most geographically interesting plays.[120] As Henry Turner has suggested, despite the lack of explicit references to the staging, "the sheer scope of its action, with its wanderings, displacements, and geopolitical subplot, ensures that the stage's spatial potential remains fully felt throughout."[121] James Shapiro has noted that this is "the first and last time that Shakespeare ever included a parallel plot or subplot in one of his tragedies."[122] Intriguingly, as well as the parallels around land and succession, the key phrase "Nothing, my lord" appears in both plots—said by Cordelia of course, in the opening scene, but also by Edmund on hiding the letter from his father.[123] Frederick Flahiff additionally suggests that "a genealogical table is a kind of map which locates its subject in time rather than space,"[124] which given the importance of lineage and inheritance underlines the cartography of the play. The economy of words generally, and concerning space and geography specifically, is an important element of Shakespeare's work. The word "earth" appears just in these five or six instances; "land" limited times in the three (Quarto) or two (Folio) closely linked senses; "territory" but once, and only in the Folio.

Why might "territory" appear so infrequently? This is, in large part, because of the uncommon nature of the word at the time, something that is perhaps not sufficiently recognized. As I have shown in detail elsewhere, the word "territory" appears relatively late in our conceptual vocabulary to think about the relation between place and power.[125] The word does not appear, for example, in the King James Bible, or its earlier translation by William Tyndale. Even its Latin etymological root, *territorium*, is an extremely rare word in the classical language, and does not have the same sense as the modern word, meaning the lands surrounding a town. This was preserved in Cawdrey's 1604 dictionary, quoted in the introduction. In that sense, the town had territory, but was not within it. Only in the late Middle Ages did *territorium* take on something like the modern sense, being first linked to the exercise of jurisdiction by Italian lawyers, and in the seventeenth century being explicitly thought in relation to sovereignty. Such territories, of course, would include towns within their limits. For Emily Leider, the word's use in *King Lear* is an intrusion, and that along with some other words in the opening scene suggests a foreign influence replaced later in the play by words of "native origin."[126] Two of those native words are, of course, "land" and "earth." Both can be traced back to Anglo-Saxon, and play important roles in the classic work of that language, the anonymous *Beowulf* poem.[127]

Nonetheless, the use of the word "territory" in *King Lear* is important, especially when viewed within the wider land politics of the play. The kingdom is viewed as a territory, which can be divided between rulers, bequeathed, and distributed. It is doubtless significant that Shakespeare portrays King Lear as dividing the British Isles into three, just as King James is uniting it through his accession to the throne of England while already king of Scotland. The theme of struggle over land runs throughout both the main plot and subplot of *King Lear*, and the play has been read as a collapse of the feudal social order, and the birth of private property.[128] Yet it is perhaps more compelling to see the play as actually showing a collapse or fall back into feudalism, where a unified state fragments into feuding aristocracy, absolutism into fractured power.[129]

While some aspects of the play's land and earth politics are historically specific, more generally there are themes that hint at a wider history. While *King Lear* may have been written in the early seventeenth century, it is a play that speaks of a pre-Roman Britain, a mythical past that was being reappropriated at the time of writing. Such a protonationalist theme anticipates some of the issues explored in *Cymbeline* (see the coda). Yet in *King Lear* Shakespeare also shows how territory functions as a crucial modern category. The kingdom is strategically important, both between the dukes and in relation to France. There are economic questions in relation to the inheritance and value of land as property; legal issues around succession and legitimacy. The spatial, calculative, and economic language displayed in Goneril's and Regan's opening speeches indicate the accounting, surveying, and cartographic techniques being developed at the time. A hint of the geophysical underpinnings of territory, the very earth, can be found here, and perhaps might be developed in terms of the elemental. Many of Shakespeare's other plays help to examine those multiple aspects of territory. This is the case often, and even sometimes especially, when the mere *words* "territory" and "territories" are absent. *Hamlet* and *Macbeth*, for example, are stories set within a wider frame of territorial conflict, set in vulnerable kingdoms with powerful neighbors. Reading these plays in the light of this question is revealing for the wider situation of their dramatic narratives. The more thorough treatment of the thematic of territory, as concept and practice, will begin with these two plays.

CHAPTER TWO

Vulnerable Territories: Regional Geopolitics in *Hamlet* and *Macbeth*

I. *HAMLET*

Both *Hamlet* and *Macbeth* concern the court politics of a usurping monarch, the consequences of the usurper's actions, and the wish for revenge. In both cases the king is murdered by the one who wishes to claim his throne (King Hamlet by his brother Claudius; King Duncan by Macbeth), and the dead man's son and heir (Hamlet and Malcolm) seek vengeance. It might therefore appear that the threats to the kingdom come from within, but both plays are also structured by external threats to the integrity of the state. In *Hamlet* the threats are Denmark's tensions with Norway, the relation between both countries and Poland, and, more peripherally, Denmark's links with England and France. In *Macbeth* it is Scotland's relations to Norway, England, and Ireland. These wider stories are often, especially in *Hamlet*, marginalized in readings and productions. It is tempting, of course, to concentrate on the internal politics of the Elsinore court, and to focus on the familial relations. But Shakespeare provided a wider framing, and a number of plot elements become clearer if this is taken into account. This chapter therefore reads the two plays through their wider, geopolitical framing.

G. R. Hibbard has suggested that *Hamlet* is distinct from other tragedies, pointing to *Othello* and *King Lear*, but we might also add *Macbeth*, which operate in a "self-contained world . . . where practically every character introduced is affected in some way by the events in which he is caught up." *Hamlet*, in distinction, is "a very open world. Figures come into it unheralded and unexpectedly, and make their exits from it unobtrusively and often unnoticed. They touch the action at some particular point, but are themselves untouched by it."[1] While this open spatiality is a good way of recognizing the wider regional setting of the play's action, it is worth noticing that the timescale of the play can be unclear. The opening part of the play has events that happen over a day or so, as does the end, while the middle of the play clearly compresses a much longer period of time.[2]

Beyond the timescale of the action on stage, the play additionally outlines an extensive backstory in the opening scene—not just the death of King Hamlet, and the quick marriage of his widow, Gertrude, to his brother; but also the wider conflict between Denmark and Norway a generation before. The antagonism between these two countries led to a battle in which King Hamlet killed King Fortinbras. Both Kings now dead, their namesake sons seek revenge: Hamlet over his uncle Claudius; Fortinbras over Denmark as a whole. While Fortinbras is a minor figure in the dramatic action of the play—mentioned a few times, and appearing only briefly twice—he is the parallel and contrasting figure to Hamlet. Other tragedies are more sparing with their backstory—we never know how Lear gained his kingdom, and only have hints of Macbeth and Othello's past.

As with so many of his plays, Shakespeare took the basic outline of the story from an earlier source. Saxo Grammaticus's *Historiae Danicae* and François de Belleforest's *Les Histoires Tragiques*, volume 5, both have versions of the story—the latter a translation and embellishment of the former; itself translated into English in *The Hystorie of Hamlet* in 1608.[3] In Saxo Denmark's King Horwendil kills King Koll of Norway, but is murdered by his brother Feng. Horwendil's son Amlethus seeks revenge. Belleforest similarly has a fight between King Horvvendille and Norway's King Collere, King Fengon as the uncle, and the son Amleth. Shakespeare probably read Belleforest, not Saxo.[4] As well as these extant sources, there is also a supposed play dating back to the 1580s, generally known as the *Ur-Hamlet*, but now considered lost. There are references to stage performances in 1587.[5] While most accounts suggest the hypothetical play was an intermediate source for Shakespeare, Eric Sams and Terri Bourus have suggested that it was actually Shakespeare's own first draft, and that the First Quarto is a version of this text.[6]

This raises a crucial textual question: there are three extant versions of this play. The First Quarto edition from 1603 is usually dismissed as a "bad Quarto," perhaps reconstructed by one of the actors involved, with missing parts, misremembered lines, and perhaps some linking passages not from Shakespeare's hand. It is about 2,100 lines, and it may tell us something of how the text was performed, even if it has been printed in a corrupted form. It may be an earlier version of the text, and this is not incompatible with the idea that it is still a corrupted version. The Second Quarto followed it into print quickly, with surviving examples printed in 1604 and 1605, and is a much longer text, almost twice the First Quarto at 3,800 lines. It was advertised as "newly imprinted and enlarged to almost as much again as it was, according to the true and perfect copy."[7] The First Folio printed a still different text, frequently similar to the Second Quarto, but with hundreds of small changes and including some 80 lines not in that text, while lacking about 200 that were.[8] Because lineation can differ, Ann Thompson and Neil Taylor suggest that word counts are more useful as a comparison: the First Quarto has 15,983 words; the Second Quarto 28,628; the Folio 27,602.[9] While some modern editions of the play begin with the Second Quarto and some with the Folio, most conflate the sources in some manner to provide a composite text, occasionally relegating variant text to appendices or marking up the text with brackets. Most stage productions tend to use a conflated, but cut, text, though there are occasional performances of the First Quarto alone. In his pioneering work on the text, John Dover Wilson notes that he was "convinced . . . that the Second Quarto was printed direct from Shakespeare's manuscript and that the Folio *Hamlet* was a text of quite inferior quality."[10] Nonetheless, he recognizes that there are several long and multiple short omissions in the Second Quarto, which must be filled in from the Folio, which must have derived by an independent route from Shakespeare's manuscript.[11] Philip Edwards suggests that some of the additional passages in the Second Quarto may be because the transcriber misread text marked for deletion.[12] While the Arden Second Series provided a composite text;[13] the Arden Third Series uses the Second Quarto as its basis, with a companion volume that includes the other two texts. I have followed the Second Quarto for most of the following discussion, though I have compared all passages to other editions.[14] These textual issues will reoccur at several crucial moments in the analysis of the play.

Norway and Poland

The play begins with the members of the night watch encountering each other at the change of guard. The cry goes out: "Who is there," to which Horatio replies, "Friends to this ground," and Marcellus adds, "And liegeman to the Dane."[15] Horatio seems to be implying a higher social status—the equality of "friend"—while Marcellus declares a relation of social deference as a "liegeman." The "ground" is significant in the play—both the site of Denmark and the space of the stage—and it reoccurs twice more in the play in crucial scenes: the "little patch of ground" over which Norway and Poland will fight; and the ground of Yorick's grave.[16] Both passages are discussed in more detail below. This relation of a social bond to a place is significant: both are identifying themselves as those who have sworn to the land of Denmark, that is, the country or territory, and its King. It is important to recognize that the term Denmark functions as a synecdoche, both as a place and as a person in the play.[17] The dead King is sometimes referred to as Denmark, just as he is in this opening scene when the watch encounter a ghostly figure that appears on the battlements. Horatio addresses it:

What art thou that usurp'st this time of night
Together with that fair and warlike form
In which the majesty of buried Denmark
Did sometimes march?[18]

Like so much else in the play, the text in the First Quarto is distinct:

What art thou that thus usurps the state
In which the majesty of buried Denmark
Did sometimes walk?[19]

Whether usurping the night or the state, this figure looks like the dead and buried King of Denmark, King Hamlet. Barnardo had earlier described the ghost as "in the same figure like the King that's dead."[20] As they repeatedly stress, it looks "like" the King.[21] It has the same shape and appearance as King Hamlet. The idea that it might actually *be* the King seems obscured by the idea that something has usurped the body, an evil spirit perhaps. Hibbard suggests Horatio takes an "orthodox Protestant position about ghosts" here.[22] Horatio goes on to note further the visible resemblance to the King in his glory days:

Such was the very armour he had on
When he the ambitious Norway combated.
So frowned he once, when in an angry parle
He smote the sledded Polacks on the ice.[23]

There are some complexities in these short lines. Norway similarly refers to the King as much as the country, so there is a confrontation between King Hamlet and the King of Norway. The first two lines describe that conflict, but do the next two lines continue that description, or refer to another episode? The construction "when . . . once," as opposed to "when . . . then," suggests two separate moments, but it is unclear. This lack of differentiation is enhanced when we read the description of the second, or second part of the, encounter. This is because "sledded Polacks" is an editorial emendation, and the phrase that reads "sleaded pollax" in both First and Second Quarto, or "sledded Pollax" in the Folio, is ambiguous.

A "poleaxe" or "pole-axe" is a weapon, a long axe or halberd, and the spellings Polax and Pollax are used for this implement in *Love's Labour's Lost* in the Quarto and Folio texts respectively—"Your lion, that holds his pole-axe sitting on a close-stool."[24] Such a weapon would end in a sharp point, like a pike, but have an axe at the end as an additional means of attack. To read the text in such a way leaves open the question of what "sledded" or "sleaded" would mean, of which the most plausible explanation seems to be leaded or studded, as in a sledgehammer.[25] Many such weapons did indeed have a hammer or studded weight on the other side of the head to counterbalance the axe. Did King Hamlet hold such a weapon, striking the ice with it during a confrontation? Alternatively, Kathleen Irace suggests that this may mean the person carrying the weapon, which would suggest that King Hamlet struck one of the King of Norway's guard.[26]

Alternatively, "sledded Pollax" can be taken to mean Poles "using sleds or sledges" on the ice.[27] This is the reading preferred by the editors of the Arden Shakespeare, in all three texts (First and Second Quartos and First Folio), adding the initial capitalization from the Folio to all versions. An inhabitant of Poland is, at a later stage in the text, described as a Polack as well as a Pole,[28] and the Arden editors note that the word Polack is not in this sense derogatory, even if today, especially in North America, it does have that connotation.[29] It is worth noting that where the First Quarto refers to people from that place, which it does in two instances, the spelling is consistently Polacke; the Second Quarto has Pollacke three times, and Pollack once; and the Folio Poleak twice and Polake once. Other instances do not match passages in the other texts. Only for the "sleaded pollax" or "sledded Pollax" do any of the texts spell the word with a final *x*.

(Incidentally, Ophelia's father, Polonius, has a name that in Latin means "of Poland"—a peculiar choice;[30] though in the First Quarto he is named Corambis, perhaps because of performances at universities, and the risk of

TABLE 1. Incidence of spellings

Act/scene/line	Q1	Q2	Folio
I.i.62	pollax	pollax	Pollax [sledded]
II.ii.63	Polacke	Pollacke	Poleak
II.ii.75	Polacke	Pollacke	Poleak
IV.iv.22	—	Pollacke	—
V.ii.360	[alt. text]	Pollack	Polake

offending someone with a similar name, though Philip Edwards wonders if "someone thought it was odd to suggest that the Danish counsellor was a Pole."[31])

Different editions interpret the term in different ways. As noted, the Arden Third Series renders it as "sledded Polacks." T. J. B. Spencer's Penguin edition prefers the weapon;[32] but Irace the people,[33] and Hibbard argues that "'He smote *the* sledded pole-axe on the ice' is un-Shakespearean and un-English. Normal usage demands *his*. . . . The use of the definite article in 'the sleaded pollax' makes it clear that here we have the direct object of 'smote'—'the Polacks.' And since those smitten are the Polacks, it follows that 'sledded' means 'equipped with sleds' or 'who travelled in sleds.'"[34] Of course, for an audience hearing the play, rather than reading it on the page, both senses would be at play at once.

Despite the spelling issues, the Polish explanation seems best: King Hamlet's military exploits are being outlined. First the combat with Norway; second the "angry parle"—turning the idea of a parley or negotiation into a combat—against the Poles. Given that Norway and Poland are later established as the major regional powers, there is a logic in the setting out of a three-way rivalry between them and Denmark at this early point. Yet while we hear no more of that possible, particular encounter, the combat with Norway is outlined in some detail, and has a particular bearing on the construction of the play as a whole. Horatio says that the ghost, "in the gross and scope of mine opinion . . . bodes some strange eruption to our state."[35] Marcellus asks what the purpose of military preparations is. As well as the watch of which Horatio and his colleagues are part, brass cannon are being prepared and forced labor is being used in the shipyards.[36] Horatio is asked if he can explain the reason, and he replies:

> That can I.
> At least the whisper goes so. Our last King,
> Whose image even but now appeared to us,
> Was as you know by Fortinbras of Norway—
> Thereto pricked on by a most emulate pride—
> Dared to the combat, in which our valiant Hamlet
> (For so this side of our known world esteemed him)
> Did slay this Fortinbras, who by a sealed compact
> Well ratified by law and heraldry
> Did forfeit with his life all these his lands
> Which he stood seized of to the conqueror;
> Against the which a moiety competent

Was gaged by our King, which had return
To the inheritance of Fortinbras
Had he been vanquisher, as by the same co-mart
And carriage of the article design
His fell to Hamlet. Now, sir, young Fortinbras,
Of unimproved mettle, hot and full,
Hath in the skirts of Norway here and there
Sharked up a list of lawless resolutes
For food and diet to some enterprise
That hath a stomach in't, which is no other,
As it doth well appear unto the state,
But to recover of us by strong hand
And terms compulsatory those foresaid lands
So by his father lost. And this, I take it,
Is the main motive of our preparations,
The source of this our watch, and the chief head
Of this post-haste and rummage in the land.[37]

Though a long speech, this is an extremely compressed narrative. Horatio knows the story, but acknowledges it as a "whisper," perhaps a rumor or a secret. Hamlet—the title character's father, the murdered king—slayed Fortinbras of Norway after being provoked into combat through jealous rivalry. Hamlet—the first time the figure is named in the play—was known across the Western Hemisphere. Quite how Horatio knows how King Hamlet was dressed—"the very armour he had on"—thirty years before is unclear.[38] While he is clearly familiar with Danish politics, he is also identified as a fellow student from Wittenberg, who admits, "I saw him once—'a was a goodly king."[39] By his death, Fortinbras forfeited lands to Hamlet—an agreement sworn, legally just, and in accordance with rules of chivalry.[40] These lands of which he was in legal possession passed, not to his son by inheritance, but to King Hamlet by conquest. They were "seized," that is, taken, but there is also a legal sense of possession of land. There is debate as to whether these were his personal lands, or those of the kingdom, but it is clear that it was not the entirety of the kingdom, because the next line suggests that the lands forfeited by Fortinbras had been balanced by a "moiety competent," an equivalent portion of land, "gaged" or wagered by Hamlet.[41] Had Hamlet fallen, these would have passed to Fortinbras; as it was, Fortinbras's land went to Hamlet.

Young Fortinbras, the dead King's son, refuses to accept this state of affairs. We discover later that this combat happened on the very day of Hamlet's birth.[42] Now, Fortinbras, wishing to prove himself, has roused in the borderlands, the outskirts, of Norway a group of men to form an army. Who are these men? Both the First and Second Quartos call them "lawless resolutes," a group without law but resolved to a purpose. The Folio, however, calls them "landless resolutes," suggesting younger sons who did not inherit land like Fortinbras himself. "Landless" is a much less common word in Shakespeare than "lawless," with the only other instance in *King John*, describing the bastard.[43] As Edwards suggests, "The idea here is not of an army of criminals but of disinherited gentry and younger sons who have nothing better to do."[44] Such men are much more likely to form a coherent army than "lawless" renegades. Fortinbras—literally "strong in arm"—threatens with a "strong hand" to "recover" the lands his father lost a generation before. It is for this purpose that Horatio and his colleagues keep watch, because of the threat to the state of Denmark. Fortinbras is the cause of the urgency and disturbance—"post-haste and rummage in the land"—and therefore the reason for the watch and the military preparations.

The "lawless" or "landless resolutes" are interesting in relation to a later episode in the play. When Hamlet and Horatio encounter the gravedigger, he is throwing skulls out of the grave: "That skull had a tongue in it, and could sing once. How the knave jowls it to the ground."[45] One of these skulls, of course, is very famous, but the others are interesting too. Hamlet comments that these might be the skulls of Cain, a politician, a courtier, and then a lawyer. In the First Quarto, he comments:

> Look you, there's another, Horatio! Why may't not be the skull of some lawyer? Methinks he should indict that fellow of an action of battery for knocking him about the pate with's shovel. Now where is your quirks and quillets now, your vouchers and double vouchers, your leases and freehold and tenements? Why, that same box there will scarce hold the conveyance of his land, and must his honour lie there?[46]

The register shifts from the lawyer suing the gravedigger for assaulting his head with a spade to the "legal quibbles or niceties" of arguments to a set of terms relating to legal property over land, with the "conveyance," a document so large it would barely fit in the grave. Perhaps "his honour" is a reference to the mode of addressing a judge as "your honour," though it is sometimes suggested that it might be a mistake for "owner."[47]

In the Second Quarto, the blurring of the lawyer-landowner is less pronounced, as the text is not so compressed:

> There's another! Why, may not that be the skull of a lawyer? Where be his quiddities now—his quillets, his cases, his tenures and his tricks? Why does he suffer this mad knave now to knock him about the sconce with a dirty shovel and will not tell him of his action of battery? Hum! This fellow might be in's time a great buyer of land, with his statutes, his recognizances, his fines, his double vouchers, his recoveries. To have his fine pate full of fine dirt? Will vouchers vouch him no more of his purchases and doubles than the length and breadth of a pair of indentures? The very conveyances of his lands will hardly lie in this box, and must the inheritor himself have no more? Ha?[48]

We move, here, from a lawyer to a lawyer's techniques to a possible assault charge to a landowner. Is this the same person, or perhaps another skull?[49] The "pate" becomes a "sconce," both words being slang for a head; and the pate reappears later as the skull filled with soil in the ground. "Quiddities" are "excessive subtle scholastic arguments concerning the *quidditas* or essence of a thing," and "quillets" both "small pieces of land" and "verbal quibbles."[50] The double meaning of "quillets" thus nicely captures the legal-land issue. "Tenures" are of course legal titles to land or other property, and "tricks" suggests sharp practices, while *Law Tricks* was the title of a 1604 play by John Day.[51] As Thompson and Taylor note, "Hamlet uses legal terminology relating to transactions in land: *statutes* were securities for debts or mortgages, *recognizances* were bonds relating to debts, *fines* and *recoveries* were legal fictions used in the *conveyancing* (the legal transfer of ownership) of land, *vouchers* and *double vouchers* were ways of securing third parties as guarantors, and *recoveries* were suits for obtaining possession."[52] A "pair of indentures," like the "indentures tripartite" of the map scene of *Henry IV, Part 1*, would be sealed by both parties, and perhaps these would be on the same sheet of paper, cut in such a way that the two or three parts, and only they, could be put back together as a means of preventing forgery.[53] This lawyer, though, has not just purchased land for others, but for himself too—perhaps also implied by "his purchases" the line before.[54] As a purchaser he is not, strictly speaking, an "inheritor," and this is why the documents are so extensive as to not fit in his small plot of burial ground.

In the First Folio, the text is further expanded:

> There's another! Why, might not that be the skull of a lawyer? Where be his quiddits now—his quillets? His cases? His tenures and his tricks? Why does he suffer this rude knave now to knock him about the sconce

> with a dirty shovel and will not tell him of his action of battery? Hum! This fellow might be in's time a great buyer of land, with his statutes, his recognizances, his fines, his double vouchers, his recoveries. Is this the fine of his fines, and the recovery of his recoveries—to have his fine pate full of fine dirt? Will his vouchers vouch him no more of his purchases, and double ones too, than the length and breadth of a pair of indentures? The very conveyances of his lands will hardly lie in this box, and must the inheritor himself have no more? Ha?[55]

"Quiddits" may be an error here—it is a word not otherwise used by Shakespeare. The knave is "rude" rather than "mad," though both have the sense of irreverent; and one line here is missing from the Quarto, perhaps because the printer's eye skipped from one instance of "recoveries" to the second. "Fine of his fines" is part of the line lost, a line that works in multiple ways—"fine" as end or conclusion, as legal document, as quality or texture, and as penalty—especially when combined with "fine pate" and "fine dirt," a sophisticated head or brain now in the soil.[56]

As Thompson and Taylor suggest, "Hamlet's interest in property and inheritance in this context seems to relate to his own position as in effect a disinherited son."[57] As Margreta de Grazia has noted, Hamlet's famous comment "My wit's diseased" would have sounded the same as "diseized," "to be illegitimately dispossessed of lands," the mirror image of his father's seizure of those lands a generation before.[58] In general, the disinherited nature of Hamlet's claim is reinforced by that of Fortinbras, though Hamlet has been unjustly robbed of his land, whereas we are told, at least by Danish sources, that Fortinbras's father lost it "by a sealed compact / Well ratified by law and heraldry."[59] As Linda Charnes notes of Hamlet and Fortinbras,

> Both fathers were prepared capriciously to gamble away their sons' patrimonies in a duel, thereby breaking the laws of entail that in Shakespeare's day bound fathers as well as sons. That King Hamlet won the wager would lead, one might think, to this story being trumpeted to general acclaim. Instead, Horatio tells Bernardo how "the whisper goes," suggesting something shameful in the enterprise. After all, the outcome of the bet might have gone the other way.[60]

Entail or "fee tail" is a legal restriction that prevents someone selling property, and therefore preventing the heir from inheriting it.[61] This point is significant: Denmark's problems may precede Claudius. There is also an important quibble about the word "hide" in the grave scene, between the animal hides that make the vellum on which laws were written, and the measure of land known as a hide.[62] A hide was the amount of land able to sustain a family, and varied depending on terrain, perhaps around a hundred acres. It became a standard term for surveys and taxation. The term is perhaps derived from the story of Dido asking for the amount of land that could be enclosed by an oxhide, and cutting this into a fine strip that enclosed enough land to found Carthage.[63]

In the next scene, we discover why Fortinbras has chosen this specific moment. King Claudius notes his brother's death—though of course does not explain it—and that his onetime sister is now his wife. This succession has apparently meant Fortinbras feels the state is vulnerable. Claudius notes:

> Now follows that you know, young Fortinbras,
> Holding a weak supposal of our worth,
> Or thinking by our late dear brother's death
> Our state to be disjoint and out of frame—
> Co-leagued with the dream of his advantage—
> He hath not failed to pester us with message
> Importing the surrender of those lands
> Lost by his father, with all bands of law,
> To our most valiant brother.[64]

This reinforces Horatio's narrative, notably stressing the "bands of law" concerning the lost lands. It adds that young Fortinbras now thinks that Denmark is militarily weak, and so he has made his demands for the return of the lands. Claudius is dismissive of Fortinbras, but wants to make sure, so sends emissaries to Norway:

> So much for him.
> Now for ourself and for this time of meeting,
> Thus much the business is: we have here writ
> To Norway, uncle of young Fortinbras—
> Who, impotent and bedrid, scarcely hears
> Of this his nephew's purpose—to suppress
> His further gait herein, in that the levies,
> The lists and full proportions, are all made
> Out of his subject; and we here dispatch
> You, good Cornelius, and you, Voltemand,
> For bearers of this greeting to old Norway,

Giving to you no further personal power
To business with the King more than the scope
Of these delated articles allow.
Farewell, and let your haste commend your duty.[65]

We discover here that, as in Denmark, the King's brother, rather than his son, has succeeded him to the throne of Norway. In Norway this makes sense—young Fortinbras was merely a child when his father died—though in Denmark it is open to question. The contrast between "old Norway" and Claudius is stark. One is bedridden and impotent; the other makes a great play of his own bedroom potency with his "sometime sister, now our Queen."[66] Claudius wants to take no chances, and to get Norway to rein in his nephew—much as he wishes to control his own. Diplomatic means—rather than any threat of force—are used to try to deflect the Norwegian invasion. The parallels between the generations are also clear. King Hamlet was clearly spurred to action, while his son will hesitate in the play to come. King Fortinbras was a noble opponent, but young Fortinbras operates in a different, and it seems, less honorable way.[67] As Spencer notes, the "personal combat" between the Kings Hamlet and Fortinbras "seems to belong to a different and older world from that of King Claudius, who is a modern politician and works through his ambassadors."[68]

The Ambassadors return in Act II, bearing a message from Norway. Voltemand reports:

Most fair return of greetings and desires.
Upon our first he sent out to suppress
His nephew's levies; which to him appeared
To be a preparation 'gainst the Polack;
But, better looked into, he truly found
It was against your highness; whereat, grieved
That so his sickness, age and impotence
Was falsely borne in hand, sends out arrests
On Fortinbras; which he in brief obeys,
Receives rebuke from Norway, and, in fine,
Makes vow before his uncle never more
To give th'assay of arms against your majesty.
Whereon old Norway, overcome with joy,
Gives him three thousand crowns in annual fee
And his commission to employ those soldiers

So levied (as before) against the Polack,
With an entreaty, herein further shown
That it might please you to give quiet pass
Through your dominions for this enterprise
On such regards of safety and allowance
As therein are set down.[69]

Thus Norway, when first alerted to Claudius's concerns, discovers that the army he thought was headed for Poland was actually a threat to Denmark. He orders Fortinbras to cease, and gets him to swear he will not attack Claudius and his state. Norway rewards Fortinbras with an annual allowance, and sends him indeed to attack the Poles. Voltemand produces a document that asks that Fortinbras be allowed safe passage through Danish lands to make this possible—a much more cautious and deferential attitude. This is a claim Claudius seems pleased to accept:

It likes us well,
And at our more considered time we'll read,
Answer and think upon this business.[70]

We hear no more of Fortinbras until Act IV, when he appears at the head of an army and sends a messenger to the Danish king. In the Second Quarto and Folio he says:

Go, Captain, from me greet the Danish King:
Tell him that, by his licence Fortinbras
Craves [Folio: claims] the conveyance of a promised march
Over his kingdom. You know the rendezvous.
If that his majesty would aught with us,
We shall express our duty in his eye;
And let him know so.[71]

The First Quarto text reads:

Captain, from us go greet the King of Denmark.
Tell him that Fortinbrasse, nephew to old Norway,
Craves a free pass and conduct over his land,
According to the articles agreed on.
You know the rendezvous: go, march away!

The agreement between Denmark and Norway allows for the passage of Norway's forces, though Fortinbras wishes to inform Denmark of this rather than simply do it. In the Second Quarto and Folio he adds that if the King wishes he will meet him face-to-face, "in his eye." There is a politeness in the words, but a veiled threat here: in the Quarto he "craves," rather than, as the Folio has it, "claims," something that has been agreed; and if the King "would aught with us" he will tell him "in his eye." As the Captain leaves, in the First Quarto the scene ends, but in the other texts Fortinbras instructs his army to "Go softly on" (Second Quarto), or "Go safely on" (Folio).[72] Either with respect, or safe in the knowledge that they will not be challenged, Fortinbras and his army exit.

In the Second Quarto, and only in that version of the text, Hamlet encounters this army as he leaves for England. He meets the Captain as he leaves to do Fortinbras's bidding, and asks, "whose powers are these?" Their dialogue establishes that they are Norway's, commanded by Fortinbras, and directed "against some part of Poland."[73] Hamlet then asks, "Goes it against the main of Poland, sir, / Or for some frontier?"[74] There are some complexities here. The "main" is the mainland, or perhaps the coast, but "frontier" is ambiguous. The *OED* suggests that this means "a fortress on the frontier," and gives it as the earliest such use; the Penguin text glosses it as "frontier-fortress."[75] The Arden Third Series, however, rightly notes that the contrast with "main" suggests it carries its more common meaning of "border or extremity."[76] Indeed, the Captain replies:

> Truly to speak, and with no addition,
> We go to gain a little patch of ground
> That hath in it no profit but the name.
> To pay five ducats—five—I would not farm it,
> Nor will it yield to Norway or the Pole
> A ranker rate, should it be sold in fee.[77]

The Captain, possibly a farmer in civilian or an earlier life, though "farm" here may mean rent or lease out, sees the target as little more than a vanity conquest. Neither the King of Norway nor that of Poland would gain much if they were to sell it. It is not clear that the Captain sees all territorial ambition as equally worthless, though his view of the merits of a struggle over this particular piece of land is clear. While Hamlet thinks the "Polack" will not defend it, the Captain notes that it is "already garrisoned"[78] by a defending army, so that both sides clearly have much at stake in its possession. Simon Critchley and Jamieson Webster describe this as "a futile territorial and religious war, between the Protestant Norwegians led by Fortinbras and the Catholic Poles."[79] Hamlet notes the men who will fight over this "plot" of land, "this straw," a trivial or worthless thing that is not even big enough to bury the dead from its contest: "which is not tomb enough and continent / To hide the slain."[80] There is some ambiguity over the number of men—initially "two thousand souls and twenty thousand ducats," and then "twenty thousand men"; the latter figure presumably taken from the figure that was estimated as the cost of the war.[81] The first lines, speaking of the two thousand men, and the cost, may actually have been spoken by the Captain *to* Hamlet; the Captain of course being in a better position to know. Hamlet then aims to repeat, but actually confuses, the figure.[82]

In an intriguing way, then, the play has a set of spatial reductions. We begin with the ground of Denmark won in the duel between King Hamlet and King Fortinbras, invoked in the opening encounter with the ghost. We move to the "little patch of ground" over which Norway and Poland will fight, operating at this crucial point in the play. Toward the end we are led to the ground of a grave, a space that Hamlet first sees emptied, then temporarily occupies when he and Laertes grapple, and then for which both are destined when they die in the duel. Hamlet anticipates this in his remarks on Alexander the Great in this grave scene: "Dost thou think Alexander looked o'this fashion i'th' earth? . . . Alexander died, Alexander was buried, Alexander returned to dust, the dust is earth, of earth we make loam, and why of that loam whereto he was converted might they not stop a beer-barrel?"[83] This reduction of territorial ambition to the earth a corpse will occupy is followed, here, by a similar claim about "Imperious Caesar,"[84] and equally found in the death of Hotspur in *Henry IV, Part 1* (see chapter 7). All these different spaces, of diverse scales, are portrayed in the dramatic action of the text, and of course in its enactment on the ground of the stage.

In his soliloquy at the end of this Second Quarto scene Hamlet describes the experience: "witness this army, of such mass and charge."[85] As he sees Fortinbras acting resolutely, willing to sacrifice his troops for this small piece of land, he implies his own actions will take a sharper turn after this scene. "O, from this time forth / My thoughts be bloody, or be nothing worth!"[86] Even so, as late as the duel with Laertes, Hamlet still seems to surrender to providence.[87] Dover Wilson suggests that the "Fortinbras scene was patently written in order to give occasion to this soliloquy."[88] But this is to see the wider geopolitics as unimportant to the story. For an example of a less crucial element, observe what happens immediately after this, with the hurried treatment of Hamlet's trip to England and his return—a point discussed below.

Fortinbras does not appear in the play again until the final scene, where shortly before his death from Laertes's blows and Claudius's poison, Hamlet hears a commotion without, and exclaims, "What warlike noise is this?"[89] The courtier Osric replies:

Young Fortinbras with conquest comes from Poland
To th'Ambassadors of England gives
This warlike volley.[90]

Fortinbras then enters at the head of an army, after "the Polack wars,"[91] alongside the English ambassadors. They meet at a scene of carnage, with the King, Queen, Hamlet, and Laertes all dead. Almost every other claimant to the Danish throne is deceased. Two geopolitical elements of the play thus collide: Fortinbras's victory in Poland, and the English messengers' news of the death of Rosencrantz and Guildenstern. Hamlet's dying words indicate that the latter will remain expected, while with Fortinbras he seals the succession:

O, I die, Horatio.
The potent poison quite o'ercrows my spirit,
I cannot live to hear the news from England,
But I do prophesy th'election lights
On Fortinbras: he has my dying voice.
So tell him, with th'occurrents, more and less
Which have solicited.—The rest is silence.[92]

Young Hamlet, born on the day old Fortinbras died at the hands of King Hamlet, thus gives his "voice" or vote to young Fortinbras. Hamlet had been named as the successor to Claudius, who had died a few lines before. Hamlet is, strictly speaking, the King at this point, and so it seems he can anoint his own successor. Fortinbras will be crowned King of Denmark unopposed in the wake of the deaths of all the other contenders.[93]

Fortinbras remarks that "Such a sight as this / Becomes the field but here shows much amiss."[94] The Ambassadors are not sure to whom they should deliver their message that the King's commandment is fulfilled, that "Rosencrantz and Guildenstern are dead."[95] Fortinbras assumes authority by military might and presumed right of succession:

For me, with sorrow I embrace my fortune.
I have some rights of memory in this kingdom
Which now to claim my vantage doth invite me.[96]

Fortinbras thus "encircles" the play:[97] his preparations against Denmark had created the need for the watch in the opening scene; his arrival to take the crown ends it. Reading the play through the Fortinbras story, both father and son, opens up the dramatic space of its action. The territorial aspects of the play are an essential setting within which the familial relations of Claudius, Gertrude, and Hamlet play out. As de Grazia puts it, "The play opens with threatened invasion and ends in military occupation"; "the play dramatizes one conflict over land after another."[98] The lands lost by his father are reclaimed; the lands won by Hamlet's father are seized. Denmark will become a territory under Norwegian dominion.

Laertes and France

While Denmark, Norway, and Poland are the three crucial countries in the regional geopolitics of the play, Germany, France, and England also play important roles. At the beginning of the play we hear that Hamlet has returned from university in Wittenberg in Saxony, and so for him the death of his father and the marriage of his mother and uncle are compressed in the narrative even more than their actual temporal proximity: "the funeral baked meats / Did coldly furnish forth the marriage tables."[99] Wittenberg was of course famous as the birthplace of Protestant thought, the place where Martin Luther taught and nailed his theses against the Catholic Church to the cathedral door. The play makes explicit reference to events following this at the Diet of Worms with the joke after Polonius's death: "A certain convocation of politic worms are e'en at him. / Your worm is your only emperor for diet."[100] Hamlet expects to return to Wittenberg, but is not allowed. In contrast, Laertes requests that the King allow him to leave the court in an early scene:

> My dread lord,
> Your leave and favour to return to France;
> From whence though willingly I came to Denmark,
> To show my duty in your coronation,
> Yet now, I must confess, that duty done,
> My thoughts and wishes bend again toward France
> And bow them to your gracious leave and pardon.[101]

On hearing that Laertes's father, Polonius, has given permission—"by laboursome petition, and at last / Upon his will I sealed my hard consent"—the King grants this wish.[102] Before he leaves, Laertes cautions Ophelia on Hamlet's limitations of choice for her:

For on his choice depends
The safety and health of this whole state.
And therefore must his choice be circumscribed
Unto the voice and yielding of that body
Whereof his is the head.[103]

The body is the nation of Denmark, the body politic, and while he will be its head, he must listen to the opinion or approval of the people that constitute it.[104]

Later, though, with Polonius dead, Laertes comes back from France and—at least in the Second Quarto and Folio—foments an uprising against Claudius. The King learns that Laertes has "in secret come from France,"[105] and he appears within moments at the head of a large crowd of followers, leading a rebellion "in a riotous head."[106] Indeed, the crowd are calling him lord, and saying, "Laertes shall be king!"[107] Claudius, however, is able to turn Laertes against Hamlet instead of himself, noting that Hamlet is the cause of his father's death—an enmity reinforced when Laertes meets the now-mad Ophelia. There are now three male characters all seeking to avenge their fathers' deaths, and the terms of their denied inheritance—in the Folio text Hamlet seems to acknowledge this in Laertes: "For by the image of my cause I see / The portraiture of his."[108] Both Laertes and Fortinbras are much more active in seeking this revenge. Laertes is praised by a mysterious Norman horseman, introducing yet another country into the play's geographies, Normandy being distinct from the France to which Laertes had been so drawn.[109]

Hamlet and England

Hamlet's encounter with Fortinbras and the Norwegian army occurs when he is on his way to England. He is heading there on the orders of the King. Initially this is proposed as a cure for his madness, and that he can be useful for Denmark: "he shall with speed to England / For the demand of our neglected tribute."[110] These hints at a relation between England and Denmark, with the former as Denmark's "faithful tributary," are largely unexplained in the text, but likely refer to the Danegeld, a tax paid to prevent Viking invasions.[111] As things develop, after the play within the play, it doubles as a means of getting Hamlet out of the way. Claudius realizes that Hamlet is a threat to him, and wants him dead. He uses Rosencrantz and Guildenstern to achieve this:

I like him not, nor stands it safe with us
To let his madness range. Therefore prepare you.

I your commission will forthwith dispatch,
And he to England shall along with you.
The terms of our estate may not endure
Hazard so near us as doth hourly grow
Out of his brows.[112]

Claudius wants to preserve the "terms of our estate"—his position or status as king and his lands. The Folio is stronger:

The terms of our estate may not endure
Hazard so dangerous as doth hourly grow
Out of his lunacies.[113]

After the killing of Polonius, Hamlet seems to embrace this departure—the repeated cries of "For England!" in Act IV, scene iii—but Claudius has another purpose:

And, England, if my love thou hold'st at aught
As my great power thereof may give thee sense,
Since yet thy cicatrice looks raw and red
After the Danish sword, and thy free awe
Pays homage to us, thou mayst not coldly set
Our sovereign process; which imports at full
By letters congruing to that effect,
The present death of Hamlet. Do it, England!
For like the hectic in my blood he rages
And thou must cure me. Till I know 'tis done,
Howe'er my haps, my joys were ne'er begin.[114]

It appears Denmark has recently inflicted some kind of victory over England, scarring it with the "cicatrice" from the "Danish sword." Claudius believes this will make them serve his "sovereign process" commanding the death of Hamlet, so to cure him of the illness caused by his presence. England, here, as with Denmark and Norway earlier, means both the king and country. Hamlet later discovers this plot when he finds the document, a "grand command," speaking of "A royal knavery, an exact command . . . My head should be struck off."[115] His suspicions were already raised because Rosencrantz and Guildenstern were given the letter, not him.[116] Hamlet switches the instruction for one that commands the killing of Rosencrantz and Guildenstern instead. He is captured by pirates but paying a ransom, negotiates successfully with the pirates, and returns to Denmark.[117] A major offstage narrative is here compressed into only a few lines of the play, because all that seems important is to give Hamlet a chance to return, though presumed dead. Nonetheless, the role of England sets up an even wider frame for the play's action. It might be seen as a convenient external power to move the narrative forward, but as with the role of France, Saxony, and the hint about Normandy it sketches a wider set of European relations aside from the Denmark-Norway-Poland ones that are so crucial to the play.

The Role of Horatio

While frequently dropped in stage and film adaptations, the role of Fortinbras and Norway more generally is a crucial geopolitical and territorial element of the play. The play is often read and performed, especially in Anglo-American contexts, in such a way as to focus on the internal struggle: internal to Hamlet and internal to Denmark.[118] As de Grazia shows, the historical parts of the text were regularly cut in early editions and adaptations, with the Ambassadors, Norway, and Fortinbras only reintroduced in 1898.[119] Even today these elements are regularly cut or marginalized. As Edwards notes, "On the stage today, *Hamlet* has become simply a personal/family tragedy."[120] One director, John Caird, suggests it is "at heart a domestic and philosophical drama, not a history play," and justifies the cuts on this basis. He argues that the more political scenes can be "the dull points" and that they "distract the audience from the metaphorical power of the world that is Denmark." He even argues they may be later additions to the text,[121] though elements of the story are there in all three versions of the text.

One film version that does include this part of the plot is Kenneth Branagh's *Hamlet*. Defending the film against its critics, Charnes suggests this may not help make it "a 'tight' film," but "this view might make sense were it not for one inescapable fact: Fortinbras' invasion is anything but superfluous. Shakespeare begins and ends the play with the problem of Norway's claims; and an Elizabethan audience worried about English succession in the last years of Elizabeth's life would have considered the threat to Denmark as a nation *at least* as important as Hamlet's state of mind."[122] Indeed, it adds to the claustrophobic nature of the majority of the action—the family drama and the personal, philosophical struggle—by having that larger frame. The struggles between King Hamlet, Claudius, and Hamlet—past, present, and putatively future kings—are operating within a single space, while larger geopolitical forces rage around them. Denmark is revealed to be a small pocket of land surrounded by dangerous enemies, reduced to being a passage route from Norway to Poland, the price of its protection from Norway itself. "Denmark's a prison," as Hamlet declares in the Folio text—in a passage missing from the Quarto editions. He agrees that the world as a whole is one "in which there are many confines, wards and dungeons—Denmark being one o'th'worst."[123] At least some of the land of the Danish court was gained by Norway's death, but it was only a "moiety" of his lands and power. Eventually the internal feuding opens up the possibility for young Fortinbras to walk into the court unopposed, and to be elected by Hamlet in his dying words.

Several of these geopolitical speeches are made by Horatio, and many of the others are made to him or at least in his presence. Hamlet himself is sometimes not present when these parts are outlined, notably in the initial background-setting story told by Horatio. Interestingly, Horatio is, alongside Hamlet himself and the Queen, one of the few parts that cannot be doubled.[124] Even though he is absent from the stage in Act II, the second half of Act III and the first half of Act IV, his presence on stage at the same time as others means he could only conceivably double Reynaldo or the Norwegian captain. For Hamlet, Horatio is a trusted friend, a link back to Wittenberg, but also an important member of the night watch and the court. He is the key figure to whom Hamlet expresses his political concerns. As he says to Horatio in an early scene, in a passage only in the Second Quarto, the celebrations of Claudius denigrate Denmark's position in the world: "This heavy-headed revel east and west / Makes us traduced and taxed of other nations."[125] One line in this extended speech nicely captures the geopolitics of the personal—the Danish reputation for drunkenness can be seen as "Oft breaking down the pales and forts of reason."[126] The "pales" are the fortifications or defenses around an enclosed park or other space, while fortresses would be part of that military strategy. But as well as a military threat—the speech is given on the battlements, just before the first appearance of the ghost to Hamlet—it also hints at the fragility of mental states.

Perhaps Horatio's otherwise connected nature is why, rather strangely, Hamlet has to tell him who Laertes is in the grave scene: the two characters had not, until this point, met on stage.[127] Despite some general misgivings about his method, Moretti's network analysis of the plot is important here. Horatio occupies a pivotal role in the connections between characters. Moretti shows that if a network of characters and speeches between them is constructed, Hamlet is of course at the core. But Horatio is not far behind. If Hamlet is removed from the network, then two other clusters of connections are clear, first the court around Claudius and second the other figures in the play. Besides Hamlet, only Horatio keeps them in touch, and without him, the network of "others" also disintegrates.[128] Horatio is the one Danish connection to Fortinbras. As Thompson and Taylor outline, "Moretti characterizes Horatio's network of ambassadors, messengers and sentinels as being rather impersonal, almost bureaucratic: it is, he suggests, as if Horatio heralds the coming of a new political reality—the replacement of the Court by the State."[129]

II. *MACBETH*

Scotland is a country at war, both with itself and with external enemies. Elements within the Scottish nobility have betrayed the king, and Norway has invaded. The key traitors are Macdonald and the Thane of Cawdor. As Bate has noted, because of court performances, Shakespeare had to take care over diplomatic issues. King James's wife was Danish, and the play may have been performed in front of both James and the Danish king. "This may explain why Norway is made Scotland's enemy in the opening battle, when it was Denmark in the *Chronicles* that were Shakespeare's source."[130] As well as the Scandinavian opponents, threats endure from England and Ireland. As Kristen Poole has noted, "*Macbeth* is arguably one of the most spatially oriented plays, as the shifting geographies of the supernatural often bring the characters to an intense self-awareness of their placement in both local and cosmic spatial settings."[131]

Macbeth, Thane of Glamis, is one of those still fighting for the Scottish crown, distinguishing himself in battle. He thus proves his loyalty to Scotland, and his defense against the threats from outside. An injured captain, fresh from the battle, reports to King Duncan and his eldest son, Malcolm, the situation, the "broil," he had witnessed.

> Doubtful it stood,
> As two spent swimmers, that do cling together,
> And choke their art. The merciless Macdonald
> (Worthy to be a rebel, for to that
> The multiplying villainies of nature
> Do swarm upon him) from the Western Isles
> Of kerns and galloglasses is supplied,
> And Fortune, on his damned quarry smiling,
> Showed like a rebel's whore.[132]

The situation was uncertain, and the armies were exhausted in their struggle. Either Macdonald's own villainy was increasing or traitors flocked to join him. The Western Isles are the Hebrides, off the coast of Scotland, perhaps extending to parts of Ireland. Troops are supplied against the King: "kerns" are light foot-soldiers; "galloglasses" axe-armed soldiers, perhaps horsemen.[133] It is notable that at the end of the play Macbeth has to rely on hired kerns to form a part of his own force.[134] The Irish root of the word "kern" also indicates Irish involvement. "Quarry" is often emended to "quarrel," as "quarry" can mean this, in the sense of either an argument or a crossbow bolt, but the word can also have the sense of "a heap of dead men, a pile of dead bodies" (*OED*).[135] The captain continues:

But all's too weak:
For brave Macbeth (well he deserves that name),
Disdaining Fortune, with his brandished steel,
Which smoked with bloody execution,
Like Valour's minion, carved out his passage,
Till he faced the slave,
Which ne'er shoke hands, nor bade farewell to him,
Till he unseamed him from the nave to th'chops,
And fixed his head upon our battlements.[136]

A Norwegian lord is at the head of the army and seeks to push ahead: "surveying vantage, / With furbished arms, and new supplies of men, / Began a fresh assault."[137] But this did not dismay Macbeth and Banquo. The captain ends his narrative, and the Thane of Ross enters, to pick up the story.[138] The King asks where he has come from:

From Fife, great king;
Where the Norwegian banners flout the sky,
And fan our people cold.
Norway himself, with terrible numbers,
Assisted by that most disloyal traitor,
The Thane of Cawdor, began a dismal conflict;
Till that Bellona's bridegroom, lapped in proof,
Confronted him with self-comparisons,
Point against point rebellious, arm 'gainst arm.
Curbing his lavish spirit: and, to conclude,
The victory fell on us . . .
That now Sweno, the Norways' King, craves composition.
Nor would we deign him burial of his men
Till he disbursed at Saint Colme's Inch
Ten thousand dollars to our general use.[139]

King Sweno therefore requests a peace treaty, a "composition," and pays a forfeit to be allowed to bury his dead—a comprehensive defeat. The King rules that Cawdor shall be executed, and Macbeth take his title: "Go pronounce his present death, / And with his former title greet Macbeth . . . What he hath lost, noble Macbeth hath won."[140] Yet there is some uncertainty here. We previously heard of the treachery of Macdonald, allied to Norway. Now we hear of the King of Norway's alliance with another traitor, the Thane of Cawdor. Does "confronted him" mean Cawdor—the most recently named person—or the King of Norway? This "him" is in conflict with someone, and it is not clear whom. Bellona was goddess of war, so her groom is a military figure of some note. Macbeth is the obvious choice, but if the other figure is Cawdor it cannot be Macbeth because he twice has lines that clearly indicate he did not know Cawdor was a traitor or that he was dead. It is not enough to suggest Cawdor's treachery came to light later: "Bellona's bridegroom" directly confronts "him" in this report.[141] It is possible that Macbeth confronts Norway, who is assisted by Cawdor, but that Macbeth did not know this, while the messenger did.

The location of the battles is also important. The first battle is in an uncertain location, but shortly afterward Macbeth and Banquo are nearing the town of Fores, or Forres, when they meet the witches for the first time.[142] There is a large obelisk known as Sweno's stone, which stands near Forres in Scotland.[143] One legend has it that this is the spot where Macbeth met the witches, and that they are imprisoned in the stone, but while this cannot serve as a basis alone, the play's location of the meeting certainly seems to be in that region. The second battle is clearly located in Fife, which is well over 100 miles from Forres. Saint Colme's Inch is Inchcolm, an island in the Firth of Forth, within a day's march of Fife. Ross's stress that he has come from Fife would be redundant if he and the messenger were talking of the same conflict. Muir suggests that "the two battles have been run together in place as well as in time. . . . But not even an audience of Scotsmen would notice the geographical difficulties."[144] Building on this, Brian Morris has suggested that "Scotland is not a presence in *Macbeth* as England is in Shakespeare's Histories. This is evident from the very faint frame of geographical reference, and Shakespeare's obvious dramatic intention not to evoke strongly any sense of place."[145] This claim, of course, runs contrary to Poole's suggestion quoted earlier, though even she notes that things are confused: "The opening of the play sets the stage for a disordered space-time. Time, space, atmospheric conditions and planetary motions become jumbled together."[146] Yet if there were two different battles, Macbeth's presence at the first, and his killing of Macdonald, allow him to be near Forres to meet the witches. The second, with the defeat of Norway and the capture of Cawdor, took place at Fife. "Bellona's bridegroom" is thus someone other than Macbeth, perhaps Macduff.[147]

The King's conferring of the title Thane of Cawdor on Macbeth, in his absence, begins the chain of succession, which we the audience, and Macbeth himself, hear about later, in the very next scene. Macbeth is hailed by the witches as, successively, Thane of Glamis—a position he already has—Thane of Cawdor and King. As Macbeth hears the witches' prophecy the audience knows that the first two positions are already his. Macbeth replies to the witches to say he knows the first: "By Finel's death, I know I am Thane of Glamis."[148] Finel was his father. Macbeth clearly does not know of Cawdor's treachery, otherwise why does he ask the witches: "But how of Cawdor? The Thane of Cawdor lives / A prosperous gentleman."[149] This is repeated when he is named as Cawdor by Ross, acting on the King's instructions; Macbeth replies, "The Thane of Cawdor lives. / Why do you dress me in borrowed robes?"[150] This idea of the clothes not fitting recurs several times in the play. Later in the same scene Banquo notes of Macbeth: "New honours come upon him, / Like our strange garments, cleave not to their mould, / But with the aid of use."[151] Cawdor, though, is a position he has earned. It is his assumption of king that is the true taking of un-entitled title. As he descends into tyranny, the metaphor returns. Cathness suggests: "He cannot buckle his distempered cause / Within the belt of rule."[152] And Angus replies: "Now does he feel his title / Hang loose about him, like a giant's robe / Upon a dwarfish thief."[153]

Macbeth's not knowing of Cawdor's treachery or death is why it makes more sense that the Captain tells of one battle, in which Macbeth defeated Macdonald; and Ross of another, in which a different figure captured Cawdor.[154] Thus Macbeth has dispatched one of the key traitors in Macdonald, and is given the title of another, the Thane of Cawdor. Of course, Macbeth, like his predecessor, will similarly be a "disloyal traitor," though he is currently loyal and has not yet deceived the King. At this point in the narrative, Angus assures him he is the rightful Thane of Cawdor, because of the previous Thane's duplicity:

> Who was the Thane, lives yet;
> But under heavy judgment bears that life
> Which he deserves to lose.
> Whether he was combined with those of Norway,
> Or did line the rebel with hidden help
> And vantage, or that with both he laboured

In his country's wreck, I know not,
But treasons capital, confessed and proved,
Have overthrown him.[155]

The confusion of Ross here is mirrored in the text. Is Cawdor part of the Norwegian force, or in support of the rebel, presumably Macdonald? The "vantage"—support or assistance—here mirrors the "surveying vantage" of the Norwegian lord earlier.[156] Ross explicitly states to Macbeth that the award of Cawdor is "an earnest of a greater honour"—that is, a foretaste or pledge, and presumably Macbeth suspects that being named Duncan's heir is next.[157] Macbeth notes that with being Glamis, and becoming Cawdor, "The greatest is behind."[158] This was anticipated in his earlier reply to the witches: "to be king / Stands not within the prospect of belief, / No more than to be Cawdor."[159]

Immediately then Macbeth imagines a way to achieve the third stage: "why do I yield to that suggestion / Whose horrid image doth unfix my hair, / And make my seated heart knock at my ribs, / Against the use of nature?"[160] But he quickly turns from this vision of a possible future: "If chance will have me King, why, Chance may crown me, / Without my stir."[161] Yet if Macbeth thinks he will quickly be named as Duncan's legitimate heir, he is to be disappointed. Duncan issues his inheritance clearly:

Sons, kinsmen, thanes,
And you whose places are the nearest, know:
We will establish our estate upon
Our eldest, Malcolm, whom we name hereafter,
The Prince of Cumberland: which honour must
Not unaccompanied, invest him only,
But signs of nobleness, like stars, shall shine
On all deservers. From hence to Inverness,
And bind us further to you.[162]

His estate is both his lands and status as king, but he suggests he will give other honors. He wants to move the party on to Inverness, Macbeth's home castle. This elevation of Malcolm to heir presents a new challenge to Macbeth:

The Prince of Cumberland: That is a step
On which I must fall down, or else o'er-leap,
For in my way it lies.[163]

Cumberland is one of the counties between England and Scotland—it is a step, yet also a border. Yet when Macbeth kills Duncan, Malcolm and his brother Donalbain flee, the first to England, the other Ireland.[164] Macduff reports that they "Are stolen away and fled; which puts upon them / Suspicion of the deed."[165] Ross thinks it likely "the sovereignty will fall upon Macbeth," which Macduff confirms: "He is already named, and gone to Scone / To be invested."[166] Macduff must already doubt things, or intend to act as a challenger to Macbeth, as he says he will not go to Scone, but to Fife.[167] Indeed, later Macbeth is told by one of the witches' apparitions to "beware Macduff / Beware the thane of Fife."[168]

At a later point, when Macbeth decides that he needs Banquo dead, he tells his wife not to worry about the process:

> Be innocent of the knowledge, dearest chuck,
> Till thou applaud the deed. Come, seeling night,
> Scarf up the tender eye of pitiful day
> And with thy bloody and invisible hand
> Cancel and tear to pieces that great bond
> Which keeps me pale.[169]

The "seeling night" is blinding, with "seeling" a term from falconry for the hooding and sewing up the eyes of the bird, although "sealing" is a legal term; "scarf up" continues the first theme of blindfolding, and "cancel . . . that great bond" the second, judicial one. "Thy bloody and invisible hand" could be the falconer's, bloody from the seeling and now unseen, or perhaps that of the legal scribe. The bond is either a tether or a legal or moral restraint, which keeps Macbeth either in a fearful pallor or confined within bounds, the double sense of "pale."[170] The meaning of "pale" as an enclosed space is crucial in Shakespeare, and is a theme that will be returned to in the conclusion to this chapter and the book's coda.

The English Connection

Malcolm, then as the rightful heir, in exile in England, lives in the court of King Edward the Confessor.[171] If, at first, Macbeth most fears internal threats against his rule, he comes to worry most about the external powers. Yet this is the case even before he has murdered Banquo:

> We hear, our bloody cousins are bestowed
> In England and in Ireland; not confessing

Their cruel parricide, filling their hearers
With strange invention. But of that tomorrow,
When therewithal we shall have cause of state
Craving us jointly.[172]

He most begins to worry about the external threat when he hears that Macduff has "fled to England."[173] We had previously been told that "the most pious Edward . . . the holy king," has been requested by Malcolm and Macduff to "wake Northumberland and warlike Siward," and now as a consequence Edward is preparing for war against Scotland.[174] Siward was the name of the Earl of Northumberland, another border county to the north of England and immediately south of Scotland. In Holinshed's *Chronicles*, Siward is Malcolm's grandfather; in *Macbeth* he is his uncle.[175]

There are some textual problems in this scene (Act III, scene vi) and its relation to the later Act IV, scene i. In the first scene, Lennox appears to be opposed to Macbeth and discusses the latter's reaction to Macduff's flight to England, while in the later scene Lennox is still serving Macbeth and tells him of this new defection, which surprises Macbeth and spurs him into action against Macduff's family.[176] This textual issue matters for various reasons, but in terms of this scene, it concerns which king is preparing for war. The unnamed "another lord" speaks only of Edward, and names him as "their king," Lennox's immediate reply "Sent he to Macduff" clearly refers to Macbeth. Thus, some editors emend "their king" to "the king,"[177] but this suggests that Macbeth knows of the English maneuvers, and so it is him preparing for war. That makes a nonsense of later scenes, where things come as a surprise.[178] Is it not possible that Lennox and the other lord do not consider Macbeth to really be a king? Or, more plausibly, is the text corrupted at this point, possibly because of the insertion of the scene involving the witches and Hecate? This is the other key textual problem of the play: the role of Thomas Middleton in the only version that is preserved, the Folio text. The intrusion of Hecate, and two songs sung by the witches, are likely his work, and there may be more.[179]

In England, Macduff and Malcolm discuss what has become of Scotland. Macduff says:

I would not be the villain that thou think'st
For the whole space that's in the tyrant's grasp,
And the rich East to boot.[180]

Malcolm describes the country in a way that may allude to Christ's passion:

I think our country sinks beneath the yoke;
It weeps, it bleeds; and each new day a gash
Is added to her wounds.[181]

Malcolm adds that "from gracious England have I offer / Of goodly thousands,"[182] and later specifies the makeup of these forces: "gracious England hath / Lent us good Siward and ten thousand men."[183] "Gracious England" is the King of England, Edward the Confessor. It is notable that "gracious" is used to qualify two kings—Duncan and Edward.[184] (The other instances of this word in the play are Seyton asking Macbeth "What's your gracious pleasure?"[185] and the messenger beginning his report of Birnam Wood moving with "Gracious my lord."[186]) The good want the external forces to return to save Scotland from itself. They are grateful for "Old Siward, with ten thousand warlike men."[187] Is England just supporting the right of Malcolm to the Scottish throne, or might there be a purpose to annexing Scotland?[188]

In this play then, because of the parlous state of Scotland, and with the support of England, Malcolm and Macduff lead a force to take the throne. This invasion of the English army and their siege of Dunsinane castle is shielded by branches cut from Birnam Wood, thus fulfilling another of the witches' prophecies:

Macbeth shall never vanquished be, until
Great Birnan wood to high Dunsinane hill
Shall come against him.[189]

Macbeth is immediately incredulous: "That will never be. / Who can impress the forest; bid the tree / Unfix his earth-bound root?"[190] But even Macbeth had recognized, only a couple of scenes before, that "Stones have been known to move and trees to speak."[191] And Macbeth hints at the use that will be made of them with the verb "impress," having a meaning of conscript into military service.[192] This is exactly how Malcolm will instruct his force:

Let every soldier hew him down a bough,
And bear't before him; thereby shall we shadow
The numbers of our host, and make discovery
Err in report of us.[193]

This is later described as "the wood began to move," and "a moving grove."[194] The geographies are shifting, and the outside world closes in on Macbeth—"and now a wood / Comes towards Dunsinane."[195]

After the victory over Macbeth, Malcolm says that one of the first acts of the new regime should be "As calling home our exiled friends abroad, / That fled the snares of watchful tyranny."[196]And what becomes of Donalbain? In the play, he disappears to Ireland and we hear nothing of him again, except that his absence is noted from his brother's side when they invade Scotland.[197] Historically, he returns to murder Malcolm's son.[198] Perhaps this is what is implied with his final lines, after his father's murder: "There's daggers in men's smiles: the near in blood, / The nearer bloody."[199]

The Question of Succession

In contrast to *Hamlet*, the geopolitical struggles and the ambitions of external forces are more central to the plot, but they similarly relate to contests at the time between Scandinavian and British powers. Both plays, too, concern the question of succession. Succession thus runs through the play, yet is frequently cut short. Banquo at the beginning of Act III recognizes that though he fears Macbeth has "played'st most foully for't," all that was promised to him has come to be.[200] He dares to think that the part of the prophecy that bore on him may be true too:

> It should not stand in thy posterity;
> But that myself should be the root and father
> Of many kings.[201]

But the succession he wishes is supplanted by another. The witches had told him that Banquo himself would not be king, but his heirs would be, and as Macbeth later looks into the cauldron he sees a line of kings, the eighth of which has a glass that shows many more. Macbeth recognizes "the spirit of Banquo" in the first, and notes that some "two-fold balls and treble sceptres carry," perhaps meaning they have been crowned twice, presumably of Scotland and England, and carry treble scepters as king of both kingdoms, as well as of Wales.[202] Shakespeare, of course, was writing at the time of King James. The accession of James, with the Union of the Crowns in 1603, meant that one person was king of both countries, not that one country was created from two. But James clearly intended this, and while he spoke of Britain, political inertia meant that it took a century before the Act of Union in 1707.[203] Balls, on this reading, are orbs. But while an orb was used in English coronations, E. B. Lyle notes that it was not for Scottish ones. However, Shakespeare may not have known this detail, or not have cared if he did. The orbs might be the mounds on the England and Scottish crowns, with mounds being "an orb of gold, often topped by a cross."[204] Whichever, Lyle rightly concedes there is "every reason to think that the 'twofold' in *Macbeth* also refers to the union of Scotland and England." Treble scepters might add Wales, and James was certainly king of all three. But he also ruled lands overseas, and treble might be France, Ireland, and Britain, not the three parts of Britain. Treble then would include Britain, which was itself double from Scotland and England.[205]

As George Walton Williams has argued, there are two narratives in the play concerning the succession. One is Macbeth's murder of Duncan; the other the attack on Banquo's line, though unsuccessful with its survival and later establishment. He goes so far as to call these "two parallel fables."[206] This is a play that has multiple chronologies running through it, both in the sense of a line of kings, but also the temporality of the action, the supernatural and the afterlife of Banquo, and the time of life. Sandra Clark and Pamela Mason think this is crucial to the play: "What is unique to *Macbeth* is the all-pervasive theme of time, the foregrounding of it in the play's last speech and the meaning of this foregrounding as an ending."[207] As Macbeth says when he hears of his wife's death,

> She should have died hereafter;
> There would have been a time for such a word.
> Tomorrow, and tomorrow, and tomorrow,
> Creeps in this petty pace from day to day,
> To the last syllable of recorded time;
> And all our yesterdays have lighted fools
> The way to dusty death. Out, out, brief candle!
> Life's but a walking shadow; a poor player,
> That struts and frets his hour upon the stage
> And then is heard no more: it is a tale
> Told by an idiot, full of sound and fury,
> Signifying nothing.[208]

It is possible to imagine *Hamlet* without the Fortinbras subplot, and some stage versions have attempted to do so. Hamlet's personal issues with the death of his father, his mother's marriage to his uncle, and the uncle's role as murderer could all remain, without the need for a backstory of King Hamlet's battle with King Fortinbras. Why Fortinbras appears at the end of the play would need some explanation, but the Ambassadors might be removed;

Hamlet need not encounter the Norwegian army (indeed, he does not in the First Quarto or Folio); and the description of the ghost need not have the whole historical narrative attached. The play would lose a lot, but it would not collapse. With *Macbeth*, things are somewhat different. Without the external threat of Norway at the beginning of the play, and the traitorous elements that join it, how would Macbeth prove himself, and how would the position of Thane of Cawdor become vacant? Without Malcolm's flight to England, the establishment there of a base to which rebels such as Macduff might fly, and the support of the English king and Siward, Lord of Northumberland, how would Macbeth be overthrown? Of course, the play could be rewritten to avoid those external elements, but it could not simply be cut to do so. Yet, paradoxically, *Hamlet* has much more detail about the external politics than *Macbeth*. In *Macbeth* enough is said to situate the story; in *Hamlet*, as in so many other aspects, the wider picture is filled out in a much fuller way. This may be a result of the cuts that are often assumed in *Macbeth*. Both plays, though, demonstrate the regional geopolitics within which their stories are set, with the vulnerability of small kingdoms to threats both at home and abroad.

Hamlet and *Macbeth* are thus both plays in which the strategic sense of territory can be found. While neither play includes the *word* "territory" or "territories," issues around territory play a significant role. Claudius's Denmark is surrounded by powerful neighbors, from Norway to Poland, with its wider relations to England, France, Normandy, and Saxony suggested to varying degrees of detail. Internal struggle over succession is only one element in its territorial precarity. Duncan's and then Macbeth's Scotland is similarly vulnerable—under attack initially by enemies within and without, from Norway to Ireland, and ultimately its rightful king can only regain his throne with the help of England. If these struggles were read anachronistically, they could be seen as a set of international relations between sovereign states. Yet even within the context of their time, both plays indicate the importance of political geographical relations, of spaces under control and under threat. Like Lear's territory, which he partitions through the division of the kingdom, these are spaces whose control is disputed by internal and external force.

Yet what sense of territory comes through in *Hamlet* and *Macbeth*? It is perhaps quite a traditional, modern sense of bounded spaces under the control of kings, or in and over which political struggle takes place. While that is certainly one sense of territory, it is a restrictive and limiting one. In the readings of both plays, I noted the use of the term "pale." Hamlet talks of "the pales and forts of reason";[209] Macbeth of "that great bond / Which keeps me pale."[210] The dual sense of fear and enclosure is strong in the latter. Pales were traditionally marked-out areas of forest, often royal hunting places. These would have a fence and ditch that allowed deer to jump in, but not back out. The term derives from the fences used to mark them out; a pale was a stake or picket, from the Latin *palus*, the root of the modern "palisade." A generation after Shakespeare's death, the political movement known as the Levellers adopted the name given to those who destroyed these barriers, either hedges or park palings.[211]

The political sense of a pale comes from this earlier use, and the term came to mean an enclosed and bounded area under the jurisdiction of some force. Examples include the English Pale of Ireland and the Pale of Calais. The Pale of Calais was the small area around the town of that name held by the English crown after the battle of Crécy in 1346 during the Hundred Years' War, and only lost in 1558, just a few years before Shakespeare's birth. The Pale of Ireland—another area held by the English, centered on Dublin—is mentioned in *The First Part of the Contention betwixt the Two Famous Houses of York and Lancaster*, published anonymously in Quarto in 1594, but probably written ca. 1588. This is a version of the play better known as Shakespeare's *Henry VI, Part 2*—either a pirated version or an early draft, or perhaps both. The text includes the lines "With troupes of Irish Kernes that uncontrolled / Doth plant themselves within the English pale."[212] This has no real parallel in the Shakespearean play that appeared in the Folio, but is likely remembered from Marlowe's *Edward II*: "The wild O'Neill, with swarms of Irish kerns, / Lives uncontrolled within the English pale."[213]

The common phrase "beyond the pale" is a generalization of the Irish use—"within the pale" was legally English jurisdiction; "beyond the pale" was not, even if still under English control. In its specific sense, then, "beyond the pale" is a political-legal question of jurisdiction, a spatial determination of the extent of rule. In the idiomatic sense in modern English, the term has a sense of beyond acceptable limits or boundaries, especially concerning morality, behavior, or judgment. While a pale in the political-geographical sense is therefore certainly a territory, this should not imply that all territories are pales, or that the pale meaning is sufficient: we cannot assume the restrictive bounded sense of pale tells us all we need to know about territory.

Hamlet and *Macbeth* are thus helpful in thinking about a strategic sense of territory, of contest and conflict over such sites. While crucial in comprehending territory, this is not nearly exhaustive. There are many other aspects of territory into which Shakespeare can offer insight. We find hints of this, for example, in *Hamlet*'s graveyard scene with its discussion of property and law. This book therefore now turns to other plays to explore some of these other issues. The next sequence of chapters looks at some of the history plays to begin this work. First, the question of rule and ownership is examined through a reading of majesty and possession in *King John*.

CHAPTER THREE

The Territories: Majesty and Possession in *King John*

Shakespeare's *The Life and Death of King John* was probably written in the 1590s, possibly 1595–96, though various years from 1592 to 1598 have been proposed.[1] Some critics date the play even earlier, which would make it one of Shakespeare's first plays, and I will return to the question of precedence to a very similar play later.[2] A date of the mid-1590s puts it after the first tetralogy of the three parts of *Henry VI* and *Richard III*, and around the same time as the first play of the second tetralogy, *Richard II*. This is also around the same time as *Edward III*, a play of disputed authorship but in which Shakespeare probably had at least a collaborative role. Like *Edward III*, and the late collaboration *Henry VIII*, *King John* is a standalone play, detached from the other plays in historical setting and dramatic presentation.

Set in the early thirteenth century, it is the earliest of Shakespeare's English history plays, though of course his Roman plays are set much earlier. Other plays—*Hamlet*, *Macbeth*, and others—take semihistorical figures and develop elaborate stories around them, but *King John* is still somewhat different. It is almost a prelude to the other English history plays, introducing the dual themes of domestic disorder and division and foreign conquest that would shape their narratives.[3] John's reign can be seen in three main periods: a dispute with the French king Philip II (Philip Augustus) over the Angevin territories in France, and the right of succession (1199–1206); a confrontation with the pope (1207–13); and the barons' rebellion and the French invasion (1213–16).[4] Shakespeare merges several elements from these periods, often compressing events from different years into the same day in his play.[5]

King John was the youngest of Henry II's five sons. The first son, William, had died aged two. The second, also called Henry, had ruled alongside his father for part of the reign as "the young king," but died before him. His third son, Richard, better known as Coeur de Lion or Lionheart, succeeded his father. A fourth son, Geoffrey, predeceased him. This is the opening tension in Shakespeare's play: should Richard, who has died returning from the Third Crusade to the Holy Land, be succeeded by his surviving brother, John, or by Geoffrey's son, Arthur.[6] (Historically, Arthur's claim was the French domains, not the crown of England.[7]) France backs the claim of Arthur, but John resolves this situation by a strategic marriage of his niece, Blanche of Spain, and Lewis, the Dauphin of France, and therefore creates an alliance with the French. John gives up lands as a generous dowry, and the French give up their support to Arthur's claim.

This truce is broken by the pope's legate, Cardinal Pandulph, who objects to the King's choice of Archbishop of Canterbury. John refuses and is excommunicated, and Pandulph threatens the French with excommunication too if they do not support the pope. The English and French fight, Arthur is captured, the order is given for his death to the King's man Hubert, but is not carried out. Arthur later dies while trying to escape. The English barons, objecting to Arthur's treatment or at least using it as an excuse, rise up against the King. In this they are aided by the French, and John has to strike a new deal with the pope in order to retain his throne. The Magna Carta is never mentioned in the play.[8]

Thus the play raises a range of questions: Who is the legitimate heir to Richard Coeur de Lion's throne? What should England rule over, and how should power be balanced between the King and feudal lords? What relation does the Crown have to the papacy?[9] John dies at the very end of the play, poisoned by a monk, and his crown passes to his son, Henry. So, despite a challenge from the French king Philip, who supports Arthur's claim; despite a dispute with Rome and Pope Innocent III; and despite a revolt of the barons within his kingdom, and a French invasion, John remains king throughout the play. John's son becomes King Henry III. Henry III is the great-grandfather of Edward III, and father, son, grandson, and great-grandson rule between them for an unbroken 161 years, though not without disputes and the dethronement of Edward II. The disputes between Edward III's grandchildren, who include Richard II and Henry IV, and the two branches of that family, the house of Lancaster and house of York, provide the material for Shakespeare's other history plays.

While the major plotline of this play concerns multiple political and territorial themes—John's threats from another potential king, supported by a foreign power; domestic turmoil from the barons; and the rivalry between temporal and spiritual power—there is another story running in parallel. This is the story of the two sons of Sir Robert Faulconbridge, and who should inherit his land. The younger, also named Robert, is a legitimate son; the elder, Philip, a bastard. They come to the King asking for judgment on legitimacy and inheritance. It becomes clear that Philip is actually the son of Richard Coeur de Lion, and he gives up his claim to land to be knighted as Sir Richard Plantagenet and to become a member of the court. (Were he not a bastard, of course, he would have a better claim to the throne than either John or Arthur. He is, though, an ahistorical character.) This relation between Robert and the Bastard, which concerns lineage and land, is, for one editor of the text, "a fine example of Shakespearean opening moves in which he ever so briefly introduces a central theme"—the question of succession, the question of a right to inheritance of land for a king—"then dramatises it in a minor key."[10] Shakespeare drops the Bastard's attraction to Blanche from the play's major source, thereby taking away part of his motive for a rivalry with France. We never hear of Robert again, but the Bastard becomes a major figure in the play, with the most lines and perhaps the best dramatic part, and his support for the young king at the end, with a speech invoking the need for England to stand together, resolves the action well.

As A. R. Braunmuller has noted, the play almost works in two halves. In the first, John is triumphant, the Bastard detached, the older generation dominates, there are strong female characters, and the lords are loyal. In the second half, John is indecisive, the Bastard confident, the younger generation dominates, there are no female characters, and the lords are disloyal.[11] Such a reading helps to make sense of what can, at times, seem quite an episodic plotline. Three themes in the play are worth further attention because they raise important questions about the concept and practice of territory. One is the question of majesty, the second the question of possession in relation to territory, and the third their relation to land.

BORROWED, BANISHED, BARE-PICKED MAJESTY

In the opening scene the play dramatizes the claim the French king is making to the English crown, on behalf of Arthur Plantagenet. The French ambassador Chatillon first describes John as "the majesty, / the borrowed majesty," with the qualified afterthought "borrowed" implying assumed or stolen, and "majesty" meaning both sovereignty and the display of such in dress and attire.[12] Chatillon believes Arthur should be king, rather than John, so "borrowed majesty" is a claim that the crown has been usurped.

John's mother, Queen Eleanor, mocks this opening: "a strange beginning: 'borrowed majesty!"[13] John claims that "strong possession and our right" are his claim to the crown, but his mother stresses that these two are not reinforcing and paired, the stress on "and," but comparative: "your strong possession much more than your right, / Or else it must go wrong with you and me."[14] When they meet, King Philip and John trade legal terms in their dispute. John challenges: "From whom has thou this great commission, France, / To draw my answer from thy articles?" Philip replies that it is God, who has made him guardian to Arthur, and he must "look into the blots and stains of right." Thus it is God "Under whose warrant I impeach thy wrong / And by whose help I mean to chastise it."[15] This leads to this exchange on usurpation:

> King John: Alack, thou dost usurp authority.
> King Philip: Excuse it is to beat usurping down.
> Eleanor: Who is it thou dost call usurper, France?
> Constance: Let me make answer: thy usurping son.[16]

The two kings, two men, challenge each other, the one accusing the other of usurping, and the other saying that it is excused to challenge it. Then the two women, John's mother and Arthur's mother, take up the challenge. All four lines, from four speakers, all with the root word "usurp." But the use of the word shifts. In the first line it is a verb: to usurp authority. In the second, a noun, a gerund, the action of usurping. In the third, a noun, a descriptor of someone, the usurper. In the fourth, a participle working as an adjective, a modifier of "son," a belittling way to describe a king.

"Majesty" in the play is generally modified with a qualifying adjective, unless someone is being directly addressed as "your majesty," "my majesty," "his majesty," and so on. The exceptions are the Bastard's speech when the powers of France and England seem about to clash: "Ha, majesty, how high thy glory towers, / When the rich blood of kings is set on fire!"; and the citizen's description of the "double majesties" of France and England.[17] But generally when majesty is invoked it is seen in a negative way, mirroring the questionable claims to it. Constance, the mother of Arthur, regularly invokes the failed nature of France's power. Fortune is "corrupted, changed," she "adulterates hourly" with John, and has made France "tread down fair respect of sovereignty / And made his majesty the bawd to theirs. / France is a bawd to Fortune and King John. / That strumpet Fortune, that usurping John!"[18] "Bawd" here, echoing the Bastard, implies France is pimping on behalf of England, creating the marriage of Blanche and Lewis. Her continual complaints receive a response. King Philip tells her that she has no right "To curse the fair proceedings of this day: / Have I not pawned to you my majesty?" But she replies that he has "beguiled me with a counterfeit / Resembling majesty, which, being touched and tried, / Proves valueless."[19]

Constance describes the French king's change of allegiance when he takes the pope's side against the English crown as "O fair return of banished majesty." Eleanor replies, "O foul revolt of French inconstancy!"[20] Constance sees Philip as returning from the banishment of his alliance with John, a trade that had destroyed the chances of her son, Arthur, who also holds, to her mind, "banished majesty"; whereas Eleanor, mirroring this line, sees this as a return to France's dissembling ways. And are we right to hear Constance's name in "inconstancy," even if in the negative?

Later in the play, when John meets Hubert after Arthur's death, he tries to deny that he had ordered the killing:

> It is the curse of kings to be attended
> By slaves that take their humours for a warrant
> To break within the bloody house of life,
> And on the winking of authority
> To understand a law, to know the meaning
> Of dangerous majesty, when perchance it frowns
> More upon humour than advised respect.[21]

John indicates the risk of his whims and careless wishes being taken as orders, the risks attached to the power of his position. It is an unfair denial of his role, but indicates the risks attached to the office. (A parallel might be the traditional story of John's father, Henry II, supposedly saying, "Who will rid me of this troublesome priest?" and this leading to the killing of Thomas Becket, Archbishop of Canterbury. Or in Shakespeare's *Richard II* where Mowbray's permanent banishment and the reduction of Bolingbroke's exile are performative utterances: "such is the breath of kings."[22]) Toward the end of this play there is another negative association, when it is described as the "bare-picked bone of majesty," a line to which I will return in context.[23]

Now, such is the precariousness of John's reign, the limits of his majesty, that in the play's fourth act he is recrowned:

> King John: Here once again we sit, once again crowned,
> And looked upon, I hope, with cheerful eyes.
> Pembroke: This 'once again,' but that your highness pleased,
> Was once superfluous. You were crowned before,
> And that high royalty was ne'er plucked off,
> The faiths of men ne'er stained with revolt;

Fresh expectation troubled not the land
With any longed-for change or better state.

Salisbury then adds, in the famous, misquoted speech:

Therefore, to be possessed with double pomp,
To guard a title that was rich before,
To gild refined gold, to paint the lily,
To throw a perfume on the violet,
To smooth the ice, or add another hue
Unto the rainbow, or with taper-light
To seek the beauteous eye of heaven to garnish.
Is wasteful and ridiculous excess.[24]

The "possessed with double pomp" and "wasteful and ridiculous excess" hint at the economic and property concerns that run throughout the play, to which I will return. Here, as Braunmuller notes, there is a complicated grammatical construction to the phrase, because "with" serves two purposes: "possessed with," where it is a "prepositional complement to the verb," and "with double pomp," where that is the thing possessed or the thing that possesses. John is made ruler again, a second ceremony, but is also "owned with excessive ceremony." The "pomp" has no object beyond the ceremonial: "crown, state, England" are not mentioned.[25] Repetition does not reinforce, but identifies the weakness.

"Double pomp" clearly jars, as does Salisbury's later "new crowned,"[26] as the King justifies himself, again falling back on a property relation, which is both active and passive: "some reasons for this double coronation / I have possessed you with and think them strong."[27] But it is not the last coronation, as in Act V the King has given up the crown to the pope, in order to receive it again. Crown here is both the literal object, and the abstraction of his sovereign power.

King John: Thus have I yielded up into your hand.
The circle of my glory.
Cardinal Pandulph: Take again
From this my hand, as holding of the pope,
Your sovereign greatness and authority.[28]

It is both the crown and the authority that Pandulph hands to John with the pope's blessing, though this is a "leasehold, a tenure."[29] Papal spiritual authority was unlimited in spatial extent, over Christian souls wherever they were. Nonetheless, even papal theorists recognized that the pope needed to delegate power to temporal rulers, whose power was spatially delimited, for political purposes. This was the doctrine of the pope's two swords, both to command, but the temporal not to be used directly.[30] While John, mirroring the dispute of Shakespeare's own time, sought to be temporal ruler alone, he is forced to acquiesce to papal authority.

Thus John has to perform the ceremony at least three times.[31] The first—sometimes staged, or sometimes preceding the opening of the play—is already threatened by the promise of France's support for Arthur. Second, having seen off the threat of Arthur, the King does it again; and third, having debased himself to the pope, who now is spiritual and temporal ruler of England, he is granted his kingship once more:

> King John: The legate of the Pope hath been with me,
> And I have made a happy peace with him,
> And he hath promised to dismiss the powers
> Led by the Dolphin.[32]
> It is left to the Bastard to denounce this surrender:
> Bastard: O inglorious league!
> Shall we upon the footing of our land
> Sent fair-play orders and make comprimise,
> Insinuation, parley, and base truce
> To arms invasive?[33]

THE TERRITORIES

Shakespeare seems to use the notions of majesty and sovereignty largely interchangeably. They both function as key operators of political rule, even if they are often qualified rather than absolute. What is that majesty, or sovereignty, over? *King John* is one of only a handful of Shakespeare's plays in which the word "territories" appears. As the analysis has shown, it is clearly not only the word that matters, and there is much in Shakespeare where what we might now call "territory" is at stake, but the words used are rather different. *Hamlet* and *Macbeth*, which never use the *word*, are both plays in which territorial conflict is a significant theme. But for a moment, since the question arises as to *use*, let me enumerate the instances, which I will come back to explain.

There are three instances in *Coriolanus*; two each in *Henry VI, Part 1*, *Henry VI, Part 2*, and in *King John*; and one each in *Richard II*, *Two Gentlemen of Verona*, and *Edward III*—though if indeed Shakespeare has a hand in this play, this is not likely one of the scenes. The word "territory," singular, only appears in *As You Like It* and the Folio text of *King Lear*.[34] In *As You Like It*, as so often, it concerns the threat of banishment: "or turn thou no more / To seek a living in our territory."[35]

In *King John*, there is one mention in the opening scene, and one in the final act. In the opening scene, it is used in the passage by the French ambassador Chatillon concerning John's "borrowed majesty." This is what he claims:

> Philip of France, in right and true behalf
> Of thy deceased brother Geoffrey's son,
> Arthur Plantagenet, lays most lawful claim
> To this fair island, and the territories:
> To Ireland, Poitiers, Anjou, Touraine, Maine,
> Desiring thee to lay aside the sword
> Which sways usurpingly these several titles,
> And put thy same into young Arthur's hand,
> Thy nephew and right royal sovereign.[36]

The sword is the sword of state, as opposed to the standard scepter, perhaps reinforcing the rule by power not right, and it rules over ("sways") titles to several parcels of land, in Chatillon's terms, without legitimacy. It is interesting that editors are frequently troubled by the text of this passage. One of the challenges for this play—with this issue, and many others—is that there is no early Quarto edition: the first publication of the text was in the First Folio of 1623.[37] There is therefore only one source text to draw upon.

In the Arden First Series, the editor, Ivor John, makes this claim:

> There is no other case of the use of "the territories" in this way by Shakespeare. One is tempted to suggest either "and *the* territories *Of* Ireland," or "*her* territories." In *The Troublesome Raign*, II. iii (ed. 1591), we have "to England, Cornwall and Wales and to their territories."[38]

The reference here is to *The Troublesome Reign of King John*, an anonymous play first published in 1591. I will return to this play, and its relation to Shakespeare's play, later in this discussion.

In a 1919 edition, edited by H. H. Furness, we find a note quoting, and endorsing, Ivor John's gloss.[39] Dover Wilson, in the Cambridge University Press edition of 1936, made the claim that he was "puzzled by the rather odd use of the word 'territories' in the sense of dependencies," similarly implying that there are no other uses of the phrase in this way by Shakespeare.[40] "Dependencies" is the only gloss given to the word "territory" in C. T. Onions's *A Shakespeare Glossary*.[41] Dover Wilson goes on to note another use in *The Troublesome Reign of King John*. In that play, Philip of France addresses Arthur, and states, "Now 'gin we broach the title of thy claim / Young Arthur in the Albion Territories."[42] Dover Wilson suggests that here "we find the words 'claim' and 'territories' once again combined, the latter in this case being used in a perfectly ordinary sense."[43]

The 1989 Oxford edition has this use of "the territories" glossed as "feudal dependencies," following the earlier Clarendon edition from 1886;[44] and in the Arden Second Series the editor, E. A. J. Honigmann, glosses it as "i.e. dependencies."[45] The Penguin edition has "dependent possessions,"[46] and the RSC edition similarly provides a note of "dominions/dependencies."[47] The most recent Cambridge edition suggests "lands within or adjacent to England."[48]

Yet as several editors note, including Dover Wilson, the term is used in *The Troublesome Reign of King John*, which most people think was one of Shakespeare's key sources for his play. Some people, though, think Shakespeare's play was the earlier, which would cause a revision of its date to 1590 or 1591, because *The Troublesome Reign* was published in 1591. It has even been suggested that Shakespeare was the author of that play too, and that the play in the Folio is his own revision of an earlier version.[49] The Second Quarto of the play, in 1611, said it was "written by W. Sh.," and the Third Quarto of 1622, "written by W. Shakespeare."[50] But given that the First Folio was 1623, both can be put down to attempts to sell the play by using his name: a version of the story by him would have been known by performance, but not in print. Ivor John, over a century ago, was dismissive: "No one who compares the *Troublesome Raign* with *King John* can for a moment entertain the idea that the former is a 'first draft' of the latter."[51]

If indeed Shakespeare's play follows it, he did care enough to rework it, not relying on other sources. Honigmann in the Arden edition, in contrast, argues that Shakespeare's play precedes *The Troublesome Reign*, and is its major source.[52] This account was so widely discredited, with *King John* seen as the successor, that Guy Hamel suggested in 1989 that the "claim now seems to me so strong as to need no further elaboration or support."[53] Yet the Cambridge edition of 1990 opened the debate up again, with some good arguments for Shakespeare's text being the earlier.[54] The authorship of *The Troublesome Reign* is also unresolved, though the most recent edition attributes it to George Peele, who died in 1596. Braunmuller puts the dilemma clearly:

If *King John* preceded *The Troublesome Reign*, Shakespeare's play must pre-date 1591; if *King John* followed *The Troublesome Reign*, then it must be regarded as Shakespeare's most important source. *The Troublesome Reign* is therefore relevant to both *King John*'s date and its sources.[55]

My instinct is to go with the later date of the play, given similarities to *Richard II*, which would suggest it follows *The Troublesome Reign*. But whichever account is believed, the plays are within a few years of each other, and *The Troublesome Reign* shows that the use of the term "territories" to describe places, perhaps belonging to other places, as opposed to lands, belonging to someone, is not without precedent.

The instance Honigmann and Ivor John select is "King to England, Cornwall, and Wales, and to their territories";[56] Dover Wilson's reference is to "Albion territories," "an expression which seems in the context to mean the continental parts of the Angevin inheritance."[57] But there are four instances of the term "territories" in that play, and one other in Shakespeare's *King John*. In *The Troublesome Reign* the uses are all possessive: "Albion territories"; "our territories"; "your territories"; "their territories"; the last possessions of places, and only indirectly of the King.[58] (The parallel passage in *The Troublesome Reign of John* to the first use in Shakespeare's *King John* is a demand for "the kingdom of England, with the lordship of Ireland, Poitiers, Anjou, Touraine, Maine."[59] Poitiers is a mistake, also found in *The Troublesome Reign*, as it should be the province, Poitou, not the town.[60])

"Territories" is also possessive in the other *King John* reference, where toward the end of the play we are told by the Bastard:

Now hear our English king;
For thus his royalty doth speak in me:
He is prepared, and reason too he should—
This apish and unmannerly approach,
This harnessed masque and unadvised revel,
This unhaired sauciness and boyish troops,
The king doth smile at; and is well prepared
To whip this dwarfish war, these pigmy arms,
From out the circle of his territories.—[61]

This sense of "territories," as the lands under the possession of a ruler, is more common in Shakespeare's use, and does not receive an editor's gloss in either of the Arden editions, in Clarendon or Oxford, in the Penguin, RSC, or the 1990 Cambridge edition.[62] But "Albion territories," "England,

Cornwall and Wales, and to their territories," and "the territories" hint at something more complicated.

The suggestion that "the territories" is unusual might appear a strange one. Given the paucity of references to "territory" and "territories" in Shakespeare we need to be careful in extrapolating anything from linguistic absence. However, the point seems to be that with the other references to territory the determiners are possessive, whereas this one gives "territories" a definite article. If we look at the other uses of the word in Shakespeare the pattern is revealing:

> There are three in *Coriolanus*—"your territories," "the Roman territories," "our territories."[63]
> In *Henry VI, Part 1*—"our territories," "Gallian territories" (i.e., French).[64]
> *Richard II*—"our territories."[65]
> *Two Gentlemen of Verona*—"my territories."[66]
> *As You Like It*—"our territory."[67]

These would seem to support the general view—all are possessive, all are lands that are properties of rulers or the Roman Republic or French kingdom. In *Edward III*—though in a scene not likely by Shakespeare's hand—we have "the territories of France alone," which, although using a genitive construction rather than a possessive pronoun, is along the same lines: "France's territories alone" could stand in its place.[68]

The most interesting semantic examples come in the remaining plays. In *Henry VI, Part 2* there is a use of "England's territories," but the more intriguing instance comes when Lord Somerset reports on the situation in France to the King: "That all your interest in those territories / Is utterly bereft you; all is lost."[69] While this may appear to be another use of "territories" in a sense of lands, or as a battlefield fought over and surrendered, the relation of interest shows that it is not simply property or a strategic sense, but the political control of, claim to, and stake in those places. This same phrasing is used of "territory" in *King Lear*, where *Lear* talks of "Interest of territory, cares of state."[70] The instance in *King John* seems clearly a reference to a list of territories beyond "this fair island," which are currently in the possession of the English crown.

As Hamel suggests,

> The curious "Territories" is glossed by its use in *The Troublesome Raigne*, where it clearly means an adjunct or subordinate possession. The distinction between England and the rest of the inheritance initially claimed by

John proves to be crucial—to my mind, central to the "historical" character of the play—as John comes to be presented less as a Plantagenet claimant to part of his father's empire and more simply as King of England. The separateness—and the greater significance—of England is indistinct in *The Troublesome Raigne*, despite the references to the "Territories," but prominent in *King John*, where it is supported by such references, all unique, as that to "English *Iohn*" (TLN [Through-line-number] 303; 2.1.10) and by the Gaunt-like picture of England as a "Water-walled Bulwarke" (TLN 320; 2.1.27).[71]

WHAT ARE THE SPECIFIC TERRITORIES?

Equally, the specific territories mentioned, and Arthur's claim to them, recur at several points in the play. John laid claim to large parts of France, which were disputed throughout the Middle Ages, and the last of which, the Pale of Calais mentioned in chapter 2, had been lost shortly before Shakespeare's birth in the reign of Queen Mary. However, Queen Elizabeth I still styled herself Queen of France, a title kept by rulers until the nineteenth century. In this play, at the beginning of Act II, France meets the Duke of Austria before the gates of the town of Angiers. It is revealed that Austria killed Richard Coeur de Lion, but Arthur is prepared to forgive his uncle's death for the support Austria will give him now. Austria reciprocates:

> Upon thy cheek lay I this zealous kiss,
> As seal to this indenture of my love:
> That to my home I will no more return,
> Till Angiers and the right thou hast in France,
> Together with that pale, that white-faced shore,
> Whose foot spurns back the ocean's roaring tides
> And coops from other lands her islanders,
> Even till that England, hedged in with the main,
> That water-walled bulwark, still secure
> And confident from foreign purposes,
> Even till that utmost corner of the west
> Salute thee for her king; till then, fair boy,
> Will I not think of home, but follow arms.[72]

The description of both the overseas possessions that are Arthur's by right, and the physical geography of England, with its cliffs and surrounding ocean, reinforces the separation. England is well fortified—"coops' meaning "encloses for protection," and "hedged" a defensive perimeter.[73] As indicated in chapter 2, and as will be discussed in more detail in the coda, "pale," as well as referencing the white cliffs, is also an archaic term for an area or jurisdiction, here glossed by Robert Smallwood as an "enclosed territory, linking with *water-walled*."[74]

A speech in the opening scene of Act II has caused a lot of confusion over the attribution of the lines. It follows a request from Austria. The Folio has the following:

> Aust: what cracker is this same that deafes our eares
> With this abundance of superfluous breath?
> King *Lewis*, determine what we shall doe strait.
> *Lew*. Women & fooles, breake off your conference.
> King *Iohn*, this is the very summe of all.[75]

"King Lewis" does not make sense, as the king is not Lewis, but Philip, and Lewis is the Dauphin. So the third line is usually emended to being a speech by King Philip, such as in Arden. In other words, it should read: "King: Lewis, determine what we shall do straight." But this would fail to respect the unindented line as well. On that reconstruction, the King gives to Lewis the Dauphin the right to state the demand. The next line then remains with Lewis—here the instruction seems clear. "King John" at the beginning of the following line is not a speech prefix, it is part of the line. Arden gives it, and the remainder of the speech, to the French king; Oxford keeps with Lewis.[76] But the Penguin edition gives the first of the disputed lines to Austria, correcting "King Lewis" to "King Philip," reading this, like "King John" two lines later, as part of the speech, making this a request for clarification, and the King gets the remaining lines. Another possibility is to give the first line to Austria, and for it to read: "King, Lewis, determine . . ."—that is, a request to both men, father and son.[77]

There is also confusion over who speaks the following lines, but they are the French side, and explain what is at stake:

> King John, this is the very sum of all:
> England and Ireland, Anjou, Touraine, Maine,
> In right of Arthur do I claim of thee.
> Wilt thou resign them and lay down thy arms?[78]

But to create the English-French alliance, John shows he is willing to part with much of this claim. France gives up Arthur's claim to the throne, and England gives up the overseas possessions as the dowry for the marriage between Lewis and Blanche:

> If that the Dolphin there, thy princely son
> Can in this book of beauty read 'I love,'
> Her dowry shall weigh equal with a queen:
> For Anjou, and fair Touraine, Maine, Poitiers,
> And all that we upon this side the sea—
> Except this city now by us beseiged.[79]

"This city" is Angiers, which John later gives to Arthur, meaning all his French possessions are gone after John gives away the following:[80]

> Volquessen, Touraine, Maine,
> Poitiers, and Anjou, these five provinces
> With her to thee and this addition more,
> Full thirty thousand marks of English coin.[81]

The overseas territories, "the territories" beside Ireland, though with Volquessen added from the initial scene, are surrendered to France. John loses land, but gains a stable kingdom; France gains land, but loses its surrogate claim on the English throne. The Bastard, for one, bemoans this: "Mad world, mad kings, mad composition! / John, to stop Arthur's title in the whole, / Hath willingly departed with a part."[82] "Composition" is an "agreement, compact," but also a compromise, trade, or purchase.[83] But this is an effective trade, one that the Lady Constance realizes will prevent France pursuing Arthur's case: "Shall Lewis have Blanche, and Blanche those provinces?" and "Lewis marry Blanche! O boy, then where are thou?"[84]

There is also a dispute about who holds ultimate political sway in these areas. Is it the king, or might there be a higher authority? John, initially, is clear. In the dispute with the pope, he tells Cardinal Pandulph:

> What earthy name to interrogatories
> Can tast the free breath of a sacred king?
> Though canst not, cardinal, devise a name
> So slight, unworthy and ridiculous,
> To charge me to an answer, as the pope.
> Tell him this tale; and from the mouth of England
> Add this much more, that no Italian priest

Shall tithe or toll in our dominions;
But as we, under heaven, are supreme head,
So under Him that great supremacy,
Where we do reign, we will alone uphold,
Without th' assistance of a mortal hand:
So tell the pope, all reverence set apart
To him and his usurped authority.[85]

No earthly—though the word here is "earthy"—power has the right to demand anything of John; "tast" is usually emended to "task," "taste," "tax," so try or test, and "interrogatories" are questions put under oath.[86] The power of the church cannot extend to tax by tithe or levy within his kingdom. As an insistence on his own divine right to rule, and his supremacy under God—"heaven" is likely a censor's emendment—and no mortal power, it is an assertion of his temporal control within his kingdom. We would now say the exercise of this power within a territory, the spatial extent of sovereignty. This was an anticipation of a late medieval idea: *rex imperator in regno suo* (the king is an emperor in his own kingdom). While that really emerges in the later thirteenth century, Innocent III himself, in the papal bull *Per venerabilem*, had noted that the King of France held supreme temporal authority: "The king therefore recognizes no superior in temporal matters."[87] Although Innocent claimed that spiritual power was higher, he declined to overrule the king on this specific case—concerning the legitimacy of a French count's offspring. By the time of the 1534 Act of Supremacy under Henry VIII, there had been a clear assertion of the primacy of the English crown in relation to Rome.[88] That was the context in which Shakespeare was writing, and John is here perhaps seen as a proto-Protestant, an earlier attempt at a Reformation.

Philip notes that John blasphemes, but John continues:

Though you, and all the kings of Christendom,
Are led so grossly by this meddling priest,
Dreading the curse that money may buy out;
And by the merit of vile gold, dross, dust,
Purchase corrupted pardon of a man,
Who in that sale sells pardon from himself,
Though you and all the rest, so grossly led,
This juggling witchcraft with revenue cherish,
Yet I alone alone do me oppose
Against the Pope, and count his friends my foes.[89]

The priest is both Pandulph and the pope, and John hints that the curse of excommunication can be overturned by bribing the pope, or salvation achieved by "merit of vile gold" rather than by faith.[90] Nonetheless, John is cursed and excommunicated by the legate, and those who oppose him are blessed: "And meritorious shall that hand be called, / Canonized and worshipped as a saint, / That takes away by any secret course / Thy hateful life."[91] The relation between England and France, so briefly created, falls apart. So, late in the play, there is an invasion of England, and to avoid this John does subjugate himself to Rome to gain support.

When John fears disorder at home, and invasion from France, he meets a messenger:

> King John: A fearful eye thou hast: where is that blood
> That I have seen inhabit in those cheeks?
> So foul a sky clears not without a storm;
> Pour down thy weather: How goes all in France?
> Messenger: From France to England. Never such a power
> For any foreign preparation
> Was levied in the body of a land.[92]

More is to come, for not only has France mobilized forces, this is possible because Queen Eleanor is dead. John realizes that this means his interests are in disorder: "How wildly then walks my estate in France!"[93] Yet the anticipated storm of the message appears at several points in the play both figuratively and literally.[94] For King John it is the rush of bad news, from which he struggles to cope: "I was amazed / Under the tide: but now can breathe again / Aloft the flood."[95] Pandulph says later, of the French invasion:

> It was my breath that blew this tempest up
> Upon your stubborn usage of the pope;
> But since you are a gentle convertite [convert],
> My tongue shall hush again this storm of war.
> And make fair weather in your blust'ring land.[96]

King Philip says that he has gathered a fleet against England:

> So by a roaring tempest on the flood
> A whole armado of convicted sail
> Is scattered and disjoined from fellowship.[97]

Like the 1588 Armada, this group of ships is "convicted" (Arden), or "convected" (Oxford), which Braunmuller glosses not as "vanquished, defeated" but as "sailing together (*convectus*)." However, the next line suggests this armada has been lost.[98] But it does not stop the French invasion, though once again the weather will intervene—a point to which I will return.

THE POLITICS OF LAND

King John, like the near-contemporary play *Richard II*, is very much a play about land, with the word recurring multiple times in the text. Both plays are highly ritualized and rhetorical, and are both entirely in verse. Along with *Henry VI, Part 1* and *Part 3* they are the only plays by Shakespeare without at least some prose; a status shared with the disputed *Edward III*.[99] Both *King John* and *Richard II*, of course, like many of Shakespeare's histories and some of his tragedies, concern the question of lineage and succession. But while *Richard II* is a play that is frequently about the economic uses of land, its use, abuse, and yield, as chapter 4 will explore in detail, *King John* has the economic aspect more in terms of property, and is much more about conquest and inheritance—issues of possession. John, of course, was nicknamed *sans terre*, "lackland," either because as a youngest child he was not expected to inherit, or perhaps later because of what he had lost. The Bastard, at the point when he gives up a claim on his inheritance to be recognized as Richard Coeur de Lion's illegitimate son, to become Sir Richard Plantagenet, parallels this relation.

This dispute between Robert Faulconbridge and the Bastard also has some similarities to and differences from the Edgar/Edmund subplot of *King Lear*.[100] While the characters are quite different, a major theme in both is the land. The Bastard alternates between knowing the reason—"if he can prove, a pops me out / At least from fair five hundred pound a year / Heaven guard my mother's honour and my land"—and pretending that he is puzzled by his younger brother's claim to inheritance, suggesting there is no reason, "except to get the land."[101] "Land" runs like a refrain through the scene, punctuating the dialogue repeatedly.[102] But as the mother was married, Philip is legitimate. As King John rules to Robert Faulconbridge: "My mother's son [i.e., King Richard] did get your father's heir; / Your father's heir [i.e., Philip] must have your father's land."[103] Robert objects that his father's will said otherwise, but this is rejected (tellingly, King John's claim to the throne is due to King Richard's will). But Philip choses instead to be knighted, discarding his inheritance for recognition as King Richard's son. He turns to Robert, and declares: "Brother by th' mother's side, give me your hand; / My father gave me honour, yours gave land."[104]

The King dismisses him in similar terms: "Go, Faulconbridge, now hast thou thy desire; / A landless knight makes thee a landed squire."[105] While tempting to see John *sans terre* as the "landless knight," it is the Bastard's decision that has created this situation, and he, on being knighted, gave up his lands. As the Bastard declares, "A foot of honour better than I was, / But many a foot of land the worse."[106]

In this play there are a set of exchanges proposed, of buying people off with titles, commodities, or lands. Honigmann suggests:

> John explores and exploits his fellow-men. He tries to find everyone's price. The Bastard is bought with a knighthood (I. i. 162), France with five provinces (II. i. 527 ff.), Hubert with promises (III. ii. 30–42), Arthur is promised lands (II. i. 551), the nobles are promised Arthur (IV. ii. 67), John tries to buy them off a second time (IV. ii. 168), buys off the pope with a nominal submission (V. i. 1), in order to buy off Lewis (V. i. 64).[107]

And property relations generally, certainly in land, but also with wider economic concerns, run through the play. Indeed, the play quite carefully marks the distinction between chattels or movables, personal property, and real estate or real property, that is immovables. This was, in part, because of distinct bodies of Roman law for different things: *res mobilis* and *res immobilis*. The American English "real estate" retains this notion, as does the French *meubles* and *immeubles*—furniture and buildings. This distinction becomes important for legal inheritance, since a will pertains to real property, and a testament to personal property.[108]

The notion of possession is important in this play. Both Philip and John use the word when making their claim on Angiers: Philip threatens to "give the signal to our rage / And stalk in blood to our possession," and John "bear possession of our person."[109] This is not just in terms of property, but also in terms of being seized by something. This figures in terms of the spiritual with Peter of Pomfret as described by the Bastard—"I find the people strangely fantasied; / Possessed with rumours, full of idle dreams,"[110] and other phrases, such as John's "If this same were a churchyard where we stand, / And thou possessed with a thousand wrongs."[111] Hubert notes that Arthur's pleas "take possession of our person";[112] John asks him, "Why seek'st thou to possess me with these fears?"[113] and at the end of the play says, "Weakness possesseth me, and I am faint."[114]

A key theme in this play's exploration of possession is the notion of "commodity," a word that appears six times in a single speech by the Bastard but nowhere else in the play, though it is a recurrent theme.[115] Commodity is profit, advantage, self-interest, as well as goods. It is

> that smooth-faced gentlemen, tickling commodity,
> Commodity, the bias of the world,
> The world, who of itself is peised well, . . .
>
> this bawd, this broker, this all-changing world.[116]

Plausible, dissembling, the "bias" being a weight in a bowling ball; but the world is "peised," that is, balanced, generally.[117] Commodity throws it off balance. As Bate notes, while "set in a feudal world where monarchs were supposed to be God's representatives on earth," the play "exposes power as a 'commodity' for which men are in hungry competition."[118]

The intertwining of feudal, dynastic relations and naked grabs for land runs through the play. Lewis makes a claim to England, even after the pope has reconciled with John, and Pandulph has told the Dauphin his troops are no longer needed: "thy threat'ning colours now wind up, / And tame the savage spirit of wild war."[119] While this takes away the justification for the invasion, Lewis replies:

> Your grace shall pardon me, I will not back:
> I am too high-born to be propertied,
> To be a secondary at control,
> Or useful serving-man and instrument,
> To any sovereign state throughout the world. . . .
>
> I, by the honour of my marriage-bed,
> After young Arthur, claim this land for mine;
> And, now it is half-conquered, must I back
> Because that John hath made his peace with Rome?[120]

Because Lewis will not back down, it is left to the Bastard to threaten a military response: John will "whip this dwarfish war, these pigmy arms, / From out the circle of *his* territories."[121] Lewis admits that the Bastard can "outscold" him, but says he will not waste his time on "such a brabbler," a quarreler.[122] They depart, seemingly to meet again in battle, which rages through the final act. But a messenger brings news to John: "the great supply / That was expected by the Dauphin here / Are wracked three nights ago on Goodwin Sands,"[123] a shoal off the coast of Kent. Nonetheless, Lewis notes that the "English measure backward their own ground / In faint retire," and that he—finally doing what Pandulph had requested—"wound our tott'ring colours clearly up, / Last in the field, and almost lords of it!"[124]

Only then does he receive the news of the loss of the ships: "And your supply, which you have wished so long, / Are cast away and sunk on Goodwin Sands."[125] That seems, now, catastrophic: Lewis seems resigned to have to try again at a later point, and takes no further part in the play; we last hear of him from Salisbury when he recounts that Pandulph has brought "offers of our peace."[126] But the Bastard also recounts that his troops have been affected by the weather:

> I tell thee Hubert, half my power this night,
> Passing these flats, are taken by the tide;
> These Lincoln Washes have devoured them;
> Myself, well mounted, hardly have escaped.[127]

He repeats the tale to the King just as he dies.[128] As in *The Troublesome Reign*, these two geophysical occurrences fuse events across several years. As Braunmuller notes, they combine "the French marine defeats at Damme and in the Channel off the Thames estuary (respectively 1213 and 1218) and making them coincide with the English losses in the Wash (1216 . . .)."[129]

Equally, the rivalry with the pope is not just concerning political control over land, and whose power should hold ultimate sway in England and its possessions. It is also over property *in* land. The church was a major landowner in feudal England, and this was one of the main sources of its wealth. In *Henry V*, the Archbishop of Canterbury persuades Henry of his right to the French crown and lands in part to prevent a raid on the church's own property (see chapter 5). Here John raids the church to pay for the defense of England and the assault on France, suggesting that "Our abbeys and our priories shall pay / This expeditious charge."[130] Later the Bastard is sent back to England ahead of John:

> Cousin, away for England! haste before:
> And ere our coming, see thou shake the bags
> Of hoarding abbots; imprisoned angels
> Set at liberty: the fat ribs of peace
> Must by the hungry now be fed upon:
> Use our commission in his utmost force.[131]

Angels are coins, worth ten shillings, with the archangel Michael on them.[132] The Bastard replies, "Bell, book, and candle shall not drive me back / When gold and silver becks me to come on."[133] Bell, book, and candle are the ritual objects of an excommunication, but these are not enough to prevent the Bastard following this "commission"—a term that means both an order or direction and a monetary gain. When he meets with the King again he notes, "How I have sped among the clergymen / The sums I have collected shall express."[134] This raid on the church's coffers, mentioned by Pandulph,[135] may explain the rationale for John's supposed poisoning by a monk.[136] It is much more explicit in *The Troublesome Reign*, which plays up the behavior of the monks and is a much more explicitly anti-Catholic play.

There are also some interesting passages that relate to *King Lear*'s "interest of territory, cares of state." When the body of Arthur is discovered, the Bastard gives a speech in which he plays on the idea that England the nation and England the body of Arthur can be seen together:

> I am amazed, methinks, and lose my way
> Among the thorns and dangers of this world
> How easy dost thou take all England up
> From forth this morsel of dead royalty!
> The life, the right and truth of all this realm
> Is fled to heaven; and England now is left
> To tug and scamble, and to part by th' teeth
> The unowed interest of proud swelling state.
> Now for the bare-picked bone of majesty
> Doth dogged war bristle his angry crest
> And snarleth in the gentle eyes of peace:
> Now powers from home and discontents at home
> Meet in one line; and vast confusion waits,
> As doth a raven on a sick-fall'n beast,
> The imminent decay of wrestled pomp.[137]

"Amazed" is stronger than the present meaning, confused in a maze, which continues through the next line of being lost in a dangerous world. Hubert carries away the body of Arthur, the "morsel of dead royalty," a soul that has "fled to heaven," but so too has "the right and truth of all this realm," given the state of law and order. England is being torn apart, as if by dogs who "tug and scamble," that is, scramble or struggle; the image is reinforced by "dogged," "snarled," and "crest," the hackles on a dog's back, a plume on military helmets, as well a royal standard.[138] "Unowed interest of proud swelling state" here is perhaps a pun on "unowned"—England has no rightful king, no order or authority, and so the lineage and property relations are breaking down. Where does allegiance go if there is no rightful owner?[139]

"Powers *from* home," that is, not *at* home, so foreign invaders or lords out of their element, unite with "discontents *at* home." As a consequence of the previous struggles, majesty is now a "bare-picked bone," stripped of the visual pomp and ceremony, devoid of the majestic eminence it once had. The struggle for power between England, France, the pope, and domestic lords has meant it has been rendered void. Notably, the state in which there is an interest is detached from the majesty, with the Bastard describing one and then turning to the other: "now for the . . ." The bare-picked bone of majesty is echoed by the raven waiting for carrion, the pomp of sovereignty about to be wrestled away from John's throne. The idea of "interest" of state anticipates both Lewis's reference to his "interest to this land,"[140] that is, a claim or title, and is used in *King Lear*'s "interest of territory, cares of state."

There are also some important parallels between the bodies of Arthur and John and the body of the English kingdom. Initially, the ravages of the kingdom are compared to those of the body of Arthur, and John's own turmoil over the convenience, and guilt over the death. John anticipates what is to come:

> My nobles leave me, and my state is braved,
> Even at my gates, with ranks of foreign powers;
> Nay, in the body of this fleshly land,
> This kingdom, this confine of blood and breath,
> Hostility and civil tumult reigns
> Between conscience and my cousin's death.[141]

As well as the literal body—fleshly, blood and breath—it is the confines of a kingdom, its land or territory with borders, that is "braved," that is, defied or violated. At the end of the play it is the body of the king and the kingdom that are mauled by invasion, whether it be foreign forces or poison.[142] Prince Henry asks John, "How fares your majesty?" asking both of the person and the status. The reply works both ways:

> Poisoned, ill fare; dead, forsook, cast off,
> And none of you will bid the winter come
> To thrust his icy fingers in my maw,
> Nor let my kingdom's rivers take their course
> Through my burned bosom, nor entreat the north
> To make his bleak winds kiss my parched lips

And comfort me with cold. I do not ask you much;
I beg cold comfort. And you are so strait
And so ungrateful, you deny me that.[143]

But when John dies, the Bastard stresses the continuity of succession and the strength of the new king. Early in the play John had said: "Peace be to France, if France in peace permit / Our just and lineal entrance to our own."[144] This is turned back by Philip's reply: "Peace be to England, if that war return / From France to England, there to live in peace. / England we love; and for that England's sake / With burden of our armour here we sweat."[145] In Philip's reply the relation of England the place and England, personified by the person he believes king, that is, Arthur, is clear: "for that England's sake" is the England he loves.[146] But in John's speech France is both France the kingdom and Philip the king. The Bastard's declaration to Henry III at the end of the play echoes this equation, the personification of rule and the equation of the king's body with the kingdom's. Henry is "the lineal state and glory of the land!"[147]

The final lines of plays are frequently spoken by the figure that will rule beyond the scope of the action. So the Bastard having the final lines may mean he is to act as regent for the young king. A mark of his plans can be measured in the defiant, and demonstrably incorrect, speech that closes the play:

O, let us pay the time but needful woe,
Since it hath been beforehand with our griefs.
This England never did, nor never shall,
Lie at the proud foot of a conqueror,
But when it first did help to wound itself.
Now these her princes are come home again
Comes the three corners of the world in arms
And we shall shock them! Nought shall make us rue
If England to itself do rest but true![148]

The appeal is to a more positive future, if England were to stick together and unite behind the new king. Of course the audience would have known something of what happens when this does not happen, from the overthrow of Richard II by Henry Bolingbroke, to the discord in his rule as Henry IV, and the Wars of the Roses that dominate the rule of Henry VI, Edward IV, and Richard III.

THE STAKES OF TERRITORY

In Jean Bodin—who coincidently died in 1596, a likely date for the composition of Shakespeare's play—the notions of majesty and sovereignty are closely fused. Bodin uses, and for some first defines, the word *souveraineté* in the French *Les six livres de la république*, first published in 1576.[149] *Majestas* is the term used in the Latin text *De republica libri sex* in 1586.[150] The Latin text is a reworking in that language, rather than a strict translation: Bodin used the opportunity to expand and develop, as well as emend, some of its claims. But on this term, the sense was strict and, for him, equivalent. As Bodin glosses, "Sovereignty is the absolute and perpetual power vested in a republic, which in Latin is termed *majestatem*. . . . We must now formulate a definition of sovereignty because no jurist or political philosopher has defined it."[151] Later he states that "as for the title 'majesty' it is clear enough that it belongs only to someone who is sovereign."[152]

It is not clear if Shakespeare read the French or Latin text, though it is certainly possible.[153] Rather the point is to stress the way the terms were seen as effectively synonyms in the late sixteenth century. The first English translation of Bodin's text was in 1606, and was a composite version of the French and Latin, along with some phrases that are not found in either.[154] It is not necessarily a good translation, but it is invaluable in tracking the way these terms came into English political debates. The translator was Richard Knolles, who was also the author of a history of the Turks that Shakespeare is likely to have used in his composition of *Othello* (see chapter 6).

In the seventeenth century, though, some writers did begin to see a distinction between majesty and sovereignty. This was especially important in the German debates about the Holy Roman Empire, and the complicated interrelation of the emperor and the kings, dukes, free cities, and other elements that made up the empire. For writers such as Knichen, Althusius, and, a little later, Leibniz, majesty was the overarching power of the emperor, a power to command obedience and demand fealty. Sovereignty was a lower level of rule, though a more effective one: it was day-to-day rule, circumscribed by geographical extent, and limited in power. In Leibniz's pithy summation—after he had taken several hundred pages to get there—to be sovereign was to be "master of a territory."[155]

But, crucially, this was not simply owner of a territory, or territories. Instead of just being the possession of a ruler, this was *now* the limit and extent of their rule. In Shakespeare's *King John* we find a minor example of the shift, a shift from a possessive pronoun to a definite article. *The Troublesome Raign* has "Albion," "our," "your," "their territories"; *King John* has the French ambassador suggest "the territories." But this assertion is disputed, politically and conceptually. As the Bastard, late in the play, asserts the King's right again, it is a right asserted over "his territories."

But just as John's majesty is precarious, so too is his grasp of the lands. The Bastard is stressing the older, feudal relation of a king to his kingdom; the French ambassador perhaps hinting at a more modern sense of territories as dependent possessions of a kingdom, or, in time, a state. This is the sense in which, for example, the Australian Northern Territory is not of the same standing as South Australia, nor the Yukon Territory to the majority of Canadian provinces, or where the Utah Territory preceded the incorporation of the state into the Union. Britain continues to have overseas dependent territories, as does France and the United States. In those instances, being a "territory" is a precondition of statehood.

Neither the English nor the French in *King John* quite see England or France as a territory themselves, but that is a sense that is being slowly asserted in political theory at the time. There are three themes, then, in this play's political geography: majesty, territories, and land. Majesty and land are obvious themes in the play, while territory may appear a minor one. But examining it in relation to the other two themes is revealing. The possession of territory, and the legal-political right to and exercise of power over it, are the topics of the next two chapters.

CHAPTER FOUR

Economic Territories: Laws, Economies, Agriculture, and Banishment in *Richard II*

THE STRANGER PATHS OF BANISHMENT

Having established that territory figures as an important theme in Shakespeare's plays, whether concerning the use of the word "territory" or "territories" in *King John* and elsewhere, or as a more general theme in plays such as *King Lear*, *Hamlet*, and *Macbeth*, this book now shifts to discuss specific aspects of the question of territory: economic and legal issues in this chapter and the next, colonial territories in chapter 6, and technical and corporeal understandings in chapters 7 and 8. Economic questions are another aspect of the examination of possession in the previous chapter. What is the relation between political control of land and its possession? What does possession entail and allow?

The focus to begin an examination of these questions, which continue through subsequent chapters, is *Richard II*.[1] From the very beginning of the play the spatial extent of rule is established through the issue of banishment. The King wants to prevent bloodshed on the land between the feuding noblemen Bolingbroke and Mowbray: "For that our kingdom's earth should not be soiled / With that dear blood which it hath fostered."[2] As a result he stops their duel and expels them. To Henry Bolingbroke, Duke of Hereford, he says:

> Therefore, we banish you our territories.
> You cousin Hereford, upon pain of life,
> Till twice five summers have enriched our fields,
> Shall not regreet our fair dominions,
> But tread the stranger paths of banishment.[3]

Richard, notably, does not expel them from England, but all the lands under his control. "Regreet" means to encounter again, and "stranger" here is an adjective meaning foreign, not a comparison between paths and dominions.[4] To Mowbray, Duke of Norfolk, he presents a harsher sentence:

> Norfolk, for thee remains a heavier doom,
> Which I with some unwillingness pronounce:
> The sly slow hours shall not determinate
> The dateless limit of thy dear exile.
> The hopeless word of 'never to return'
> Breathe I against thee, upon pain of life.[5]

As Charles Forker has noted, Richard II's own words have a powerful performative aspect here: "As a king by divine right, Richard assumes that his words have an almost supernatural power to enact what they refer to."[6] As Richard says, "We were not born to sue"—that is, request or entreat—"but to command."[7] Thus the final lines above: "The hopeless word of 'never to return' / Breathe I against thee, upon pain of life."[8]

Thus the sovereign power of the king is clearly established, and with a spatial determination: he expels from the territories he controls elements that risk their stability. The exile of Mowbray and Bolingbroke is the immediate punishment, because of the feud between them, but Richard is concerned by what may follow.[9] The key thing the King wants to prevent is they do not plot against him or the land. Perhaps their enmity to each other will be forgotten in their shared fate. He realizes that, in banishment, they are no longer within his spatial control, and his former subjects are no longer bound to him as king. He therefore asks them to make their final vow—not to "embrace each other's love in banishment"—not to him but to God: "Swear by the duty that you owe to God— / Our therein we banish with yourselves— / To keep to oath that we administer."[10] He closes by the instruction not "To plot, contrive or complot any ill / 'Gainst us, our state, our subjects or our land."[11]

The invocation of "state" might appear anachronistic, but what is meant by it here is the throne, with a link to the two other elements of rule—subjects and land, what we would now call the population and the territory. The term "state" figures with multiple meanings in the play. Gurr notes that in its meaning as the throne, or "chair of state," it appears as "state of law" (the high position of the king), the "state and crown" that must be resigned to Bolingbroke, and the "sacred state" of Richard's divine position, which Henry takes over.[12] In some of these phrases the meaning is also clearly the king's *status*, as it is in multiple uses of "thy state" or "my state" in the play;[13] and this meaning recurs when the king's status is most debased, such as in the instance of the antic figure of death "scoffing his state"; when it is so reduced that "state" has become "a peasant"; and when Richard counsels his queen to "think our former state a happy dream."[14] There is also the phrase "state and profit," which is used as part of the accusation against Richard. Most of these passages are discussed in more detail below. But it is clear that in none does Shakespeare impart the more modern sense of a political apparatus separate from both the governor and the governed. The closest, perhaps, is the Queen's remark to the gardener that "They will talk of state, for everyone doth so / Against a change; woe is forerun with woe."[15] By this she means that with the transition of power evident, woe tidings before woe events, people will talk of politics, affairs of state.[16]

The power of the king is clear in this scene, showing how Richard still holds sway within his kingdom. Bolingbroke's father, John of Gaunt, is Richard's uncle, and perhaps because of this family relation, Richard shortens Bolingbroke's sentence. Responding to Gaunt's obvious grief he reduces his term:

> Thy sad aspect
> Hath from the number of his banished years
> Plucked four away. Six frozen winters spent,
> Return with welcome home from banishment.[17]

Bolingbroke himself recognizes the performative aspect of this speech:

> How long a time lies in one little word!
> Four lagging winters and four wanton springs
> End in a word; such is the breath of kings.[18]

Nonetheless Gaunt is not satisfied, because he realizes that while the exile has been shortened by four years, his own life will not last that long:

> But little vantage shall I reap thereby, . . .
>
> My inch of taper will be burnt and done,
> And blindfold Death not let me see my son.[19]

The King suggests that he has "many years to live," but Gaunt replies that "not a minute . . . that thou canst give."[20] The limits of the King's power are evidenced here: he can reduce sentences and take life, but not grant it. This was, of course, Foucault's point: the privilege of "sovereign power was the right of life and death."[21] But this was really "the right to *make* die or *let* live [faire *mourir ou de* laisser *vivre*]." Only with biopolitics did it become the "power to *make* live or *disallow* it in death [faire *vivre ou de* rejeter *dans la mort*]."[22]

The banishment theme throughout the play is linked to notions of pilgrimage. One example is Bolingbroke's speech before the duel:

> Lord Marshal, let me kiss my sovereign's hand
> And bow my knee before his majesty.
> For Mowbray and myself are like two men
> That vow a long and weary pilgrimage;
> Then let us take a ceremonious leave
> And loving farewell of our several friends.[23]

The pilgrimage then is the threat of death in the duel, but later in the same scene, now exiled, he speaks of his "enforced pilgrimage."[24] Equally Mowbray's challenge to Bolingbroke had been that he would "meet him were I tied to run afoot / Even to the frozen ridges of the Alps, / Or any other ground inhabitable / Where ever Englishman durst set his foot."[25] "Inhabitable" here means not habitable, uninhabitable: an anticipation of a fight to the death or of the exile to come.[26] Now Bolingbroke speaks of the sentence passed upon him:

> Then England's ground, farewell! Sweet soil, adieu—
> My mother and my nurse that bear me yet!
> Where'er I wander, boast of this I can,
> Though banished, yet a true-born Englishman.[27]

The issue of pilgrimage makes a return at the end of the play, discussed in the conclusion below.

REALM, NATION, FARM

With this internal dispute between Mowbray and Bolingbroke seemingly resolved, Richard then turns his attention to the rebellion in Ireland. Rather than a threat to his rule within his territories, such as the feud between Bolingbroke and Mowbray, this is a threat of a more significant breakaway of one of those territories. His new concern is how to provide the resources to deal with this:

> We will ourself in person to this war
> And, for our coffers with too great a court
> And liberal largesse are grown somewhat light,
> We are enforced to farm our royal realm,
> The revenue whereof shall furnish us
> For our affairs in hand. If that come short,
> Our substitutes at home shall have blank charters
> Whereto, when they shall know what men are rich,
> They shall subscribe them for large sums of gold,
> And send them after to supply our wants;
> For we will make for Ireland presently.[28]

Richard recognizes that the way he has been conducting his court and dispensing favors has left his funds low. In order to attend to the situation in Ireland, a new source of revenue is needed. The idea of farming the realm is important: this is a play filled with the language of soil, land, earth, and ground. Forker suggests that this "roots the motif of patriotism in the pervasive imagery of earth, land and ground."[29] Forker uses Richard Altick's analysis, which shows that "the three words occur a total of 71 times."[30] For Forker, this is part of a wider linguistic context, where "nature imagery suffuses *Richard II*, projecting a landscape of potential fruitfulness."[31]

Patriotism and attachment to place are certainly significant themes in this play, but there is a strong land politics at stake beyond this, and an economic rather than merely a natural sense. We might say that there is a geopolitical economy at work in the play. The King is condemned for selling rights and precisely for farming the realm.[32] "Farm" means to farm out, to lease to tenants ("let this land by lease"[33]). He will thereby get a stable rent, but the revenue yielded by the land and its resources go to those that pay the rent.[34] His deputies left behind will have "blank charters" to extract resources. Forker glosses these as "writs authorizing agents of the crown to extort revenues in unspecified amounts from the affluent."[35] The charters would have the names and amounts inscribed (subscribed) in them, and the "blank" aspect was that the charge was initially unknown.[36] Farming the royal realm may also mean leasing out the offices of being king; "his 'royalties,' his prerogatives and responsibilities."[37] But Richard thinks that there may be easier ways to obtain money—simply seizing it from others. One of these is John of Gaunt, Bolingbroke's father, who, as the opening scenes indicated, realized he did not have long left to live. Richard's shift in focus is clear when he hears of Gaunt's illness:

Now put it, God, in the physician's mind
To help him to his grave immediately!
The lining of his coffers shall make coats
To deck our soldiers for these Irish wars.
Come, gentlemen, let's all go visit him.
Pray God we may make haste and come too late![38]

Yet, while dying, Gaunt has the life left to make his powerful and famous speech where he both praises the land and deplores what it has become:

Methinks I am a prophet new inspired,
And thus, expiring, do foretell of him.
His rash fierce blaze of riot cannot last,
For violent fires soon burn out themselves;
Small showers last long but sudden storms are short;
He tires betimes that spurs too fast betimes;
With eager feeding food doth choke the feeder.
Light vanity, insatiate cormorant,
Consuming means, soon preys upon itself.
This royal throne of kings, this sceptred isle,
This earth of majesty, this seat of Mars,
This other Eden, demi-paradise,
This fortress built by Nature for herself
Against infection and the hand of war,
This happy breed of men, this little world,
This precious stone set in the silver sea,
Which serves it in the office of a wall
Or as a moat defensive to a house
Against the envy of less happier lands,
This blessed plot, this earth, this realm, this England,
This nurse, this teeming wound of royal kings,
Feared by their breed and famous by their birth,
Renowned for their deeds as far from home,
For Christian service and true chivalry,
As is the sepulchre in stubborn Jewry
Of the world's ransom, blessed Mary's son,
This land of such dear souls, this dear dear land,
Dear for her reputation through the world,
Is now leased out—I die pronouncing it—

Like to a tenement or pelting farm.
England, bound in with the triumphant sea,
Whose rocky shore beats back the envious siege
Of wat'ry Neptune, is now bound in with shame,
With inky blots and rotten parchment bonds.
That England that was wont to conquer others
Hath made a shameful conquest of itself.
Ah would the scandal vanish with my life,
How happy then were my ensuing death![39]

The speech has, rightly, been seen as a protonationalist invocation of English identity, perhaps more in keeping with the sense at the time of writing rather than the historical moment being portrayed. Yet the speech equally speaks of a specific geography to this England. The "sceptred isle," the maritime nature of the kingdom, and the physical geography of its setting are all significant themes. While England is not itself an "isle," and the relation to Scotland and Wales is obscured, the sense of an island nation is clearly stressed. The physical situation becomes a military setting, as a fortress, helping to keep the island free from the dangers of the wider world—infection and war. The later use of "moat," and its "rocky shore," stress the defensive properties, here against invaders and the god of the sea. The double use of "bound in with," first as a geographical guard, and then as statement of its decline, is equally powerful. "Plot," here, is both site and map, drama or strategy—a sense that is scaled up to earth, realm, and kingdom. The triple invocation of "dear" implies both what is loved, and what is valued; a bound is a spatial limit as well as a restraint.[40] A tenement is an estate with a tenant farmer, a "pelting farm" is a small, paltry farm—nothing to do with animal pelts.[41] The line echoes the anonymous play *Woodstock: A Moral History*, usually dated between 1591 and 1595, and sometimes called *King Richard II, Part 1*. In that play it is Richard describing his own role: "Become a landlord to this warlike realm, / Rent out our kingdom like a pelting farm."[42] In *Woodstock* the division of land is much more explicit: "Reach me the map, we may allot their portions, and part the realm among them equally."[43] As Sullivan notes, "Map and survey produce a landscape of absolute property. . . . If survey and map make possible the cartographic construction of the nation, the use of the same also allows the management of the constituent part of the nation in a way that is understood as detrimental to it both at the local level and as a whole."[44]

The "inky blots" are the charters used to seal these deals. Gaunt then explicitly taunts the King: "Thy death-bed is no lesser than thy land, /

Wherein thou liest in reputation sick."[45] He continues, with the punning criticism of Richard's obsequious followers:

A thousand flatterers sit within thy crown,
Whose compass is no bigger than thy head;
And yet, encaged in so small a verge,
The waste is no whit lesser than thy land.[46]

If the compass is the circumference, then the real jurisdiction of the King extends no further than his body, within the space enclosed by his crown; while a verge is a "limit, demarcation, border," or, again, a compass. Forker adds that for verge "wordplay on two more technical terms is perhaps also intended: the twelve-mile sphere of jurisdiction around the King's court assigned to the Lord Marshal; a measure of land of some fifteen to thirty acres."[47] Waste too has multiple meanings.[48] Forker notes that it may relate to "head," as the homonym "waist"; but more specifically the word means extravagance, wasteland, or "damage done to a rented property by a tenant, either deliberately or through neglect."[49] And finally Gaunt returns to the criticism of the economic extraction from the land of the kingdom:

Why cousin, wert thou regent of the world,
It were a shame to let this land by lease;
But for thy world enjoying but this land,
Is it not more than shame to shame it so?
Landlord of England are thou now, not king.
Thy state of law is bondslave to the law.[50]

The use of "landlord" is particularly striking. As Peter Ure has noted, the term is used five times in *Woodstock*. *Woodstock* is a possible source for Shakespeare, and Ure thinks that this word proves he was the borrower. As well as the passage cited above, a key formulation is "And thou no king, but landlord now become / To this great state that terrored christendom."[51] For Ure, "it is of course more likely that Shakespeare remembered the word because it is repeated so often in the other play than that the author of *Woodstock* expanded the single reference in *Richard II* into so abundant a treatment in his own work."[52] The "state of law" is Richard's status as king, the lawgiver of the country, and yet because of these economic trades, he is now himself bound by the law. He is beneath, rather than above, the law.[53]

Yet while Gaunt's criticism is powerful, and his points well founded, this neither delays his death nor dissuades Richard from his course. When Richard hears of the death of Gaunt he is merciless:

> The ripest fruit first falls, and so doth he.
> His time in spent; our pilgrimage must be.
> So much for that. Now for our Irish wars:
> We must supplant those rough rug-headed kerns,
> Which live like venom where no venom else
> But only they have privilege to live.
> And, for these great affairs do ask some charge,
> Towards our assistance we do seize to us
> The plate, coin, revenues and moveables
> Whereof our uncle Gaunt did stand possessed.[54]

This will become a recurrent theme through the play. Richard dispossesses not just Gaunt, but also the exiled Bolingbroke of the lands and title he inherits from Gaunt, as well as the particular kinds of property listed here.[55] In this he creates a grievance far more serious than the banishment he laid on Bolingbroke, because he is breaking the line of inheritance.

The Duke of York counsels him on precisely this point. He tells him not to remove "the royalties and rights of banished Hereford," his "charters and his customary rights," because of the precedent: "for how are thou a king / But by fair sequence and succession?"[56] This is an important point: Richard is not just dispossessing one man, in his "robbing of the banished Duke," but effectively challenging the very right of inheritance.[57] York continues:

> If you do wrongfully seize Hereford's rights,
> Call in the letters patents that he hath
> By his attorneys-general to sue
> His livery and deny his offered homage,
> You pluck a thousand daggers on your head. [58]

Shakespeare closely follows Holinshed's *Chronicles* here. The reference is to the particular legal proceedings that would be used to take the lands and goods. Both "letters patents" and "sue his livery" are references to suits Bolingbroke would make through lawyers to inherit the lands from his father, which would go to the King unless he could prove he was a legitimate heir, and of age. In inheriting, he was supposed to make homage to the King.

Presumably because he is in exile he is unable to do so, and so Richard denies him. Richard thus seizes the rights, calls in—that is, revokes—the letters patent used by the lawyers to sue his livery—to claim delivery of his lands. Thus the lands would revert to the Crown.[59]

The King will have none of this caution: "Think what you will, we seize into our hands / His plate, his goods, his money and his lands."[60] As Forker puts it, this is "Richard's most criminal act, apart from destroying Gloucester and farming the realm . . . an egregious violation of the cherished law of inheritance on which the royal title itself depends."[61] Other noblemen in the play recognize this. Shortly after Gaunt dies, Northumberland declares, "Well, lords the Duke of Lancaster is dead." Ross comments, "And living, too, for now his son is duke." Willoughby notes, "Barely in title, not in revenues"; to which Northumberland responds, "Richly in both, if Justice had her right."[62] Richard thus creates a legitimate grievance in Bolingbroke, beyond mere exile. As a result, Bolingbroke returns to England at the head of an army. He does this while Richard is abroad seeking to pacify the Irish rebellion. This is part of the reason of course Richard needed the funds he raised by such illegitimate means.

The nobles then voice the generalized concerns:

Ross: The commons hath he pilled with grievous taxes,
And quite lost their hearts. The nobles hath he fined
For ancient quarrels, and quite lost their hearts.
Willoughby: And daily new exactions devised,
As blanks, benevolences, and I wot not what.[63]

What is striking here is that both the nobles and the commons have been raided for money, the commons stripped or peeled, the notion akin to pillage; the nobles through a range of means, including taxes and fines. "Blanks" here are the blank charters; "benevolences" are forced loans.[64] While we tend to hear only from the nobles directly, the garden scene gives a sense of the King's difficult relations with the commons, and here and elsewhere we hear that reported. We are told for instance by York that the King has offended them: "the commons . . . are cold" to him and may revolt against him on Bolingbroke's side.[65]

Ross: The Earl of Wiltshire hath the realm in farm.
Willoughby: The King's grown bankrupt like a broken man.
Northumberland: Reproach and dissolution hangeth over him.

Ross: He hath not money for these Irish wars,
His burdenous taxations notwithstanding,
But by the robbing of the banished Duke.[66]

EARTH AND THE KING'S TWO BODIES

The mirror image between what Richard seeks to achieve in Ireland and what he will lose at home is striking. As York says to the Queen,

Your husband, he is gone to save far off,
Whilst others come to make him lose at home.
Here am I left to underprop his land,
Who, weak with age, cannot support myself.[67]

Richard wants to save Ireland, that is to protect and preserve Ireland under English rule, while the threat from others means he is likely to lose what he currently has over England itself.[68]

Strikingly, on his return from Ireland, Richard praises his land through a sequence of uses of the language of earth:

Needs must I like it well. I weep for joy
To stand upon my kingdom once again.
Dear earth, I do salute thee with my hand,
Though rebels wound thee with their horses' hoofs. . . .

So weeping, smiling, greet I thee, my earth, . . .

Feed not thy sovereign's foe, my gentle earth . . .

This earth shall have a feeling, and these stones
Prove armed soldiers, ere her native king
Shall falter under foul rebellion's arms.[69]

Several commentators have drawn parallels between Richard's sense of the earth of his kingdom, and his physical, corporeal connection to it, and the notion of the *corpus mysticum*. The *corpus mysticum* was crucial to the medieval notion of "the King's Two Bodies," in Ernst Kantorowicz's phrase. The bodies were the King's own vulnerable and weak physical body, and the eternal body politic.[70] This notion was crucial to the Tudor political imagination, and shaped ideas of monarchy. The physical, natural body was human, and so could become sick, injured, and die, while the body politic was effectively the state, and would endure beyond the death of the individual. As King Richard II himself says in Shakespeare's text, "thus play I in one person many people."[71] Indeed, Kantorowicz's influential study has a detailed reading of *Richard II* on precisely this point: Shakespeare "has made it not only the symbol, but indeed the very substance and essence of one of his greatest plays: *The Tragedy of Richard II* is the tragedy of the King's Two Bodies."[72] Kantorowicz shows how Richard is confronted with the split between the two bodies, for as his own body suffers from torment it is increasingly detached from the body politic.

The break between the two bodies is vividly evident when Richard regards himself in a mirror after his deposition, only to throw the mirror to the floor so that it is "cracked in an hundred shivers."[73] While for most kings their own mortality is balanced by the idea that the monarchy will continue, Richard realizes that his own body is becoming detached from the kingly ideal. Kantorowicz shows how this actually magnifies his physical suffering, and the pain of his shedding the trapping of kingship, and with them the unity of the corporeal and mystical bodies.

In Ruth Nevo's reading Richard's relation to the earth is in contrast to John of Gaunt's praise of the land:

> Richard, unlike Gaunt, is invoking an ancient, sacramental magic. It is prenational, a-historical; it is the sacred, animistic bond between king and land—the *corpus mysticum* which includes and transcends both political kingdom and physical earth, as kingship includes and transcends both the king's eternal "body politic" and his personal, natural self. . . .
>
> Gaunt's patriotism is anachronistic, far more Elizabethan than feudal, is less to the purpose than the perception that Richard, by contrast, is drawing upon constitutional and legal doctrine, and quasi-erotic sentiment to supply the strength and confidence which "worldly men" derive from the exercise of political arts. Richard's identification of himself with his kingship and with the land he is part of and possesses constitutes the inherited and as yet untried conception of himself to which he retreats at the first crisis.[74]

What is intriguing in this is that Gaunt, the elder figure, who might be seen as invoking a lost world, is actually the more modern. In a previous section I have indicated the protonationalist overtones of his speech, which speak both of a national identity to the people and of the island geographies of the nation, but the play is intriguing in setting that up in opposition to the other view. There are tensions here: on the one hand, Richard has a primal relation to the kingdom's earth; on the other, he sees it as a commodity to be monetized. It is Richard who is suggesting a feudal view of land as resource, while still holding to an ancestral bond, rather than the kingdom as personification of the nation. Bolingbroke will be the inheritor of the sense of the kingdom as resource, even if Richard sought to dispossess him of his father's lands. For Forker,

> Richard's essentially feudal world, a world of oaths and codes and honour, of titles and fixed identities, of ritual solemnity and ceremonial beauty, puts heavy stress on the seriousness and potency of words. Bolingbroke, who challenges and overturns that world, brings to bear a more modern, relativistic, sceptical and less comely understanding of how meaning is generated.[75]

Kantorowicz's hypothesis concerning the king's two bodies also helps to resolve a tension in Gaunt's accusation, since Richard is supposed to be landlord; yet one of his crimes is that of a tenant. How is it that Richard is treating the kingdom as wasteful tenant, while also its landlord? Klinck suggests that "the paradox might be explained by saying that Richard, who in his 'Body politic' is lord paramount, is in his body natural a tenant who holds or occupies the 'Dignity royal' and its appurtenances; as tenant, he commits waste upon that which he holds, the consequence being that he is deprived prematurely of that holding."[76]

BOLINGBROKE'S RETURN

Strikingly, for a man whose tie to the land is emotive rather than narrowly economic, Bolingbroke's name frequently invokes places. As Forker notes,

> Unlike Richard, his adversary is known in the play by a variety of names—Derby, Hereford, Bolingbroke, Lancaster and finally Henry IV. The variation suggests his more uncertain and shifting identity as subject, rebel, usurper and monarch in a way that ports with his more relativistic and pragmatic attitude towards language.[77]

But it also stresses his attachment to the specific sites, owned, inherited, lost, and regained. His relation to the land is clear when he returns from exile early, back in a forbidden place before the time has elapsed. As the Duke of York says to him,

Even in condition of the worst degree,
In gross rebellion and detested treason.
Thou art a banished man, and here art come,
Before the expiration of thy time,
In braving arms against thy sovereign.[78]

For York, his very presence in the country, before the expiration of the sentence, is treason against the Crown. But Bolingbroke has a clear reply:

As I was banished, I was banished Hereford;
But as I come, I come for Lancaster. . . .

Will you permit that I shall stand condemned
A wandering vagabond, my rights and royalties
Plucked from my arms perforce and give away
To upstart unthrifts? Wherefore was I born?[79]

Bolingbroke thus claims he is justified in returning, saying he will see the King and "Even at his feet to lay my arms and power / Provided that my banishment repealed / And lands restored again be freely granted."[80] There are, now, two elements to his appeal: not just the lands, but also the original sentence. The alternative is bloodshed, blood from the people of England into the very land itself:

I'll the advantage of my power
And lay the summer's dust with showers of blood
Rained from the wounds of slaughtered Englishmen—
The which how far off from the mind of Bolingbroke
It is such crimson tempest should bedrench
The fresh green lap of fair King Richard's land
My stooping duty tenderly shall show.
Go signify as much, while we here march
Upon the grassy carpet of this plain.[81]

Yet whatever his justification for appealing the loss of his land, Bolingbroke's mere presence in the land is treason. Richard states as much:

Tell Bolingbroke—for you methinks he stands—
That every stride he makes upon my land
Is dangerous treason. He is come to open

The purple testament of bleeding war;
But ere the crown he looks for live in peace,
Ten thousand bloody crowns of mothers' sons
Shall ill become the flower of England's face,
Change the complexion of her maid-pale peace
To scarlet indignation, and bedew
Her pastor's grass with faithful English blood.[82]

While Bolingbroke claims to only be after his own title and lands, he ends up with Richard's crown and kingdom as well. As Forker suggests, "Publicly, Bolingbroke claims only to seek what is legally his—his hereditary lands and title. And Richard makes it unnecessary for him to claim more by agreeing to relinquish the crown before it has been formally demanded."[83] The indirect route to the crown is an intriguing aspect of Bolingbroke's story in the play. Richard tries to reconcile himself to this changing state of affairs: "Say, is my kingdom lost? Why, 'twas my care; / And what loss is it to be rid of care?"[84] Yet this loss is of course forced by the presence of Bolingbroke and his forces in the kingdom: "So high above his limits swells the rage / Of Bolingbroke, covering your fearful land / With hard bright steel and hearts harder than steel."[85] "Steel," here, refers back to Richard's own suggestion that his position was impregnable:

Not all the water in the rough rude sea
Can wash the balm off from an anointed king;
The breath of worldly men cannot depose
The deputy elected by the Lord.
For every man that Bolingbroke hath pressed
To lift shrewd steel against our golden crown,
God for his Richard hath in heavenly pay
A glorious angel. Then if angels fight,
Weak men must fall; for heaven still guards the right.[86]

The idea of angels posed against soldiers is compelling, just as "the breath of worldly men" is impotent compared to "the breath of kings." Yet Richard knows that his position is, in truth, entirely dependent on earthly support. An angel is, of course, a coin, so the economy of his position is clear here too. As his position erodes, he blames others including the Earl of Wiltshire, Bagot, Bushy, and Green, not knowing all but Bagot are already dead: "That they have let the dangerous enemy / Measure our confines with such peaceful steps?"[87] Forker glosses "measure our confines" as "travel through (literally, 'around the boundaries of') my territory."[88] Richard accuses them of making peace with Bolingbroke, of being Judases, and having "spotted souls," that is, stained with treason.[89] It is left to Scroop to reveal that they have not betrayed Richard, but been executed: "Their peace is made / With heads and not with hands."[90] Richard now bemoans his position, in some of the most beautiful verse of the play:

> Let's talk of graves, of worms and epitaphs,
> Make dust our paper and with rainy eyes
> Write sorrow on the bosom of the earth.
> Let's choose executors and talk of wills.
> And yet not so, for what can we bequeath
> Save our deposed bodies to the ground?
> Our lands, our lives and all are Bolingbroke's,
> And nothing can we call our own but death
> And that small model of the barren earth
> Which serves as paste and cover to our bones.
> For God's sake let us sit upon the ground
> And tell sad stories of the death of kings—
> How some have been deposed, some slain in war,
> Some haunted by the ghosts they have deposed,
> Some poisoned by their wives, some sleeping killed
> All murdered. For within the hollow crown
> That rounds the mortal temples of a king
> Keeps death his court; and there the antic sits,
> Scoffing his state and grinning at his pomp.[91]

Much of this is readily transparent, though with the idea that all has gone to Bolingbroke, Richard now thinks he is left only the barren earth that will enclose his mortal remains. The reduction of the space of his kingdom to the earth, ground, or grave parallels the shrinking scales of *Hamlet*'s dramatic action. "Model" either means the flesh as a smaller version of the world, or model as mold—paste being a pastry cover over a meat pie.[92] The "sad stories" is perhaps a reference to a book that would have been known by Shakespeare's audience, *The Mirror for Magistrates*.[93] Of course, it could not have been known by the historical Richard because it long postdates his life, and he, along with Mowbray, Thomas of Woodstock, and other characters in Shakespeare's plays, including Glendower, Henry Percy, and Jack

Cade appear within it. The ghosts appear, of course, to Richard III as well as to Macbeth; and Richard himself will haunt Bolingbroke when he is Henry IV. The antic, the jester, laughs at the King's state and pomp—his status and splendor. Now, compared to the power of the opening scene, we clearly see Richard's late linguistic impotence:

> O God, O God, that e'er this tongue of mine
> That laid the sentence of dread banishment
> On yon proud man should take it off again
> With words of sooth![94]

He also lists the trades he is now prepared to make—an attempted exchange economy:

> I'll give my jewels for a set of beads,
> My gorgeous palace for a hermitage,
> My gay apparel for an almsman's gown,
> My figured goblets for a dish of wood,
> My sceptre for a palmer's walking staff,
> My subjects for a pair of carved saints
> And my large kingdom for a little grave,
> A little, little grave, an obscure grave;
> Or I'll be buried in the King's highway,
> Some way of common trade, where subjects' feet
> May hourly trample on their sovereign's head;[95] . . .
>
> We'll make foul weather with despised tears;
> Our sighs and they shall lodge the summer corn
> And make a dearth in this revolting land.[96]

The holy nature of Richard comes through here, where, reduced from being a divinely appointed king, he imagines himself as a holy beggar saying prayers for those who give him alms, a pilgrim or monk with simple robe, a begging bowl, staff, and wooden figures of holy figures. It is the removal of the apparel of kingship, the breaking of the bond between his physical body and the *corpus mysticum*. The reduction of the kingdom to the grave is stressed once more. And then, in one of their final exchanges, Richard demurs when Bolingbroke kneels: "Fair cousin, you debase your princely knee / To make the base earth proud with kissing it."[97] Bolingbroke states, "My gracious lord, I come but for mine own," to which King Richard replies,

"Your own is yours, and I am yours and all."[98] While claiming to be pursuing only his right to his father's lands, Bolingbroke attains the kingdom as a whole.

THE CATERPILLARS OF THE COMMONWEALTH

Bolingbroke uses the agricultural language of pests or parasites to describe those who have farmed the land on Richard's behalf: "Bushy, Bagot and their complices, / The caterpillars of the commonwealth, / Which I have sworn to weed and pluck away."[99] This continues when he condemns them to death—a premature and presumptive assumption of the sovereign right:[100] "Whilst you have fed upon my signories, / Disparked my parks and felled my forest woods."[101] "Signories" are manor and estates, the lands he should have inherited from his father, and Forker suggests that "disparked my parks" means "'converted my parklands to uses other than keeping game and hunting.' Less aristocratic uses such as timbering and farming seem to be implied as well as destruction of the walls or fences that traditionally enclose game parks."[102] This may be another instance of tenant behavior, constituting waste.[103] Most crucially, however, he is already asserting power over space and its use.

This is then played out in the garden scene:[104]

> Gardener: Go thou, and like an executioner,
> Cut off the heads of too fast-growing sprays
> That look too lofty in our commonwealth.
> All must be even in our government.
> You thus employed, I will go root away
> The noisome weeds, which without profit suck
> The soil's fertility from wholesome flowers.
> 1 Man: Why should we in the compass of a pale
> Keep law and form and due proportion,
> Showing, as in a model, our firm estate,
> When our sea-walled garden, the whole land,
> Is full of weeds, her fairest flowers choked up,
> Her fruit trees all unpruned, her hedges ruined,
> Her knots disordered and her wholesome herbs
> Swarming with caterpillars?[105]

The "fast-growing sprays" are new branches, a none-too-subtle reference to the names of the men just executed. They were too "lofty" and needed to be cut off. Though quickly they are also likened to weeds that need to be removed from the "sea-walled garden" that is the island of Britain. Within "the compass of a pale" is another instance of "pale" in the sense of an enclosed space, here a walled garden, but the sense is of the English pale around Dublin that Richard had sent troops to defend, that is, a bounded legal territory. Violence is needed to keep things in order. The gardeners then, while clearly against Richard, are closer to Bolingbroke than his father, John of Gaunt. While Gaunt had praised the kingdom in these terms, it is Bolingbroke who used force to maintain and repair it.[106] "Caterpillars," of course, echoes Bolingbroke's own description, and we then hear the gardener explicitly say that the weeds "are plucked up, root and all, by Bolingbroke— / I mean the Earl of Wiltshire, Bushy, Green."[107]

The Gardener tells us that they are dead and that Bolingbroke

> Hath seized the wasteful King. O, what pity is it
> That he had not so trimmed and dressed his land
> As we this garden! We at time of year
> Do wound the bark, the skin of our fruit trees,
> Lest, being over-proud in sap and blood,
> With too much riches it confound itself.
> Had he done so to great and growing men,
> They might have lived to bear and he to taste
> Their fruits of duty. Superfluous branches
> We lop away that bearing boughs may live.
> Had he done so, himself had borne the crown,
> Which waste of idle hours hath quite thrown down.[108]

The "wasteful king" is of course a reference back to earlier notions of waste; and the need to prune, cut, and discard is clear in keeping a well-managed kingdom, as much as a garden. Richard has failed to do this; the hope is that Bolingbroke will attend to such matters. His execution of the King's supporters—already exercising a royal prerogative—is an example of Bolingbroke seizing this opportunity.

TIRED MAJESTY

Returning to one of the opening images, the Bishop of Carlisle is arrested for a speech that echoes Gaunt's key discussion earlier in the play: this one in support of Richard; that against. For Stanley Wells, this is the key tension in the play: while Richard is a bad king, more suffering will come from his overthrow.[109] Carlisle says that if Bolingbroke is crowned king, anticipating the situation in the *Henry IV* plays,

> The blood of English shall manure the ground,
> And future ages groan for this foul act.
> Peace shall go to sleep with Turks and infidels,
> And in this seat of peace tumultuous wars
> Shall kin with kin and kind with kind confound.
> Disorder, horror, fear and mutiny
> Shall here inhabit, and this land be called
> The field of Golgotha and dead men's skulls.
> O, if you raise this house against this house,
> It will the woefullest division prove
> That ever fell upon this cursed earth.[110]

In the longer version of the scene, which appeared first in the Fourth Quarto of 1608, Richard is brought before the assembled lords to cement this transfer of power. (The textual question here will be discussed in the conclusion.) The Duke of York says to the King:

> You do that office of thine own good will
> Which tired majesty did make thee offer—
> The resignation of thy state and crown
> To Henry Bolingbroke.[111]

"Tired majesty" is, here, the stress or responsibility of being king, as well as the tarnished status of this, and Richard resigns his kingly status, his rank, and his throne.[112] As Richard says to Bolingbroke,

> No mark me how I will undo myself:
> I give this heavy weight from off my head,
> And this unwieldy sceptre from my hand,
> The pride of kingly sway from out my heart;
> With mine own tears I wash away my balm,
> With mine own hands I give away my crown,
> With mine own tongue deny my sacred state,
> With mine own breath release all dutious oaths.
> All pomp and majesty I do foreswear;

My manors, rents, revenues I forgo;
My acts, decrees and statutes I deny.[113]

He ends by suggesting, "Long mayst thou live in Richard's seat to sit, / And soon lie Richard in an earthy pit!"[114]

This, though, is still not enough. He may give up all the trappings of power, be they wealth, ritual and legal. In particular, we should note the surrender of his "manors, rents, revenues." He is not simply giving up the land, but the economic use and yield it provides. But Northumberland suggests that the deposed King also needs to sign confessions to various crimes:

No more, but that you read
These accusations, and these grievous crimes
Committed by your person and your followers
Against the state and profit of this land,
That, by confessing them, the souls of men
May deem that you are worthily deposed.[115]

"State and profit" is, Forker suggests, a hendiadys, the meaning being "profitable state" or "ordered prosperity" of the kingdom.[116] Richard resists, suggesting that his tear-filled eyes mean that he cannot see to read, and that he has already said enough:

For I have given here my soul's consent
T'undeck the pompous body of a king,
Made Glory base and Sovereignty a slave
Proud Majesty a subject, State a peasant.[117]

While Northumberland continues to insist, even Bolingbroke eventually says they should desist.[118] Northumberland notes that "the commons will not then be satisfied," and Richard says he will "read enough" to ensure they are.[119] The additional Fourth Quarto and Folio scene ends with Richard's imprisonment, and both texts converge with Bolingbroke making plans for his coronation.[120] The coronation, notably, happens offstage, and Bolingbroke reappears on stage as King Henry IV later in the final act.

THE RETURN OF BANISHMENT

In her farewell scene with Richard, the Queen talks of the "rebellious earth" of the kingdom,[121] and suggests that the King should still have the strength to resist, as a dying lion "wounds the earth."[122] The King suggests, in the final instance of this key word, that "I wasted time, and now doth Time waste me / For now Time made me his numb'ring clock."[123] Richard, while initially prisoner, does eventually die, killed by Exton. Bolingbroke's role in this death is equivocal: both desired but not explicitly sanctioned. The blood that had been for so long deferred is finally spilt on England's earth:

> That hand shall burn in never-quenching fire
> That staggers thus my person. Exton, thy fierce hand
> Hath with the King's blood stained the King's own land.
> Mount, mount, my soul! Thy seat is up on high,
> Whilst my gross flesh sinks downward here to die.[124]

One of the first challenges to the newly crowned King Henry IV, the former Bolingbroke, comes from his cousin, Aumerle. A plot against Henry is uncovered, and his mother, the Duchess of York, begs for Aumerle's forgiveness. Henry says, "I pardon him, as God shall pardon me."[125] This is a telling line—the new king now has that performative power, but still recognizes his own sin, which needs to be repaid. The Duchess cements this equation: "A god on earth thou art!"[126] But these lines are spoken before the death of Richard. In two of the final lines of the play, Henry suggests it is the crime of Richard's death, not his overthrow, that must be paid in penance. He looks for redemption through geography, through the voluntary exile of pilgrimage: "I'll make a voyage to the Holy Land / To wash this blood off from my guilty hand."[127]

At the beginning of the deposition scene Bolingbroke had decided that the banishment of Mowbray, Duke of Norfolk, should also be repealed. While he was his enemy, Bolingbroke says that he will be "restored again / To all his lands and signories"—"signories" being estates.[128] Bolingbroke thus wants to do for Mowbray what Richard denied to him. But this is not to be. From the Bishop of Carlisle we hear that Mowbray served in the Crusades and died in Venice, giving "his body to that pleasant country's earth."[129] Thus, finally, the play invokes the return of the idea of banishment, of pilgrimage, with which it began.

The issue of this trip to the Holy Land recurs in both parts of *Henry IV*. In *Part 1*, right at the start of the play the king says that "short-winded accents of new broils," that is, conflicts within the kingdom, will prevent his promised pilgrimage to the Holy Land—"to be commenced in strands afar remote," strands being places or shores or lands.[130] He mentions the possibility of a crusade, but repeats his reason for a delay twice more in this opening scene.[131]

In *Part 2*, he again expresses his wish that "were these inward wars once out of hand, / We would, dear lords, unto the Holy Land."[132] The civil situation never becomes stable enough that his wish becomes possible. Henry, ironically, will die in the Jerusalem chamber in the palace of Westminster at the end of *Henry IV, Part 2*. As he falls sick, he asks Warwick the name of the room in which he first swooned. He learns that it is called Jerusalem:[133]

> Laud be to God, even there my life must end.
> It hath been prophesied to me many years
> I should not die but in Jerusalem,
> Which vainly I supposed the Holy Land,
> But bear me to that chamber: there I'll lie.
> In that Jerusalem shall Harry die.[134]

Richard II is thus a crucial play concerning the political economies of land and the political geographies of banishment. There were important contemporary resonances to attempts to depose Elizabeth—there is a story of *Richard II* being played to an audience at the Globe the day before the failed attempt to depose the queen by the Earl of Essex, on February 7, 1601.[135] The queen was said to have drawn parallels between her own reign and that of Richard, saying to her keeper of the Tower, William Lambarde: "I am Richard II, know ye not that."[136] This does not mean that Shakespeare wrote the play to support deposition, or that the performers intended to support the rebellion, but the parallel is striking.[137] As Gurr notes,

> It was boldly adroit of Shakespeare, when Elizabeth had forbidden any discussion of her likely successor, to write a play showing how the door had first been opened to disgorge the horrors of a long and bloody civil war when no legitimate heir could be found. The historical parallel was a warning. Later he used a similar tactic with *King Lear*, when he set out the questions inherent in James's wish to unify his two kingdoms by showing its opposite: what could happen to a divided state.[138]

It is important to note that the crucial deposition scene that shows the confrontation between Richard and Henry Bolingbroke, and the transfer of the crown from one to the other, was not in the first three quartos of the play. It appeared first in the Fourth Quarto of 1608 and then in a better text in the First Folio of 1623. It is an addition to an already long scene, known as the Parliament scene.[139] It was presumably used in at least some performances, including the 1601 version at the Globe, otherwise its political value would have been lost.[140] The first half of the scene includes the confrontation between Aumerle and Bagot, Fitzwalter, Harry Percy, and other lords; and the arrest of the Bishop of Carlisle for treason. In the shorter version of the scene the Duke of York tells Bolingbroke that he is adopted as Richard's heir—already anticipated in an earlier scene—and that he has yielded "his high sceptre . . . / to the possession of thy royal hand." Protocol, though forced, appears to be followed. Bolingbroke can legitimately succeed Richard, "descending now from him."[141] The irony is that Richard and Bolingbroke were born in the same year, 1367, with Richard only three months older. In the longer version of the text there is a confrontation directly between Richard and Bolingbroke.[142] The debate concerns whether the rest of this scene was written with the rest of the play, around 1595, and censored from print versions, or written later.[143] Politically it makes sense that it did not appear in print until after Elizabeth's death.

Richard II thus opens up historical, economic, agricultural, and legal elements around the question of land, and, by extension, territory. Looking forward, Bolingbroke does not resolve this tension himself when he himself is king. As he says to his lords, having ascertained that they have read the letters about the situation in the country:

> Then you perceive the body of our kingdom
> How foul it is, what rank diseases grow,
> And with what danger, near the heart of it.[144]

Here the two bodies of the king seem to be confused, the sickness of the king's physical body bleeding into the eternal body of the state, or the ills of the state taking somatic form. Warwick replies on behalf of the lords:

> It is but as a body yet distempered,
> Which to his former strength may be restored
> With good advice and little medicine.[145]

Some of these issues are played out in the last play of Shakespeare's second tetralogy, in the legal disputes about the land of France in *Henry V*. That will be discussed in the next chapter. Two further striking passages from the *Henry IV* plays will be discussed in chapter 7, "Measuring Territories."

Without ever entirely resolving the issue, *Richard II* raises a host of questions about the relation between political rule and land. Kings can exclude people from their territories, and exercise majesty within them, but they do not own the territory as a simple possession of land. The kingdom is not theirs to do with entirely as they please. The rights of landowners within a kingdom must equally be respected. The king's relation to his territory is not simply that of a landowner to his land, at a greater scale. Possession of land allows the derivation of economic yield in revenue or rent; sovereignty is an overarching power that nonetheless comes with limitations. Economic factors are therefore crucial in understanding the relation between politics and land, but this play makes it clear they are not the only aspect. Richard's mystical invocation of the earth and Gaunt's patriotic appeal to the nation show two other ways of imaging the relation. One points back; the other forward. Yet legal questions, hinted at in this play, are also crucial.

CHAPTER FIVE

Legal Territories: Conquest and Contest in *Henry V* and *Edward III*

HENRY V

Henry V is one of Shakespeare's most famous histories, well known for its patriotic fervor and the comedy of the scenes with Katherine. While it talks of battles, we see little, perhaps because Shakespeare was stung by Ben Jonson's criticism of "three rusty swords" portraying "York and Lancaster's long jars" in the Wars of the Roses, presumably as portrayed in the *Henry VI* and *Richard III* plays.[1] *Henry V* tells of the young king and his war against France, famously portrayed by Laurence Olivier in the 1944 film shortly before Britain launched another invasion of the Continent. A somewhat more cynical and unvarnished version was directed by Kenneth Branagh in 1989, and most versions today show at least some of the play's ambiguity.[2] One of the striking things about both film versions is that they display battles that are not in the play, but are only mentioned in the speeches by the Chorus.[3] The play exists in two early printed versions: a much shorter Quarto edition of the text printed in 1600, and the version in the First Folio of 1623. The Quarto, while the first to be printed, represents a later version of the text, perhaps produced for a touring company, but transmitted to us in a version that is possibly a memorial reconstruction.[4] The Quarto, among other substantial cuts, removes the first scene and some of the second—the key parts of the text to be analyzed here. As Taylor has noted, one of the other effects of the Quarto's cuts "is to remove almost every difficulty in the way of an unambiguously patriotic interpretation of Henry and his war."[5]

That debate about the play—prowar or a cynical take on war—has dominated critical assessments, but the play has fascinating geographies too. In the opening prologue, the Chorus asks if this "cockpit"—the stage within the theater—can enclose the "vasty fields of France"—the territories and the battlefields over which, and on which, the fighting will take place.[6] While the spatiality of the stage is certainly one aspect of this speech, the geographies being imagined within it are also significant. Fields is a recurrent theme through the play—both the fields of France in an agricultural sense and the battlefields where England and France will decide their fate.[7] They are imagined as the terrain through which they march to reach the battle; the "bloody field" where valor may be won; the strategic position of forces; the battle itself ("this glorious and well-foughten field"); and the burial ground for those who will die in war.[8] The fields are the site and stake of the contestation, the object and terrain of their struggle. The Chorus continues to say:

> Suppose within the girdle on these walls
> Are now confined two mighty monarchies,
> Whose high upreared and abutting fronts
> The perilous narrow ocean parts asunder.[9]

The imagined geographies of conflict, with the steep cliff-faces of the two countries split by the narrow Channel, are of course more appropriate for England's Dover cliffs than France's shores. Much could perhaps be made of the military tactics and the space of the conflict, from the "churlish turf of France"[10] to "those that leave their valiant bones in France / Dying like men, though buried in your dunghills."[11] But there are legal as well as military battles to fight, and they take place at home as well as abroad.

The Salic Law

The play's opening scene, following the Chorus, only appears in the Folio edition of the text. In it the Bishop of Ely and the Archbishop of Canterbury discuss the political situation of the time, and the position of the church. The scene perhaps is at odds with what might be expected given the Chorus's introduction to a tale of political action: one editor has even suggested that "we almost seem to be in the wrong play."[12] Yet the opening scenes, after the Chorus prologue, need to be slow and restrained in order to build to the confrontation with the Dauphin at the end of the first act.[13] The bishop and archbishop are here to discuss the threat to their "temporal lands."[14] These are lands that the church owns directly as landlord, not by nature of their spiritual power but because of their earthly wealth. The struggle between church and king over such lands recurs through European history of this period.[15] Shakespeare had explored it before. It was a significant theme in *King John*, for example, with the Bastard sent to raid the monasteries at the same time that the King is in dispute with Pope Innocent III and his legate, Cardinal Pandulph, over jurisdiction in the English kingdom (see chapter 3). Later in this play the King assents to the hanging of Bardolph, a friend from his youth, for the crime of stealing from a church, though this is obviously of a much smaller scale.[16] More generally, the question of ownership of land between a monarch and a landlord is a central theme in *Richard II*, as the previous chapter explored. But here the legal questions dominate the economic ones.

At the outset of *Henry V* the discussion is occasioned by a bill passing through Parliament that will tax the church's temporal lands, with the idea being that the revenue raised from them can go directly to the Crown, in part to support its wars and arm and feed the necessary armies. It is a bill, we hear, that is the same as one proposed in "th'eleventh year of the last king's reign"—that is, 1413, in the time of Henry IV—but which failed because of the "scambling and unquiet time" in which it was first put forth: "scambling" meaning contentious or disordered.[17] How will the church be able to "resist it now?"[18] Canterbury notes that

> It must be thought on. If it pass against us
> We lose the better half of our possession:
> For all the temporal lands which men devout
> By testament have given to the Church
> Would they strip from us, being valued thus:
> As much as would maintain, to the King's honour,
> Full fifteen earls and fifteen hundred knights,
> Six thousand and two hundred good esquires,
> And to relief of lazars and weak age,
> Of indigent faint souls past corporal toil,
> A hundred almshouses right well supplied,
> And to the coffers of the King beside,
> A thousand pound by th'year. Thus runs the bill.[19]

Ely replies, "This would drink deep"; to which Canterbury retorts, " 'Twould drink the cup and all."[20] These, more than half of what the church possesses, are lands that have been given to the church in the "testaments" of devout men, which is a slight legal slip, because "testament" pertains to personal rather than real, that is, landed, property.[21] The value of the land has been reckoned, and as well as the specific support envisioned, would leave enough capital to produce interest yearly. Shakespeare gives the likely figure gained by interest; his historical source in Holinshed says "twenty thousand pounds" as capital. "Thus runs the bill" plays both on the legal sense of the bill and the sum, the total yield from this dispossession. The church objects to the support for the aristocracy and the army, but also to sick and diseased persons and those too old to be able to work and support themselves. Lazars are, specifically, lepers, though the term has perhaps a broader sense of the ill; "weak age" are the elderly, "indigent" meaning poor or otherwise needy. Economic questions again, then, but here more explicitly situated within a legal framework.

The key legal struggle of the bulk of the play, however, is not between the church and state, but between the Kingdom of England and the Kingdom of France. In this opening scene we discover that Canterbury is trying to divert the King from taking lands from the church by suggesting the conquest of France. Canterbury remarks that "The King is full of grace and fair regard," and Ely endorses this, suggesting that he is "a true lover of the holy Church."[22] They go into a short digression about how his previous behavior did not suggest this was likely—which for the audience of course harks back to his actions in the two parts of *Henry IV*—but that he is now a reformed and religious character. Ely pulls the question back to the key matter at hand: "How now for mitigation of this bill, / Urged by the Commons? Doth his majesty / Incline to it, or no?"[23]

Canterbury replies that he has a plan:

> He seems indifferent,
> Or rather swaying more upon our part
> Than cherishing th'exhibitors against us.
> For I have made an offer to his majesty,
> Upon our spiritual convocation,
> And in regard of causes now in hand
> Which I have opened to his grace at large,
> As touching France, to give a greater sum
> Than ever at one time the clergy yet
> Did to his predecessors part withal.[24]

"Indifferent" here has the sense of impartial rather than unconcerned; "swaying" implies vacillating but inclining; while the "exhibitors" are those who have sponsored the legislation.[25] Canterbury suggests that the King may be persuaded by this case, but that there was not enough time to go into all the details. He hints that more needs to be explained concerning

> The severals and unhidden passages
> Of his true titles to some certain dukedoms,

And generally to the crown and seat of France,
Derived from Edward, his great-grandfather.[26]

"Severals" here has the sense of specifics or particulars; "unhidden passages" suggests that the lineages are clear. Some claims to dukedoms in France have been made already, while the larger claim to France as a whole is still to be decided. Now while legal questions are behind this, it ends up being just such a war as needs the kind of financial support that the bill had been designed to provide. This question of France is pressing: the French ambassador has already come to the court, and it appears that this arrival was what broke off the earlier exposition by the Archbishop. But the King delays the ambassador's audience until he has heard more from the legal, that is, his religious, advisers: "we would be resolved, / Before we hear him, of some things of weight / That task our thoughts concerning us and France."[27] The dying Henry IV had suggested as much to his son: "Be it thy course to busy giddy minds / With foreign quarrels, that action hence borne out / May waste the memory of the former days."[28] This is advice the new king quickly follows. As his brother suspects: "I will lay odds that, ere this year expire, / We bear our civil swords and native fire / As far as France. I heard a bird so sing."[29]

The legal question concerns the Salic law, so the King asks Canterbury to explain it:

My learned lord, we pray you to proceed
And justly and religiously unfold
Why the law Salic that they have in France
Or should or should not bar us in our claim.[30]

The King clearly knows the potential, but affects caution. Is he simply looking for an excuse to go to war, or genuinely wanting to be on solid legal ground? He asks the Archbishop to explain the situation, but that he should not "fashion, wrest or bow your reading" or employ sophistry or invent false claims or otherwise violate the truth.[31] He underlines this by noting the risk and the likelihood of many deaths, and asks him to weigh his words carefully:

For God doth know how many now in health
Shall drop their blood in approbation
Of what your reverence shall incite us to.
Therefore take heed how you impawn our person,

How you awake our sleeping sword of war:
We charge you in the name of God take heed.[32]

Canterbury replies:

Then hear me, gracious sovereign, and you peers
That owe your selves, your lives and services
To this imperial throne. There is no bar
To make against your highness' claim to France
But this which they produce from Pharamond:
In terram Salicam mulieres ne succedant,
'No woman shall succeed in Salic land':
Which Salic land the French unjustly gloze
To be this realm of France, and Pharamond
The founder of this law and female bar.
Yet their own authors faithfully affirm
That the land Salic is in Germany,
Between the floods of Sala and of Elbe.[33]

Thus he claims that it is only the Salic Law that might prevent a legitimate claim by Henry. "Gloze" is to interpret, to gloss, that is, to comment on the text and expand on its meaning and significance. Pharamond was an ancestor of the first Frankish kings. The logic and proscription of the Salic Law is not in doubt, yet the location that it applies to is open to question. France declares it applies to its own territory, and therefore the inheritance through the English king Edward III cannot be allowed, because that is matrilineal, thus preventing Henry of England from laying claim to it. Edward III was great-grandson of Philip III of France, and great-grandfather to Henry V. This was a descent through the female line, with Edward II's wife Isabella being granddaughter to Philip. She was the daughter of Philip IV, son of Philip III. All of Philip IV's sons succeeded him as king, but none produced male heirs.[34] If the descent could not pass down the female line, then the existing French king, Charles VI, who was descended from Philip III's other son, Charles, the Count of Valois, was legitimate in holding the crown and lands. The French defense, then, required the Salic Law to apply to France. The Archbishop claims the law applies to lands further to the east, in modern-day Germany, between the Sala and the Elbe, the region known as Meissen, and therefore Henry can claim a legitimate right to the crown of France, a kingdom located in a place where the Salic Law does not apply. It is not then simply an interpretation of the law that is at stake, or a tracing of the genealogy, but the spatial determination of that jurisdictional question.

Canterbury then expounds at length on the genealogy, and the complicated and interwoven ancestries of European nobility. It has been described as a speech "unrivalled for tedium throughout Shakespeare's works";[35] "interminable";[36] one "which no one could possibly hear without indifference."[37] Like the discussion of the bill to take the church's temporal lands, the Salic Law material in the play is taken almost word for word from Holinshed, who provided all the basis and background needed, and has only minor metrical adjustment from Shakespeare.[38] (These passages, incidentally, are one of the most compelling arguments for the Quarto being a corrupted version of the text used for the Folio: because the Folio follows very closely from its source, and the Quarto's "nonsense" follows from it in distorted and poorly remembered form.[39]) The geographical clincher, for Canterbury, is that Charles the Great, that is, Charlemagne, did not "seat the French / Beyond the river Sala" until 805, which is some time after Pharamond's rule and the Salic proclamation. In the theater, humor is often generated from his long speech, full of legal niceties and complicated lineages, which he may claim is "as clear as is the summer's sun," but which usually is said at such speed as to make it incomprehensible. At its conclusion the King replies, simply wanting a clear yes or no answer: "May I with right and conscience make this claim?"[40]

Dry it may be, but this scene is fundamental to the plot, and extremely serious.[41] Apart from anything else, the Archbishop's speech is filled with phrases concerning deposition and usurpation by King Pepin and Hugh Capet that hint that, as son of the usurper Henry IV, Henry's claim to the English throne is suspect, let alone to the French.[42] It therefore provides an important link back to earlier plays, as well as providing the basis for the rest of this play. Did Shakespeare intend to satirize the convoluted argument, or was he at this point obsessed with historical accuracy and rather tone-deaf to how it might play to an audience?[43] *If* played for comedy and to create bewilderment, the audience might be forgiven for siding with Henry here: what really is the point and is at stake? Yet, as T. W. Craik notes, "it hardly needs saying that this speech about the Salic law is to be taken as seriously by a theatre audience as it is by the stage one, and that the King's subsequent line . . . is not expressing bewilderment or suspicion but reinforcing the 'conjuration' that he laid upon Canterbury to unfold the case 'justly and religiously.'"[44] Taylor adds that the difference between Shakespeare's time and ours is crucial:

> In defence of the Archbishop's speech it has been noted that Shakespeare's audience would have been interested in the Salic law (as we

> are not) and accustomed to listening to long and intellectually complex sermons (as we are not); it has also been observed that, although such dynastic justifications of military conquest seem to us wholly trumped-up and chimerical, the legitimacy of a king's (or a nobleman's) inheritance was the foundation of the entire political and social system of medieval and Renaissance Europe. But such partial apologies do not go far enough: the speech's reputation for tedium has become such a critical commonplace that reports of its obscurity have been greatly exaggerated.[45]

Indeed, as Taylor concludes his analysis, "the speech is therefore not the nonpareil of specious convolution some critics claim: but it is undeniably dry, unpoetic, and complex."[46] As I have suggested here, there are fundamental legal geographies in these opening scenes, concerning both the relation of church and state, and the lines of inheritance that structure succession and claim. The Archbishop provides Henry with the encouragement he wishes, suggesting that the book of Numbers provides the scriptural basis for inheritance to pass to the daughter, and saying:

> Gracious lord,
> Stand for your own, unwind your bloody flag,
> Look back into your mighty ancestors.
> Go, my dread lord, to your great-grandsire's tomb,
> From whom you claim; invoke his warlike spirit,
> And your great-uncle's, Edward the Black Prince,
> Who on the French ground played a tragedy,
> Making defeat on the full power of France,[47]

It is an effective piece of rhetoric, linking the legal and patriotic claim to the lineage, both of birthright and of martial courage. "The French ground" means both the French lands, or territory, and also the terrain of the battle itself; "making defeat" means defeated France, not defeated by France.[48] But there are complexities glossed over here. Edward III and the Black Prince are held up as male exemplars, which quickly displaces the female inheritance that gives Henry his actual claim.[49] In addition Henry holds the crown because his father, Henry IV, had seized it from Richard II, son of Edward the Black Prince, now seen as an ideal. The link to the Black Prince's father, Edward III, especially as this too relates to the Salic Law, will be examined later in this chapter.

Canterbury is not the only one advocating war against France. The Bishop of Ely lends his support, suggesting a "thrice-puissant" basis: "you are their

heir, you sit upon their throne, / The blood and courage that renowned them / Runs in your veins."[50] Descent, possession, and valor give him a threefold basis of strength as support for his claim. Encouragement also comes from the Dukes of Exeter and Westmorland. Blood is a common noun in their speeches too, invoking the ties of inheritance, the martial spirit, and the consequences of war. As Exeter says,

> Your brother kings and monarchs of the earth
> Do all expect that you should rouse yourself
> As did the former lions of your blood.[51]

We hear that the hearts of the nobles and subjects of England have already reached France; the King is petitioned to let their bodies join them "pavilioned in the fields of France."[52] Canterbury almost gives away his real purpose when he suggests that this action will be supported by the aid of the "spiritualty," that is, the church, which will raise more for the King than the clergy ever did for his ancestors.[53]

Against the Scot

The King is clearly convinced, and almost ready to bring in the French ambassador to commence terms, yet it is striking that he has a more cautious approach to these questions than anyone else. He does not want to bring the French confrontation to a head so quickly that he neglects an enemy closer to home:

> We must not only arm t'invade the French,
> But lay down our proportions to defend
> Against the Scot, who will make road upon us
> With all advantages.[54]

Necessary protection must thus be made for the north of England, and the King asks for an estimate of what will be needed and how to distribute his military forces. If the Scots have a favorable opportunity, with the English army sent to France, he fears they will take advantage and invade. Canterbury suggests that the troops garrisoned in the north, and likely under the command of the Duke of Westmorland, be used to protect the lands of England: "They of those marches, gracious sovereign, / Shall be a wall sufficient to defend / Our inland from the pilfering borderers."[55] Here the geographical terms are clear. The marches are the borderlands, the frontier zone between England and Scotland. The wall obviously invokes Hadrian's Wall as well as the protection of a well-armed force. The inland, that is, the internal parts of the kingdom, needs protection from marauding Scots. The King replies:

> We do not mean the coursing snatchers only,
> But fear the main intendment of the Scot,
> Who hath been still a giddy neighbour to us.
> For you shall read that my great-grandfather
> Never went with his forces into France
> But that the Scot on his unfurnished kingdom
> Came pouring, like the tide into a breach,
> With ample and brim fullness of his force,
> Galling the gleaned land with hot assays,
> Girdling with grievous siege castles and towns;
> That England, being empty of defence,
> Hath shook and trembled at th'ill neighbourhood.[56]

The King then suggests it is more than just "coursing snatchers," or opportunistic raids that he is concerned about, but "the main intendment of the Scot," the chief purpose of a major invasion. "Coursing" is the sport of hunting a hare with dogs; the "snatch" is the quarry itself.[57] Since the time of Edward III the Scot has exploited the undefended kingdom whenever forces are sent to France. Canterbury fills in the detail with the capture of the Scottish king David II at the battle of Neville's Cross in Durham in 1346.[58] If England is "unfurnished," that is, unequipped, it may be attacked; "galling the gleaned land" suggests stripping the land bare by its raids, "assays," or assaults.[59] Westmorland, or one of the lords, or perhaps the Bishop of Ely, because the speech heading is unclear and different between Quarto and Folio, adds his advice to the narrative:

> But there's a saying, very old and true:
> If that you will France win,
> Then with Scotland first begin.
> For once the eagle England being in prey,
> To her unguarded nest the weasel Scot
> Comes sneaking, and so sucks her princely eggs,
> Playing the mouse in absence of the cat,
> To tear and havoc more than she can eat.[60]

To resolve this we are told, in a somewhat tangled mess of corporeal and animal metaphors, that "the cat must stay at home"; the hand can "fight abroad" as long as the "advised head defends at home," and that like the different tasks of honeybees in a hive the political community must share responsibilities in different ways.[61] Thus the King is advised to take only a fraction, one-quarter, of his forces to France, and to leave three times as many forces at home to defend the kingdom.[62] It is notable that this play is almost entirely directed toward the English-French conflict, leaving domestic turmoil aside. Henry's army is, famously, made up of all the British nations—Jamy from Scotland, Fluellen from Wales, MacMorris from Ireland, and Gower from England.[63] The national aspects will be returned to in the coda. In addition, many of these soldiers are small landowners, yeoman, who are asked to show "the mettle of your pasture."[64]

Taking the advice of his bishops and nobles, the King calls for the French ambassadors, sent by the King's son, the Dauphin.[65] As Craik notes,

> It is dramatically important that in this scene where everyone is urging him to emulate his valiant ancestors the King should seem not passive but active, and not hot-headed but level-headed; thus it is he who raises the topic of defence against Scottish invaders, and when that matter has been settled he orders "the messengers sent from the Dauphin" to be admitted.[66]

He stresses that "we are well resolved," and that by God's help and that of his supporters ("the noble sinews of our power") he will either take France ("bend it to our awe") or destroy it ("break it all to pieces").[67] It is therefore clear, even before he sees the ambassadors, that his mind is set for war. The ambassador, however, provides him with a clear pretext. Stating that the Dauphin dismisses his claim to "certain dukedoms in the right / Of your great predecessor King Edward the Third," he says, "you cannot revel into dukedoms there," and presents him with a treasure in their place: "in lieu of this / Desires you let the dukedoms that you claim / Hear no more of you."[68] The treasure, swiftly revealed, is tennis balls,[69] the point of which is later revealed when they are labeled Paris-balls, the game having been introduced from France to England. Henry responds forcefully:

> We are glad the Dauphin is so pleasant with us.
> His present and your pains we thank you for.
> When we have matched our rackets to these balls,

We will in France, by God's grace, play a set
Shall strike his father's crown into the hazard.[70]

The stage is thus set for the invasion of France, and the battles and courtship for which the play is best known. The balls are changed to "gun-stones."[71] All England seems to be behind this, sacrificing at home to support the war abroad. As the Chorus notes in the prologue to Act II, "they sell the pasture now to buy the horse / Following the mirror of all Christian kings."[72] The "mirror" of princes was a long-established medieval practice of guidebooks or examples for rulers: following ones by the likes of Aegidius Romanus, Thomas Aquinas, and Erasmus, Machiavelli's *Il Principe* was one of the last such texts.

Courtship and Legal Articles

At the end of the play, following his military success, when the King is wooing Katherine, she declares that she cannot love the enemy of France. Henry replies, "in loving me you should love the friend of France; for I love France so well that I will not part with a village of it; I will have it all mine: and Kate, when France is mine, and I am yours, then yours is France, and you are mine."[73] Claire McEachern suggests that "Henry's verbal confusion—of pronouns, of political and sexual desires, indeed of territorial boundaries—forges an accommodation between power and affect, political possession and sexual possession, which accommodates the body to hegemony, and vice versa."[74] This is most striking in the sexually explicit exchange between Henry and the Duke of Burgundy after Katherine's agreement.[75] Even the earlier scene with her maid Alice had been about the words for different parts of her body—a common-enough initial language lesson, certainly, but a linguistic prelude to physical control by the English.[76] This is particularly strong when Katherine and Alice find humor in the way that English words for at least one body part (the foot) and a garment (a "gown" [*coun*]) sound obscene in French.[77] Katherine's body is already sexualized even before she has met Henry.[78] The language-lesson scene immediately follows the speech at the gates of Harfleur, where Henry threatens mistreatment of civilians, including destruction, murder, and rape:

If I begin the battery once again,
I will not leave the half-achieved Harfleur
Till in her ashes she lie buried.
The gates of mercy shall be all shut up,

> And the fleshed soldier, rough and hard of heart,
> In liberty of bloody hand shall range
> With conscience wide as hell, mowing like grass
> Your fresh fair virgins and your flowering infants.
> What is it then to me if impious war,
> Arrayed in flames like to the prince of fiends,
> Do with his smirched complexion all fell feats
> Enlinked to waste and desolation?
> What is't to me, when you yourselves are cause,
> If your pure maidens fall into the hand
> Of hot and forcing violation?
> What rein can hold licentious wickedness
> When down the hill he holds his fierce career?
> We may as bootless spend our vain command
> Upon th'enraged soldiers in their spoil
> As sent precepts to the leviathan
> To come ashore.[79]

He continues, holding the people responsible if his troops are let loose on the town:

> The blind and bloody soldier with foul hand
> Defile the locks of your shrill-shrieking daughters,
> Your fathers taken by the silver beards,
> And their most reverend heads dashed to the walls,
> Your naked infants spitted on pikes,
> Whiles the mad mothers with their howls confused
> Do break the clouds, as did the wives of Jewry
> At Herod's bloody-hunting slaughtermen.[80]

The town surrenders, and we are left wondering if this was mere bluff or genuine prediction. One of the crucial things concerning this passage is that most of it does not appear in the Quarto, and in that text it moves directly from "Till in her ashes she lie buried. / The gates of mercy shall be all shut up" to the closing lines of his speech: "What say you? Will you yield and this avoid? / Or, guilty in defence, be thus destroyed?"[81] The Quarto removes the threat of war crimes, though not Henry's later order for the execution of prisoners at Agincourt.[82] Contemporary readers have focused on these actions in various critical ways. Stressing the interconnection between the conquest of France and the conquest of Katherine, Karen Newman has argued that "in these notorious lines, the expansionist aims of the nation-state are worked out on and through the woman's body."[83] The same point is made by McEachern: "The equation of woman and territory, with its insistence upon the body-as-politic, develops in the juxtaposition of scenes iii and iv of Act III."[84]

A similar reinforcement comes in the relation between the battle of Agincourt and the conversation between Henry and Katherine. The legal tone of the beginning of the play therefore returns at the end, because the peace to be sealed between France and England depends on a number of "articles," first of which is the possession of Katherine as his queen: "She is our capital demand, comprised / Within the fore-rank of our articles."[85] The courtship scene comes here, and as serious a message as this, speaking of the conquest and unification of the kingdoms through war and marriage, quickly turns again to humor as the King tries to explain this message in French.[86] When Kate assents, so too does the French king: "The King hath granted every article: / His daughter first, and in the sequel all, / According to their firm proposed natures."[87] The Chorus hypothesizes much earlier that if Katherine had been offered, along with some "petty and unprofitable dukedoms" as dowry, this would not be sufficient.[88] She is only enough when it comes with the French crown. Henry is thus proclaimed to be the King of England and the heir to France, in papers signed by the French king in French and Latin.[89] Thus Henry will succeed the present king of France, rather than his son, the Dauphin. The French king wishes that this will bring an end to discord, telling Henry to take his daughter as his wife:

> Take her, fair son, and from her blood raise up
> Issue to me, that the contending kingdoms
> Of France and England, whose very shores look pale
> With envy of each other's happiness,
> May cease their hatred, and this dear conjunction
> Plant neighbourhood and Christian-like accord
> In their sweet bosoms, that never war advance
> His bleeding sword 'twixt England and fair France.[90]

Yet these wishes are unfulfilled. In the final lines of the play the Chorus hints at what is to come in the historical narrative. The infant son of Henry and Katherine will take the throne as Henry VI on the death of his father at the age of thirty-five, again fighting in France, which the Chorus labels "the world's best garden."[91] The Chorus notes that while Henry VI was "in infant bands crowned King / of France and England," the multiple hands at play in his government only succeeded to manage "that they lost France and made his England bleed."[92] In referring to the audience and saying that "oft our stage hath shown" these events concerning the times of Henry VI, Shakespeare ties the end of his second great tetralogy, in order of being written, to the beginning of the first.[93] But it also hints at the resolution of that earlier tetralogy. Katherine's subsequent marriage, to Owen Tudor, produces the son Edmund Tudor, and his son, Henry VII, is first of the Tudor line of kings and queens. Henry VII, of course, is the Earl of Richmond in *Richard III*, who brings the Wars of the Roses to an end with his defeat of Richard at Bosworth Field.

The Politics of Succession

This question of succession is not just of historical interest. It had a direct parallel in the time Shakespeare was writing, in the late years of Elizabeth's reign. As Gurr suggests,

> By the 1540s one major lesson from the long dynastic conflict in England called the War of the Roses was evident: primogeniture, inheritance by the eldest son, was no secure guarantee of a clear title to the succession. From the time when the childless Richard II's cousin claimed his crown down to the gamble between two equal claimants at the battle of Bosworth, the question of title to rule came under pressure from the various strong right arms that contested it.[94]

This is one of the reasons why the discussion of the Salic Law and inheritance in the opening scenes is such a serious issue. Given contemporary questions about succession, Elizabeth's status on the English throne as a woman and the future King James's claim to it, as Taylor suggests, "parody of the Archbishop's argument would have been, to say the least, uncharacteristically indiscreet."[95] Because Elizabeth did not have an heir, there had long been issues about succession. If Henry's claim to France is legitimate, on the basis of the Salic Law applying elsewhere and matrilineal succession being allowed, then the Stuart claim to England is also legitimate.[96] Gurr indicates the context:

> When in 1578 she started dallying with the possibility of marrying the French king's brother, France's Salic Law denying succession through the female line emerged as a boiling issue. A combination of childlessness and what John Knox called the "monstrous regiment of women"

> had called primogeniture into question in terms of gender. Now the threat of Elizabeth marrying a French prince, and the likelihood that her children would suffer the imposition of Salic Law to block their chances of inheritance, threatened England's dynastic security.[97]

The full title of Knox's tract was "The First Blast of the Trumpet Against the Monstrous Regiment of Women."[98] "Monstrous regiment" sounds different to our modern ears than it would have at the time. "Monstrous" means unnatural, and "regiment" is government, rule, or reign. God, on Knox's reading of the Bible, did not allow women "to bear rule, superiority, dominion or empire above any realm, nation or city."[99] The specific queens Knox had in mind were Mary, Queen of Scots and Mary I of England, who were both Catholics. His argument was therefore not just against women, but that of a Protestant against Catholic monarchs, which backfired when Elizabeth, a Protestant queen, took the throne. Gurr goes on to discuss a pamphlet by John Stubbs, *The Discoverie of a Gaping Gulf whereinto England is like to be Swallowed*.[100] Gurr suggests that

> Stubbs's pamphlet did help to lodge the Salic Law firmly in English conscience. It became a well-known grievance. . . . Of Henry V's three claims to France, specified by Stubbs as descent, conquest and covenant, the first was validated by the denial of Salic Law. Agincourt assured the second. And the third was strengthened by Shelley's Case [an inheritance case between 1579 and 1581] with its endorsement of the right to bequeath a title by will, as the French King Charles did when he signed the Treaty of Troyes, rather than by the old law of primogeniture. Shakespeare's play gives equal emphasis, one after the other, to the three claims.[101]

In the reading of the opening scenes, we can see how the Archbishop works through the details of the genealogical basis of descent. The conquest of France through military power is the main dramatic thrust of the play, from Harfleur to Agincourt. Then, in the final act, the covenant is established through the marriage to Katherine and the passing of the French line to his heirs. As Gurr adds, "Undue emphasis on the doubtful nature of Henry's claim to the English crown, which itself was the basis for his claim to the French, would have redoubled the other doubts that are discreetly implied in the first act."[102] The national and the personal intertwine, with legal debates about the legitimacy of each.

EDWARD III

This book's introduction mentioned the disputed authorship of *Edward III*, a history play that is slowly gaining acceptance into the canon of Shakespeare's plays. It did not appear in the First Folio, and the Quartos are anonymous. It has begun to appear in *Complete Works*, and separate critical editions have been produced by Yale University Press, Cambridge University Press, and the Arden Shakespeare, with other editions forthcoming.[103] Lacking indisputable historical and bibliographical evidence, various attempts have been made to resolve the authorship question. Many of these have been on linguistic evidence, now enhanced by technological developments. Most analyses recognize that the play is the work of at least two hands, or one author working at two different times. Inna Koskenniemi noted as far back as 1964 that her "considerations tend to support the theory, based on internal evidence of style, vocabulary and general structure, that Shakespeare wrote at least some parts of *Edward III*."[104] Of the recent editors, Sams is categorical that it is entirely by Shakespeare, and the bulk of his critical apparatus is devoted to demonstrating this point; Giorgi Melchiori is more measured. In recent years, computer evidence has been used to study this question, looking at incidence of particular words or constructions.[105] One of the most detailed studies does not prove it was by Shakespeare, but says that is possible, and if so, then both parts are likely by him:

> It is concluded that, as far as the rare-word vocabulary of *Edward III* goes, it is compatible with authorship by Shakespeare at an early stage in his dramatic career. Both part A and part B are regarded as his work, through probably written at different times.[106]

Interestingly, whether by Shakespeare or not, there are a few phrases taken from or very close to his sonnets. The most important is the line "Lilies that fester smell far worse than weeds,"[107] which comes from Sonnet 94, described by Melchiori as Shakespeare's "meditation on the ethics of power."[108] In *Edward III* it is used as a description of the king's corruption. Ultimately I am not especially interested in the authorship question, but rather in the text itself, which is dated by the First Quarto of 1596, and thus can be usefully juxtaposed to Shakespeare's early histories—*Henry VI*, *Richard III*, and the transition to *King John* and *Richard II*. My analysis will be cautious: while the play provides a rare instance of the word "territories," this is in one of the scenes least likely to have been written by

Shakespeare. As some useful discussions have noted, the play is worthy of analysis whether or not Shakespeare is its author, in part or whole.[109]

The action of *Edward III* follows from Marlowe's *Edward II*, in which the king is overthrown and then killed, and his son crowned in his place.[110] While the child is the nominal ruler, political power clearly rests with Roger Mortimer, who also takes Edward II's wife, Isabel. As Proudfoot notes, *Edward III* "avoided Edward's early years, perhaps because Marlowe had handled them already, more certainly because Isabel's liaison with Mortimer was an undesirable element in a play in which her role was to afford her son an unimpeachable title to the throne of France."[111] David Bevington and Eric Rasmussen draw the parallels between Marlowe's play, an uncontested Shakespeare history, and this disputed play: "In suffering, Edward [II] is compassionate and dauntless of mind, like Shakespeare's Richard II. His son, the young King Edward III, brings Edward's tormentors to account and gives every promise of greatness on the throne."[112] As well as the direct influence of *Edward II*, it is widely recognized that *Edward III* owes a debt to Marlowe's other work. As Proudfoot notes, it was "written by a poet heavily under the influence of *Tamburlaine*."[113]

There are strong thematic links between *Edward III* and *Henry V*, especially in the genealogical claim to France and the second half concerning the wars there, but also to other history plays.[114] Indeed, Edward III as a person is frequently mentioned in Shakespeare's other history plays, more so than any other English king who is not a character in those plays. In part this is because of his historical lineage.[115] Edward was the last strong and undisputed king; the early death of his son, Edward the Black Prince, began conflict and decline. Edward was therefore grandfather to the Black Prince's son Richard II; he was also the father of John of Gaunt, and so grandfather to Henry IV, great-grandfather to Henry V, and great-great-grandfather to Henry VI—the house of Lancaster. Another of his sons, Edmund, Duke of York, created the house of York, whose descendants included Edward IV, the uncrowned Edward V, and Richard III. Only with Henry VII, great-great-grandson of John of Gaunt (albeit through a complicated route down the female line, with a declaration of his great-grandfather's retrospective legitimacy), who married Edward IV's daughter Elizabeth of York, were these two houses reunited as the beginning of the Tudor dynasty. Chronologically between *Edward III* and *Richard II* there is another play, *Thomas of Woodstock*, sometimes known as *King Richard II, Part 1*, but that play is almost certainly not by Shakespeare. It was briefly discussed in chapter 4. But in this sequence of plays between Marlowe's *Edward II* and Shakespeare's *Richard III*, 178 years of reign are dramatized.

The other reason Edward III features so much is due to the conquests in France. *Edward III* has some important discussions of political geography, which closely relate to *Henry V* and point to the losses in *Henry VI, Part 1*. Edward responds to the King of France's offer of the dukedom of Guyenne, relayed by his messenger the Duke of Lorraine:

> Dare he command a fealty in me?
> Tell him the crown he usurps is mine,
> And where he sets his foot he ought to kneel:
> 'Tis not a petty dukedom that I claim,
> But all the whole dominions of the realm,
> Which if with grudging he refuse to yield,
> I'll take away those borrowed plumes of his
> And send him naked to the wilderness.[116]

The idea that the French king John is demanding a feudal obligation from Edward is offensive, more so because the position from which he claims it is not rightfully his. The claim of Edward to the French throne, like his great-grandson Henry's, cannot be bought off with a single dukedom, because he wants the entirety of the realm, all its provinces and territories. The idea of "borrowed plumes" relates to similar claims in *King John* ("borrowed majesty") and *Henry V* ("borrowed glories"), and the use of plumes or feathers to invoke social status in the first two parts of *Henry VI*.[117] Similarly the idea of usurpation has echoes in many of Shakespeare's plays.[118] Edward boasts, "Shall the large limit of fair Bretagne / By me be overthrown"[119] and lays claim to "the whole dominiums of the realm of France,"[120] but this is thrown back at him when he actually lands on the Continent:

> Edward, know that John, the true King of France,
> Musing thou shouldst encroach upon his land
> And, in thy tyrannous proceeding, slay
> His faithful subjects, and subvert his towns,
> Spits in thy face, and the manner following
> Upbraids thee with thine arrogant intrusion.[121]

("Musing" here is complaining.) The geopolitical intrigue brings in a major third party, the emperor. The Earl of Derby and Lord Audley discuss the message:

> As good as we desire: the emperor
> Hath yielded to his highness friendly aid,

And makes our kings lieutenant-general
In all his lands and large dominions.
Then *via* for the spacious bounds of France![122]

Via here is the Italian "come on" or "onward"—a recommendation of the next step. "Spacious bounds" means expansive territories, and "spacious" is a very early use of the word in this sense: used a few lines earlier in the play—"a spacious field of reasons"—while the *OED* (2b) gives *Titus Andronicus* as the first recorded use.[123] The French army, however, is reinforced by the army of the King of Bohemia, and John declares that he is well equipped to resist Edward:

At sea we are as puissant as the force
Of Agamemnon in the haven of Troy;
By land, with Xerxes we compare of strength,
Whose soldiers drank up rivers in their thirst.[124]

Later in that scene he says to his youngest son: "Now boy, thou hearest what thundering terror 'tis / To buckle for a kingdom's sovereignty."[125] "Buckle" is to wear armor, therefore to fight, to contest.

Additionally, as in *Henry V*, the attempted conquest of France must be balanced by defenses at home. While Edward thinks it is acceptable for him to break his alliance with France, he wants to check on the alliance with Scotland. Montague reports that it is

Cracked and disserved, my renowned lord:
The treacherous king no sooner was informed
Of your withdrawing of your army back,
But straight, forgetting of his former oath,
He made invasion of the bordering towns:
Berwick is won, Newcastle spoiled and lost,
And now the tyrant hath begirt with siege
The castle of Roxborough, where enclosed
The Countess Salisbury is like to perish.[126]

As well as showing the territorial vulnerability of northern England, this also adds a personal dimension: the King's relation with the Countess, daughter of the Earl of Warwick and wife to an absent husband, is a major part of the play. His turn northward is not just to deal with Scotland, but to protect a woman he covets. The Countess resists Edward's advances, just as France resists his attempts at conquest. Edward, of course, already has a wife and a kingdom, and there are strong parallels between his pursuit of additional ones. Edward seems willing to do almost anything, including murder, to gain her, but is ultimately denied. The Countess draws parallels between her virtue and the geographies of conquest:

> Let not thy presence, like the April sun,
> Flatter our earth, and suddenly be done;
> More happy do not make our outward wall
> Than thou wilt grace our inner house withal.
> Our house, my liege, is like a country swain,
> Whose habit rude and manners blunt and plain
> Presageth nought, yet inly beautified
> With bounty's riches and fair hidden pride.
> For where the golden ore doth buried lie,
> The ground, undecked with nature's tapestry,
> Seems barren, sere, unfertile, fruitless, dry;
> And where the upper turf of earth doth boast
> His pride, perfumes, and parti-coloured cost,
> Delve there and find this issue and their pride
> To spring from ordure and corruption's side.
> But, to make up my all too long compare,
> These ragged walls no testimony are
> What is within, but like a cloak doth hide
> From weather's waste the undergarnished pride.[127]

This dense speech, with its panoply of natural and agricultural imagery, is ambiguous in many ways. The Countess is asking the King not to take the plain walls of the house as a reflection on what lies within. The castle is likened to a countryman who is poor on the outside, and so may appear nothing, but is rich within. "Parti-coloured" is variegated, meaning that while there is usually an abundance of riches above the soil, on the "upper turf of earth," even in this barren place there are multiple mineral riches to be found beneath the surface. Yet these stem from "ordure and corruption's side"—excrement and rotting corpses. Does the Countess suggest that in reverse her own beauty should not obscure the risks of pursuing her? "Unfertile" and "undecked" are both the first recorded uses of the words,[128] as is "undergarnished," meaning the sumptuous garments worn beneath a cloak to prevent the weather's spoil or damage.

Just as his pursuit of the Countess shows his impoverished moral nature, the military hero of the play is not the eponymous Edward, but his son, Edward the Black Prince. It is here that the one instance of the word "territories" appears in the play. After the battle, in the diplomatic discussions, he asks that all the challenges he has faced were multiplied:

> I wish were now redoubled twentyfold,
> So that hereafter ages, when they read
> The painful traffic of my tender youth,
> Might thereby be inflamed with such resolve
> As not the territories of France alone,
> But likewise Spain, Turkey, and what countries else
> That justly would provoke fair England's ire,
> Might at their presence tremble and retire.[129]

While this scene is in an act that is not so likely to be by Shakespeare's hand, it is still interesting for the grammatical construction. Using a genitive, it is the equivalent of France's territories—that is, belonging to the French king—not French territories, or the French territory. The latter senses would postdate at least the action of the play, if not its composition.

Henry V and *Edward III* thus raise some important legal questions about the question of territory. Foremost in *Henry V* is the relation between church and state, especially the wealth of the church's temporal lands. Yet while Richard II had simply seized John of Gaunt's property, here there is the question of a bill to allow taxation of wealth. Even a king, seemingly, must operate within a legal framework. The question of inheritance is raised, not simply in terms of property, but in terms of legal claim to succession. The law does not just apply to ownership of land, but its rule, and crucially with the Salic Law the question is *where* the law applies: a geography to the law as well as a law over the geography. There are also issues concerning just conduct in war. The legal ownership of a body—both the body of a nation and the body of a woman, what McEachern called the body-as-politic[130]—features in both plays, Katherine in *Henry V* and the Countess of Salisbury in *Edward III*. The sense of territory that emerges from these and the other history plays is a more complicated, multifaceted one than the simple sense of a bounded area under jurisdiction. The political-economic and the political-legal work as additional aspects alongside the political-strategic.

Some of these issues concern the territory of England; some, external territories. Yet in the English history plays the external territories under debate are those of contiguous lands or near neighbors—Scotland, Ireland, and France. Expansionist though England's territorial ambitions clearly are, they are geographically limited. While in Shakespeare's time the British Isles came to be increasingly united, the loss of the Pale of Calais in 1558 meant that overseas ambitions moved further afield, to the Americas, India, and, in time, Africa. Having discussed economic and legal aspects of the territorial, the book now turns to a theme that connects the two: the colonial.

CHAPTER SIX

Colonial Territories: From *The Tempest* to the Eastern Mediterranean

THE TEMPEST

The Tempest is the most obvious place to begin an examination of how Shakespeare thinks about colonial territories. It is a play in which the question of inside and outside, possession and dispossession of territory or land, take a crucial role. Philip Edwards has gone so far as to suggest that "rights to territory are a fundamental question in *The Tempest*," and that the play is more concerned with this issue than is *Henry V* or *Cymbeline*.[1] Four key figures are important in understanding these concerns in the play: Prospero and Caliban, naturally, but also Antonio, the usurper, and Gonzalo, the adviser.

Having lost the dukedom of Milan to his brother Antonio, Prospero is both dispossessed and a colonizer. Of his own dispossession, Prospero is eloquent:

> To have no screen between this part he played
> And him he played it for, he needs with be
> Absolute Milan. Me, poor man, my library
> Was dukedom large enough. Of temporal royalties
> He thinks me now incapable; confederates,
> So dry he was for sway, wi'th' King of Naples
> To give him annual tribute, do him homage,
> Subject his coronet to his crown, and bend
> The dukedom yet unbowed (alas poor Milan)
> To most ignoble stooping.[2]

Antonio wanted no distinction between his own role and the position he aspired to, occupied by Prospero. He needed to be absolute ruler of Milan.

Prospero suggests he—"poor man"—had otherworldly concerns, focused on study and learning. (We hear later how Antonio had paid off Prospero: "Knowing I loved my books, he furnished me / From mine own library with volumes that / I prize above my dukedom."[3]) "Temporal royalties" are secular, as opposed to spiritual, powers, which may be a reference to political power compared to that of the church, or compared to his intellectual or magical ability; "incapable" means that he is both unable to exercise them and unable to pass them on: the standard lineage has been broken. So thirsty for power, Antonio entered into an alliance with the King of Naples, to whom he pays allegiance, and has bound his rule (the coronet) to the larger kingdom (the crown). A previously proud and superior Milan now bows to Naples—Naples of course being a larger kingdom with more maritime power. Later in the play Antonio tries to manipulate Sebastian, brother of the King of Naples, Alonso. They plot the death of the King, and Sebastian's succession to the throne, the payoff for Antonio being the end of the tribute.[4]

There is of course another dispossession in the play. Caliban is the son of the "foul witch Sycorax" from Algiers, who had been sentenced to death for sorcery, but had it commuted to banishment for an unsaid reason, possibly because she was pregnant.[5] Prospero recounts to the spirit Ariel how "the blue-eyed hag was hither brought with child / And here was left by th' sailors."[6] Sycorax, like Prospero, had arrived on the island through no choice of her own, and took it over. Caliban was born shortly afterward, and Sycorax made Ariel her servant.[7] While Prospero thus acknowledges that Caliban's birth on the island was before he arrived, he does everything he can to diminish this birthright. After Sycorax died, Prospero contends that "Then was this island / (Save for the son that she did litter here, / A freckled whelp, hag-born) not honoured with / A human shape."[8] Ariel responds in a more humanizing way: "Yes, Caliban, her son,"[9] but Prospero seeks to deny the humanity. While he too uses the word "son," he uses the verb "litter," more commonly used of animal births than human; "freckled whelp" suggests a canine pup; "hag-born" again demeans the mother. He only concedes that Caliban shares the *shape* of humans. Shortly after, Prospero calls Caliban "Thou poisonous slave, got by the devil himself / Upon thy wicked dam";[10] in this context "dam" is a derogatory word for "mother."

Caliban is thus the dispossessed, and directly claims this of Prospero. "This island's mine by Sycorax, my mother, / Which thou tak'st from me."[11] As Stephen Orgel notes, he need not make the claim on the basis of inheritance, because he could make it simply by prior possession.[12] Caliban recounts that at first Prospero treated him with kindness, and that he responded by love and showing him "all the qualities o'th'isle: / The fresh springs, brine pits, barren places and fertile."[13] Now, Caliban alone is "all the subjects that you [Prospero] have, / Which first was mine own king; and here you sty me / In this hard rock, whiles though do keep from me / The rest o'th' island."[14] Initially he was his own master, a king, ruler of and able to roam over the whole island, but now he is imprisoned within a specific site, a slave, the only subject of a new master, reduced to an animal—"sty" again implies a bestial dwelling. Prospero responds that he is only imprisoned because of what he has done, which we learn is an attempted rape of Prospero's daughter Miranda. Caliban dreams of colonizing the island for himself, wishing this had been done, with the idea of how he would have "people else / This isle with Calibans."[15] Shakespeare uses the phrase "violate / The honour of my child."[16] Rape, of course, while today having primarily a sexual sense, comes from the Latin word *rapere* meaning to seize, to abduct, to capture. *Stuprum*—defilement, dishonor, disgrace—is the more common Latin word for "rape" in the modern, sexual sense. Abduction, theft, and sexual violence were all meanings in circulation at this time, and Shakespeare had of course explored these issues in his poem *The Rape of Lucrece*.[17] For Simon Estok, "the intended rape reveals Caliban's desire to fill the geographical space of the island and the vaginal space of Miranda by force."[18] The rape of Caliban's island through Prospero's actions is paralleled by Caliban's attempted rape of Prospero's daughter.

Caliban continually stresses his dispossession: "As I told thee before, I am subject to a tyrant, / A sorcerer, that by his cunning hath / Cheated me of the island";[19] and "I say, by sorcery he got this isle. / From me he got it."[20] Caliban makes this case to try to persuade the butler, Stephano, and the jester, Trinculo, of his right to the island, and to get them to help him reseize it. Caliban's identity is complicated, with a debate about whether he is African—the play being set off the coast of North Africa somewhere in the Mediterranean—or Native American.[21] The discussion hinges on what has been called the play's "confusing" or "ambiguous geography."[22] It seems more plausible to follow a broadly African line: "The play's most obvious African connection is the island location: if plotted literally, it must have been within a hundred or so miles from a line between Naples and Tunis."[23] This is based on both the shipwreck of a boat making a journey between those two locations, and Sycorax's banishment from Algiers. Daniel Brayton takes exception to this as symptomatic of "positivistic calculations" that see this location as "an empirical matter."[24] In a sense the location does not matter; it is indubitably a colonial set of encounters, and the play is able to evoke both a Mediterranean and New World setting without the need to resolve it as one or the other. Whether or not Annabel Patterson's suggested etymology is correct, her claim that "Caliban's name (an anagram of 'cannibal') and his education permanently implicate him within the full colonial discourse of reclamation, demarcation and territory-formation" rings true.[25] Caliban's initial welcome of Prospero is paralleled by that he shows to Stephano and Trinculo: "I'll show thee every fertile inch o'th' island, / And I will kiss thy foot. I prithee, be my god. / . . . I'll swear myself thy subject."[26] For Virginia Vaughan and Alden Vaughan, Caliban is "in tune with nature and lord of the island until overthrown by Prospero and later corrupted by Stephano and Trinculo."[27] It is the latter that may perhaps be most important in the long run, but the former is the spur to his immediate grievance.

Stephano and Trinculo had been among those shipwrecked on the island by the tempest, which had also brought Prospero's brother Antonio. When the Duke and his colleagues are in the storm at sea, his old counselor Gonzalo speaks for them all:

> Now would I give a thousand furlongs of sea for an acre of barren ground—long heath, brown furze, anything. The wills above be done, but I would fain die a dry death.[28]

We thus have a three key figures in the play with relations to territory: the one who is outside territory who creates it anew, truly a place he can be master of the domain (Prospero); the one who is dispossessed and enslaved but dreams of a biological colonization of his own (Caliban); and the one who is dispossessed but then himself loses all (Antonio). All three figures have colonizing tendencies; all three have moments when they are set outside of territory.

There is an unusual interlude in the play when Iris and Ceres appear in Prospero's masque. Their speeches are profoundly geographical. First, Iris:

> Ceres, most bounteous lady, thy rich leas
> Of wheat, rye, barley, vetches, oats and peas;
> Thy turfy mountains where live nibbling sheep,
> And flat meads thatched with stover them to keep;
> Thy banks with pioned and twilled brims
> Which spongy April at thy hest betrims
> To make cold nymphs chaste crowns and thy broomgroves
> Whose shadow the dismissed bachelor loves,
> Being lass-lorn; thy pole-clipped vineyard,
> And thy sea-marge, sterile and rocky-hard,

Where thou thyself dost air—the queen o'th' sky,
Whose watery arch and messenger am I,
Bids thee leave these, and with her sovereign grace,
Here on this grass-plot, in this very place,
To come and sport. Her peacocks fly amain.
Approach, rich Ceres, her to entertain.[29]

"Pioned" is dug; over which poles are "twilled," that is, laid out to form a hedge or barrier. It refers "to a man-made boundary, the hedge, which was used to enclose sheep, determine ownership, and protect grazed pasture from elemental assaults."[30] It is worth noting that hedges predate enclosure, and, as Oliver Rackham explains, can result from three processes: planting; as "ghosts of woods that have been grubbed out leaving their edges as field boundaries"; or ones that "have developed naturally at the edges of fields. Whenever a fence, ditch, or earthern bank is neglected for a few years, not too far from a source of tree seed, a hedge will result."[31]

In this play the reply by Ceres is not short of related imagery either: "my bosky acres and my unshrubbed down, . . . my proud earth."[32] Her fuller speech to this masque is similarly geographical:

Earth's increase, foison plenty,
Barns and garners never empty.
Vines with clustering bunches growing,
Plants with goodly burden bowing;
Spring come to you at the farthest,
In the very end of harvest.
Scarcity and want shall shun you,
Ceres' blessing so is on you.[33]

As Vaughan and Vaughan have noted, "The Roman goddess Ceres represented the fecundity of the cultivated earth. Wheat and barley were sacred to her. She presided over the labours of ploughing, tilling, planting and harvesting, and was known as a maternal fertility goddess."[34] What Prospero's masque accomplishes is the idea of agri-culture, of improving the land more than those who merely lived in it, of—in the language of a modern colonial project—making the desert bloom. Indeed, in the play the lord Adrian says this place "seems to be desert," which most obviously means deserted, and relates the idea of a place that is barren as well as un- or underinhabited.[35]

Such a place, for the European men washed up on its shores, seems to invite imperial ambitions. Gonzalo wishes he had "plantation of this isle," with its clear colonial connotation of the plantations first in Ireland and then in the New World, while Antonio and Sebastian (brother to the King) respond with more agricultural ideas: "He'd sow't with nettle-seed. / Or docks, or mallows."[36] But Gonzalo has loftier ideals:

> I'th' commonwealth I would by contraries
> Execute all things, for no kind of traffic
> Would I admit; no name of magistrate;
> Letters should not be known; riches, poverty
> And use of service, none; contract, succession,
> Bourn, bound of land, tilth, vineyard—none;
> No use of metal, corn, or wine, or oil;
> No occupation, all men idle, all;
> And women, too, but innocent and pure;
> No sovereignty . . .
>
> All things in common should produce
> Without sweat or endeavour; treason, felony,
> Sword, pike, knife, gun, or need of any engine
> Would I not have; but nature should bring forth
> Of its own kind all foison, all abundance,
> To feed my innocent people. . . .
>
> I would with such perfection govern, sir
> T'excel the Golden Age.[37]

Despite Antonio and Sebastian's interruptions, and often modern editors' accusations of hyperbole, Gonzalo is actually outlining a near-utopian commonwealth.

A key inspiration for this speech is Michel de Montaigne's essay "Of the Cannibals," itself based on classical sources including Ovid, and the idea of a golden age of early man. [38] There are many aspects to be noted. The commonwealth or body politic will not be ruled by the market or commerce ("traffic"); no bureaucracy or written records ("letters"); no discrepancy of wealth or indentured servitude; no inheritance; no "bourn" (a boundary or limit) or "bound of land"—a critique of enclosure; no agricultural working of the soil; no occupation—a multifaceted term that has a literal sense of seized presence, and employment, but also marital cohabitation; no sovereignty; but instead shared property, an absence of ills, producing abundance and happiness. But if this is a commonwealth without internal divisions, it is still hard to imagine it without boundaries on the outside, to imagine it entirely outside territory.

In the end of the play, although Prospero regains his kingdom, this will be short-lived. Having no male heir, his inheritance will pass to his daughter Miranda.[39] And Miranda, through Prospero's own matchmaking, is now betrothed to Ferdinand, son of the King of Naples. While this is both romantic and strategically-politically appropriate, as Butler notes, "In the next generation Milan will become a possession of Naples. Even as Prospero resumes his power, he loses it. There is no way that he can protect Milanese independence beyond his lifetime."[40] The King, earlier in the play and thinking his son lost in the shipwreck, described him as "mine heir / Of Naples and of Milan."[41] If Ferdinand is dead, then the next heir is Claribel, his daughter. It was Claribel's wedding to the king of Tunis that gave the King and Duke the reason for their sea voyage, on which they end up shipwrecked on the island. The King here imagines that his kingdom will pass to Tunis; with Ferdinand's survival he not only preserves his inheritance, but, with Ferdinand's marriage to Miranda, creates the lineage that brings Milan and Naples together. Prospero's dukedom will be incorporated into a larger empire.

The Tempest is thus an important play in showing Shakespeare's interest in colonialization, and the seizure and shaping of territories beyond the British Isles or *Hamlet*'s Denmark. From English expansion in Wales, Scotland, Ireland, and France, as explored in the history plays, or the domestic and regional turmoil of *Macbeth* or *Hamlet*, many of Shakespeare's plays explore what might be called the near abroad.[42] However, at his time of writing, European states were in the process of expanding far beyond their immediate neighbors. *The Tempest* is the most striking place in which Shakespeare explores the practices of enclosure, possession and dispossession, land transformation, and its government in a colonial setting.

THE CYPRUS WARS IN *OTHELLO*

Yet *The Tempest* is not the sole play with colonial interest. *Othello* is largely set in Cyprus, an island that is contested between Venice and the Ottoman Turks. Othello himself has made his reputation as a soldier in "the Cyprus wars,"[43] and it is a threat to Cyprus that leads him back there with Desdemona and Iago. Shakespeare's source for the story, Giovanni Battista Giraldi Cinthio's 1565 novella, *Un Capitano Moro*, from his *Gli Hecatommithi*, does not name the main character.[44] Shakespeare's choice of the name is intriguing. As Bate suggests, it is similar enough to Ottoman, especially when pronounced as Otello, and to the founder of that empire Othman, to suggest "his origin in the Ottoman territories, against which he is now fighting."[45]

Let us leave aside the relation between Othello and Desdemona, the wider racial politics that undercut his relations with her father and Venetian society, and the plot Iago weaves, and focus instead on the wider geopolitical frame.[46] As Tom McAlindon rightly notes, the play's connection to contemporary political issues has been understated in many accounts of its construction. It has been, he suggests, seen as "a claustrophobic personal tragedy divorced from the political and cosmological contexts that lend grandeur of implication to the other tragedies."[47] Yet there is a very clear context established. Right at the start of the play we hear of Turkish invasion plans, and indeed these interrupt the narrower interpersonal feud. It may be that Shakespeare knew Knolles's 1603 study, *The Generall Historie of the Turkes*, which has an account of the Turkish attacks on Cyprus.[48] Cassio—himself an outsider too, as a Florentine—tells Othello of the events, and that the Duke has summoned him:

> Something from Cyprus, as I may divine;
> It is a business of some heat. The galleys
> Have sent a dozen sequent messengers
> This very night, at one another's heels,
> And many of the consuls, raised and met,
> Are at the duke's already. You have been hotly called for.[49]

The play's third scene is then focused on a discussion of this imminent attack on Cyprus. As Mark Rose has indicated, this scene is like the waking of Desdemona's father, "in each case an authority—Brabantio, the Venetian council—is roused in the middle of the night to deal with confusing reports of a barbarian's assaults on a precious dependent."[50] The multiple successive messengers mentioned by Cassio are indeed sending reports, but they contradict each other. The Duke and his Senators are first trying to ascertain the numbers:

> Duke: There is no composition in these news
> That gives them credit.
> Senator 1: Indeed, they are disproportioned.
> My letters say a hundred and seven galleys.
> Duke: And mine, a hundred forty.

Senator 2: And mine, two hundred.
But though they jump not on a just account—
As in these cases, where the aim reports,
'Tis oft with difference—yet do they all confirm
A Turkish fleet, and bearing up to Cyprus.
Duke: Nay, it is possible enough to judgment:
I do not so secure me in the error
But the main article I do approve
In fearful sense.[51]

There is inconsistency in the reports they have received, with different estimations of the size of the force. But while Brabantio reacts intemperately, there is a cold, calculative sense to the council's assessment, and W. Watkiss Lloyd has suggested it "strikes a tone of dispassionate appreciation of evidence and opinion that dominates all the succeeding scenes of agitation and disorders."[52] Not only the numbers, but the terms speak of this calculative rationality—"composition," "credit," "disproportioned," "account," "difference." The use of "composition" to mean "consistency, congruity" is, according to Michael Neill, "the only example of this use in *OED*."[53] This sense is mirrored in Iago's subsequent claim "I stand accountant for as great a sin."[54]

But the second senator insists that whatever the difficulties in counting, there is one thing that should be clear—there is a Turkish fleet bound to attack Cyprus. Then there is geographical confusion: a sailor enters who indicates that "The Turkish preparation makes for Rhodes" instead,[55] but this is immediately doubted by the first senator:

This cannot be,
By no assay of reason: 'tis a pageant,
To keep us in false gaze. When we consider
Th'importancy of Cyprus to the Turk,
And let ourselves again but understand
That, as it more concerns the Turk than Rhodes
So may he with more facile question bear it,
For that it stands not in such warlike brace,
But altogether lacks th'abilities
That Rhodes is dressed in. If we make thought of this
We must not think the Turk is so unskilful
To leave that latest which concerns him first,

Neglecting an attempt of ease and gain
To wake and wage a danger profitless.[56]

Themes of deception, display, and falsehood are established here, anticipating themes that the play will treat in its major plot. The Senator is clear that this is a ruse, and the Duke agrees. Both will be reinforced in this view when another messenger suggests the fleet has turned toward Cyprus:

Duke: Nay, in all confidence, he's not for Rhodes.
Officer: Here is more news.
Messenger: The Ottomites, reverend and gracious,
Steering with due course toward the isle of Rhodes,
Have there injointed with an after fleet—
1 Senator: Ay, so I thought; how many, as you guess?
Messenger: Of thirty sail; and now they do re-stem
Their backward course, bearing with frank appearance
Their purposes toward Cyprus.[57]

The Turks have reinforced their numbers with a second flotilla, and have turned back toward Cyprus. As Harold Goddard has noted, this deceit has parallels in the wider deception in the play: "It would be prosaic to put the analogy on all fours. But who can miss it? The Turk is apparently taking one course that under cover of it he may take an entirely different one. Iago is about to do the same. 'A pageant To keep us in false gaze.' What better description could we ask of his plot?" He continues to suggest that the senator's speech "We must not think the Turk is so unskilful . . ." surely applies as much to Iago as to the external threat.[58]

The unrealistic quick succession of these messengers and their conflicting reports can be overlooked: what is important is the account that the Turks are challenging Venetian supremacy across the eastern Mediterranean. Cyprus is the key focus, hence the need to call for Othello to meet this challenge. When he appears the Duke immediately charges: "Valiant Othello, we must straight employ you / Against the general enemy Ottoman."[59] This scene of military preparation is interrupted by the complaint of Brabantio. The Duke had already told Othello of his commission when he notices Brabantio has entered alongside him. Brabantio pushes the case, but he is defeated by Desdemona's pleading on Othello's behalf, and her share of the responsibility, as well as the clear need that Venice has for Othello in this moment of crisis. Indeed, were it not for the pressing strategic concern, it is possible that Brabantio would have pushed his suit harder, or at least found a more receptive hearing.[60] Even he eventually recognizes the immediate, political issue, though parallels it with the theft of his daughter: "So let the Turk of Cyprus us beguile, / We lose it not so long as we can smile."[61]

In one of his speeches in his defense, Othello tells his own story, mentioning "most disastrous chances, / Of moving accidents by flood and field, / Of hair-breadth scapes i'th' imminent deadly breach."[62] These unfortunate events, lucky escapes, near misses, and so on have happened "by flood and field," in sea and land battles, with a breach having a sense of a violation or rift, both a penetrated fortification and a ruptured hull. This dual land and sea sense is important. The maritime aspect is the most clearly stated, and as Clemen has noted, "In the whole play the sea has an important rôle—as scene, background, and as Othello's own vital element."[63] However Othello's speech continues to talk about land, physical terrain:

> Wherein of antres vast and deserts idle
> Rough quarries, rocks and hills whose heads touch heaven.[64]

"Antres" are caves or caverns, from the Latin *antra*; "vast" means empty as well as expansive; "idle," desolate or barren. "Rough quarries" are natural features of rugged mountainsides rather than man-made excavations. This and the subsequent descriptions come from Pliny's *Natural History* and Elizabethan travel accounts.[65] Sometimes the details are striking: when Iago suggests to Othello he may change his mind, he replies to the geography of his self as much as that of the world, mentioning the Black Sea, the Sea of Marmara, and the Dardanelles Strait:

> Never, Iago: Like to the Pontic sea
> Whose icy current and compulsive course
> Ne'er feels retiring ebb, but keeps due on
> To the Propontic and the Hellespont:
> Even so my bloody thoughts, with violent pace
> Shall ne'er look back, ne'er ebb to humble love,
> Till that a capable and wide revenge
> Swallow them up. Now, by yond marble heaven
> In the due reverence of a sacred vow
> I here enrage my words.[66]

Thus the opening scenes of the play establish both the major plot concerning Othello and his relationship to Desdemona, along with Iago's role in revealing this to her father, and his jealousy of Cassio and the Moor; and the wider geopolitical frame for the story. Cinthio's *Un Capitano Moro* does not set up that wider context. As McAlindon has indicated, in doing so, Shakespeare connects the story to a recent concern about the Ottoman Empire, the 1570 fleet that sailed from Constantinople via Rhodes to attack Cyprus.[67] As the Duke notes, "The Turk with a most mighty preparation makes for Cyprus. Othello the fortitude of the place is best known to you."[68]

The actual military contestation is abbreviated in the narrative with the second act opening with Montano and Gentlemen observing the battle at sea and quickly reporting on Othello's success.

> Montano: What from the cape can you discern at sea?
> 1 Gentlemen: Nothing at all, it is a high-wrought flood:
> I cannot 'twixt the haven and the main
> Descry a sail.
> Montano: Methinks the wind hath spoke aloud at land,
> A fuller blast ne'er shook our battlements
> If it hath ruffianed so upon the sea
> What ribs of oak, when mountains melt on them,
> Can hold the mortise? What shall we hear of this?
> 2 Gentleman: A segregation of the Turkish fleet:
> For do but stand upon the foaming shore,
> The chidden billow seems to pelt the clouds,
> The wind-shaked surge, with high and monstrous mane,
> Seems to cast water on the burning bear
> And quench the guards of th'-ever-fired pole.
> I never did like molestation view
> On the enchafed flood.
> Montano: If that the Turkish fleet
> Be not ensheltered and embayed, they are drowned.
> It is impossible to bear it out.[69]

The sea is roughened by the wind, tossing the ships around, with mountains as huge waves, as is the "high and monstrous mane," the sea being likened to some wild beast. The Turkish fleet is split up by the storm, breaking some ships up. The Folio's "Foaming shore" is "banning shore" in the Quarto, with "banning" possibly meaning "chiding, cursing,"[70] or "forbidding the encroachment of the waves,"[71] though the Folio's reading is also possible. The "guards of th'-ever-fired pole" are two stars known as the Guardians, part of the Little Bear constellation.[72] "Enchafed" is angry, and the suggestion is that unless the Turkish fleet was docked, it must be destroyed. Quickly this is confirmed, as a minor character reports on the storm:

> News, lads! our wars are done!
> The desperate tempest hath so banged the Turks,
> That their designment halts. A noble ship of Venice
> Hath seen a grievous wreck and sufferance
> On most part of their fleet.[73]

It was Cassio's ship that witnessed this destruction, and he adds to the report, speaking of "Tempests themselves, high seas and howling winds, / The guttered rocks and congregated sands."[74]

Thus a much wider world is established in these opening scenes, but surprisingly we hear little of it after the defeat of the Turkish fleet. Aside from a few place references, the rest of the play concentrates on interpersonal relations. Nonetheless, as Neill indicates, "the framework of momentous public events lends gravity to what might otherwise seem a mere domestic tragedy."[75] Honigmann suggests that this geographical imbalance might have something to do with the specific places:

> Shakespeare saw Venice as part of his own world, but not so Cyprus. At the far end of the Mediterranean, Cyprus had passed into Turkish rule and, unlike Venice, seems not to have interested Shakespeare as a place, for it might be almost any one of many islands fortified by the Venetians. The play is quite specific about the Mediterranean background, mentioning Spain, Barbary, Egypt, Rhodes, the Pontic and Propontic seas, Mauretania, Jerusalem, Aleppo, Arabian trees, etc. and it reminds us some two dozen times of its location in Cyprus without really adding to the sketchy information given by Cinthio—namely, that the Venetians maintained forces there and sent out the Moor as their commander.[76]

After the defeat of the Turkish fleet, with the main characters now in Cyprus, the wider setting seems to recede, and the tragedy becomes one of interpersonal relations. Yet has the entire Turkish fleet really been destroyed, or is this another ruse? Are some of the ships perhaps concealed, awaiting a more opportune moment? As McAlindon suggests, Shakespeare's contemporary audience would know that the Turkish threat had not receded, and that the Turks had been successful in their 1570 attack, still occupying the island in the early seventeenth century, when the play was written and first performed.[77] Venice had surrendered Cyprus in the 1573 peace treaty. King James wrote a poem celebrating a 1571 naval victory over the Turks at Lepanto in 1591, and it was republished in 1603 after he acceded to the English throne. Emrys Jones suggests it may have been the spur to Shakespeare's play.[78] Despite this defeat the Turks continued to dominate the eastern Mediterranean, and the loss of Cyprus marked the decline of Venetian influence.[79] Othello might confirm that "our wars are done, the Turks are drowned,"[80] but the figure of the Turk remains an important focus of the play. As Goddard insists,

> Reread the play with sharp attention to the parts in which war figures, pondering particularly every allusion to the Turks—there are many of them—and it is inescapable that what Shakespeare is bent on is an insinuation into the underconsciousness of the reader of an analogy between Iago and the Turk. Indeed, in one passage Iago openly makes the identification himself. Desdemona has dubbed him "slanderer" for his strictures upon women. "Nay, it is true," retorts Iago, "or else I am a Turk." But it is not true. And so he is a Turk.[81]

Bate comments on these same lines. There is "deep irony" here, "for it *is* Iago who does the Turkish work of destroying the Christian community. . . . Christian Iago is a functional Turk."[82]

There is, though, a clear contrast between Venice and Cyprus. In Venice, there are established norms of political and domestic behavior, and Othello's marriage to Desdemona is, at least theoretically, open to challenge. Brabantio clearly has the grounds for believing this, until the Turkish attack forces attention elsewhere. Neill stresses that Cyprus is rather different. It "is an embattled military outpost," where, in contrast to Venice, "no such predictable sanctions obtain," a place where "no one is at home," an island that "belongs to the stormy domain of the passions; and its distance from Venice is symbolically marked by the ferocious tempest that opens the second act."[83] Crucially, as Norman Sanders suggests, Cyprus is the ideal setting for the domestic drama—Desdemona is alienated from her family and Venetian friends by her actions, and now geographically isolated on the island. The military setting puts Iago in his element, and Othello is distracted from his military purpose by the presence of his wife, and vice versa.[84] The interruption of their first night together with the barrack-room fight leads Othello to liken his soldiers to the enemy: "Are we turned Turks? And to ourselves do that / Which heaven hath forbid the Ottomites? For Christian shame, put by this barbarous brawl."[85] The Turk, the Ottomite, the barbarian, set up as a contrast to the Christian. And yet, as *Othello* shows, this is far from a clear distinction. The barbarian in this play is a native of Barbary, a Berber, a description that is used of Othello, when Iago describes Othello as a "Barbary horse"; and where he talks of a vow between "an erring Barbarian and a super-subtle Venetian."[86] Perhaps the most intriguing and unexplained reference is to Desdemona's mother's maid, Barbary, who taught her the willow song.[87] The Mediterranean focus of this play is as much its southern shores as its eastern islands.

THE AQUATIC GEOGRAPHIES OF *PERICLES*

The eastern Mediterranean settings of *Pericles, Prince of Tyre* are crucial to both descriptive color and narrative structure. This is a play that was not included in the First Folio, but is now acknowledged to have at least large parts written by Shakespeare.[88] It was included in an appendix to the second printing of the Third Folio in 1664 and is the only one of those additions to be generally accepted.[89] It has been compellingly argued that the travel elements in the first part were from another hand, probably George Wilkins, while the story of Pericles's daughter Marina and the return home was written by Shakespeare.[90] The text we have is based on a Quarto version, first published in 1609, but this is corrupted and presents various interpretative difficulties.[91] There is also a narrative version published by Wilkins as *The Painful Adventures of Pericles* in 1608 that some editions use to correct or fill in text.[92] The story is loosely based on the tale of Apollonius of Tyre, filtered through the version presented by Chaucer's contemporary John Gower in his collection *Confessio Amantis*.[93] The play's dramatic episodes are introduced by a Chorus, named Gower, in tribute to this debt.

In the play that bears his name, Pericles is continually on the move. It is therefore more episodic than most of Shakespeare's plays, and Pericles is neither geographically nor mentally in control of his own destiny. It is as much a spiritual or symbolic as a spatial journey, and yet "psychologically as well as geographically the pattern of his travels slips out of his control."[94] While he chooses some of his destinations, he is shipwrecked, has to go into port because of a storm, escapes by sea, and returns to the sea before finally returning home to replace his father as ruler of Tyre (now in Lebanon). This is both a geographical return and a political one. As he leaves an early problem, one character comments, "He scaped the land to perish at the seas."[95] While he does not die, he certainly seems more at home in the water. The play also focuses on his loss and rediscovery of his wife, Thaisa, and daughter Marina, over a fourteen-year gap in the play. While those years are not dramatized, he has been wandering all the time. These and the earlier travels mean the location of the play continually changes between open sea, islands, and coastlines. Antioch in Syria is the first mentioned place,[96] and he visits other parts of the Levant, modern-day Turkey, and places in Greece and North Africa. These eastern Mediterranean sites have a double focus. On the one hand, they relate to places in late antiquity; on the other, they are of relevance to the seventeenth-century struggle between Western Europe and the Ottoman Empire for strategic control of this region. As Suzanne Gossett suggests, "The play moves all around the Mediterranean basin, using as settings locations that signify differently when read within the classical framework inherited from the Apollonius tale and when situated in the mental world of the politically alert Jacobean theatregoer."[97]

Richard Halpern and Linda McJannet have analyzed this historical setting in the late Hellenistic world, showing how the political actors are those who are struggling for effective control over what used to be Alexander the Great's empire. These issues shape the political and cultural geography of the play.[98] But Gossett highlights that "at the same time they were areas of current contestation, with the Turk for political power and with other nations for mercantile dominance."[99] So Shakespeare and Wilkins are linking the struggles after the fall of Alexander's empire in the Near East—the historical setting of the play—to contemporary struggles for control of this region, broadly the eastern Mediterranean or the Levant. The time of the play's writing is disputed, but it is likely to be the early seventeenth century, probably after 1606 and certainly before 1608.[100] Whatever its date, this was a time new maps of the world were being produced, of which one of the most famous is Ortelius's *Theatrum Orbis Terrarum*. Crucially this atlas did not just provide a map of the New World, but also historical maps, such as one of the Roman Empire. As Gossett points out, the eastern Mediterranean part "shows most of the play's locations: Tyre, Antioch, Tarsus, Ephesus and Lesbos, of which Mytilene was the capital. . . . Missing is only Pentapolis."[101] Gossett adds that there is some basis for thinking Pentapolis is a reference to Cyrene, although there were at least seven places called Pentapolis—the name means "five cities"—in the ancient world.[102] The moment of the play's writing may have been around the time that the first permanent English settlement in the Americas, Jamestown, was founded in 1607:

> Pericles appeared at a time of constant debate about the purposes of travel. . . . Jamestown, the first permanent English settlement in America, was successfully founded in 1607. The claim of sovereignty beyond

national boundaries, Cormack shows, extended the king's authority and jurisdiction, creating a transmarine empire. Yet travel was a hazard, voyaging always potentially "transgressive," permitting "deeply compromising relationships with the exotic" (Gillies).[103]

In *At the Bottom of Shakespeare's Ocean*, Mentz has suggested that *Pericles* should be seen alongside *The Tempest* and *Henry V* as an example of colonialism.[104] Mentz argues that "Pericles is a colonizer. . . . Pericles's protracted travels comprise a series of attempts to expand his kingdom's sway. He seeks an alliance through marriage with powerful Antioch, cultivates Tarsus with gifts of corn, seals an alliance with Pentapolis through marriage, and then, when he needs help, returns to Tarsus to redeem his debt."[105] It is a play filled with maritime imagery and that of navigation. One example would be the early claim "As I am son and servant to your will, / To compass such a boundless happiness."[106] "Compass" here is to enclose, to encompass—a deliberate contrast to the idea of a boundless, limitless emotion.

For Estok, *Pericles* is "spatialized and mapped to provide at times a residence for monstrosity and at others an escape route from it," and "competing geographies flash through the play like a surreal slide show."[107] Yet this is not just colonization of land, for the sea matters too, both as a medium of travel and as a place to be controlled in its own right. Pericles imagines that fish have a "watery empire,"[108] which Mentz reads as "reinforcing his focus on territorial expansion."[109] The expansion does occur across the sea—a variety of locations not far from the shores of the Mediterranean. Ortelius's map is of course of the Roman Empire, an empire that expanded not only by land, but also by sea, hence the Romans' name for the Mediterranean: *mare internum* or *mare nostrum*; an internal sea, a Roman lake. So there is a mix of issues here in terms of the purposes of travel, and the extension of political authority and jurisdiction, especially across water. Noncontiguous or saltwater colonization characterized the British Empire, whereas China or Russia tended to colonize adjoining lands. The Roman Empire did both, around all of the Mediterranean, and also by land. Britain was itself a colonial project.

One chorus by Gower highlights these very issues:

Thus time we waste and long leagues make short,
Sail seas in cockles, have and wish by for't,
Making to take our imagination
From bourn to bourn, region to region.
By you being pardoned, we commit no crime

To use one language in each several clime
Where our scenes seem to live.[110]

Cockleshells were traditionally the vessel for supernatural journeys, which can be achieved simply by desire. The geographical travel, from place to place, frontier to frontier, is mirrored in the play's changing language.[111] The key shift in the text, of course, is from the part believed to be by Wilkins to that by Shakespeare. The opening scene of that second part has a wonderful speech from Pericles seeking to control the storm:

The god of this great vast, rebuke these surges
Which wash both heaven and hell, and thou that hast
Upon the winds command, bind them in brass,
Having called them from the deep. O, still
Thy deafening dreadful thunders; gently quench
Thy nimble sulphurous flashes![112]

The sea is, of course, crucial to Pericles's familial relations. His daughter is born during a storm at sea, hence her name Marina,[113] but her mother seems to die in childbirth. Pericles is forced by the sailors to abandon her body to the sea:[114]

Th'unfriendly elements
Forgot thee utterly, nor have I time
To give thee hallowed to thy grave, but straight
Must cast thee, scarcely coffined, in the ooze,
Where, for a monument upon thy bones
And aye-remaining lamps, the belching whale
And humming water must o'erwhelm thy corpse,
Lying with simple shells.[115]

Pericles asks the ship's master which coast they are near, which is said to be Tarsus. He asks that the course be changed to there instead of Tyre so he can leave the baby with Cleon, the governor of that city, and his wife, Dionyza.[116] This is a redemption of the help he previously gave that place, when he fed the starving town with his corn,[117] although he is to be cruelly betrayed when Dionyza plans to murder Marina because she grows up to be more beautiful than their own daughter.[118] Cleon, for one, regrets this: "Were I chief lord of all this spacious world / I'd give it to undo the deed."[119]

There are also historical and contemporary parallels in the pirates who abduct Marina—both rescuing her from the planned murder, and selling her into prostitution.[120] Pirates can be traced historically—Pompey the Great was asked by the Roman Republic to rid the Mediterranean of pirates, which was one of the bases of his power, his *imperium*—but were also a concern at Shakespeare's own time. As Gossett indicates,

> Similarly complex in their origins are the pirates who abduct and rescue Marina. They are both a literary trope, with forebears in Greek prose romances and New Comedy, and a pressing governmental concern. Piracy in the Mediterranean was especially severe towards the end of Elizabeth's reign and in the first years of James's. Pirates could be identified with the view of the contemporary Mediterranean as "other" or reprobate.[121]

Piracy is significant because pirates are outside a land of jurisdiction, but work by using the sea as a lawless space.[122] Travel at this time still had the transgressive potential of encounters with the exotic, the other. "Travel might be dangerous, frightening and potentially corrupting, but for the English, at the beginning of centuries of imperial expansion, it was irresistible."[123]

There is a mirroring of loss and redemption through the sea too. Pericles meets his wife by chance through shipwreck and finds her again through sea voyages, just as he leaves his child and finds her again through his travels. As Thaisa recognizes, he came to meet her "only by misfortune of the seas / Bereft of ships and men cast on this shore—."[124] In their reconciliation scene, he returns again to maritime imagery:

> O Helicanus, strike me, honoured sir,
> Give me a gash, put me to present pain,
> Lest this great sea of joys rushing upon me
> O'erbear the shores of my mortality
> And drown me with their sweetness.[125]

Pericles, then, as Gower says in a late chorus, spends much of his time "thwarting the wayward seas."[126] "Thwarting" here means both crossing and challenging, "wayward" both unpredictable and hostile.[127] It is as apt a description of this play's colonial, aquatic geographies as any other.[128]

ANTONY AND CLEOPATRA AND THE ROMAN *IMPERIUM*

The other key eastern Mediterranean play is *Antony and Cleopatra*. While there are pronounced geographical and territorial issues in *Coriolanus*, discussed in chapter 8, *Antony and Cleopatra* is the most colonial of Shakespeare's Roman plays. However, the ways that its geographies are configured highlight two different aspects alongside the colonial. One is the contrast between sea and land, and the second is the way that understandings of measure and division structure both. As well as being an integral part of this chapter, the discussion of this play also forms a link to the discussion of technologies of measuring in chapter 7.

Unlike some of Shakespeare's other love stories, there is a wider geopolitical struggle in *Antony and Cleopatra*, and the key players are more central to it than in *Troilus and Cressida*. These interpersonal affairs are not just between the title characters, but also between Antony and Octavius, Julius Caesar's nephew and adoptive son and later the emperor Augustus. Octavius is known as Caesar or Octavius Caesar in the play, though to avoid confusion Octavius will generally be used here. Octavius says that he wishes for something that would bind him and Antony together, given the clash in their characters ("our conditions / so differing in their acts").[129] "Yet, if I knew / What hoop would hold us staunch, from edge to edge / O'th'world I would pursue it."[130] The solution, proposed by Agrippa, is that Antony marry Octavia, Octavius's sister—"to join our kingdoms and our hearts."[131] The fallout between brothers-in-law, between Julius Caesar's heir and his closest military lieutenant, is essential to the story. As John Wilders suggests, *Antony and Cleopatra* is "a play about international politics, a public as well as private drama in which Antony and Octavius compete for mastery over the Roman Empire which, at the time, extended from Britain in the west to what is now Turkey in the east, and the battles in which this contest was fought out occupy much of the third and fourth acts."[132] While *Much Ado About Nothing* can easily pass over the military context, and even some of the tragedies have their political aspects neglected or cut in stage productions, in *Antony and Cleopatra* the geopolitical struggle is "not simply a background against which the love tragedy is played out but an inseparable part of it."[133]

The Roman Empire extended at this time from Britain to Turkey, and into large areas of the African continent. The key figures of Antony and Cleopatra, along with Octavius, Lepidus, and Sextus Pompeius, are central both to the dramatic struggle and to the wider political one. Their actions cross the known world, and the fate of that world is tangled up in their interpersonal affairs. One of Antony's followers, Philo, sets the terms in the play's opening lines: "this dotage of our general's / O'erflows the measure."[134] He means that Antony's infatuation with Cleopatra exceeds the limit, and he is turning his attention away from the world and its wars to a different focus. Philo goes on to say that Antony is the "triple pillar of the world," one of three key Roman leaders, but he is "transformed / Into a strumpet's fool."[135] The "triple pillar" is a crucial term, because after the defeat of Julius Caesar's assassins the Roman world was divided into three by the triumvirate—Antony, Octavius, and Lepidus.[136] Elsewhere we hear of "these three-world sharers,"[137] and when Cleopatra describes Antony as "This demi-Atlas of this earth, the arm / And burgonet of men!"[138] she is seemingly forgetting Lepidus. A burgonet is a helmet, and so he a complete soldier with both an offensive arm and a defensive helmet.[139] Of course, Antony and Octavius are by far the most important of the triumvirate. On seeing Lepidus carousing, Enobarbus says: "There's a strong fellow . . . / 'A bears the third part of the world, man."[140] Menas replies: "The third part then he is drunk."[141] Antony continues the watery metaphor (a key theme to be explored later): "These quicksands, Lepidus, / Keep off them, for you sink."[142]

As Maurice Charney has noted, world words and symbols represent "the most general pattern of imagery in the play."[143] Philo's opening speech suggests it is a world that is divided, measured, conquered, and contested. The language of measure and extent is marked throughout the play. When Cleopatra asks Antony for the extent of his love, she asks him to "tell me how much."[144] "Tell" can mean to inform or explain, but also to count, to reckon, like a bank teller. Antony takes it in this last sense, but denies all calculative measures: "There's beggary in the love that can be reckoned," though then Cleopatra replies to say she will "set a bourn how far to be beloved."[145] Antony suggests that she will need to find a "new heaven, new earth" to encompass his love.[146] But as news comes from Rome, he disdains worldly concerns, suggesting that if his support is removed, so too will much of Rome's power disappear:

> Let Rome in Tiber melt, and the wide arch
> Of the ranged empire fall! Here is my space!
> Kingdoms are clay! Our dungy earth alike
> Feed beast as man. The nobleness of life
> Is to do thus, when such a mutual pair
> And such a twain can do't, in which I bind,
> On pain of punishment, the world to weet
> We stand up peerless.[147]

The arch is linked to the triple pillar of the world, and "ranged" is "ordered, with connotations of an army arranged in ranks."[148] Cleopatra is his empire, his space, the extent of what he wishes to command and possess. "Weet" is to recognize or know.

These divisions continue when Antony claims that with his sword he has "quartered" the world.[149] As he says to his new wife, Octavia: "The world and my great office will sometimes / Divide me from your bosom."[150] Divisions are not just by action or choice, but by instruments of measure: "Read not my blemishes in the world's report. / I have not kept my square, but that to come / Shall all be done by th' rule."[151] Here, the square or set square and the rule or ruler, the tools of carpentry or geometry, are used in relation to principles of conduct and government—a theme that will be returned to in chapter 7.

There are a range of colonial issues in terms of territories ruled by Rome. Yet this is not a stable empire:

> Labienus—
> This is stiff news—hath with his Parthian force
> Extended Asia. From Euphrates
> His conquering banner shook, from Syria
> To Lydia, and to Ionia,[152]

Labienus had defected to the Parthians, a people occupying lands in modern-day Iran and Iraq, and Antony himself had been defeated by them.[153] Some of these lands were those divided between Cleopatra and her brother Ptolemy by Antony, making Cleopatra ruler of far more than just Egypt:

> Unto her
> He gave the establishment of Egypt; made her
> Of lower Syria, Cyprus, Lydia,
> Absolute Queen.
> His sons he there proclaimed the kings of kings:
> Great Media, Parthia and Armenia
> He gave to Alexander; to Ptolemy he assigned
> Syria, Cilicia and Phoenicia.[154]

One of the challenges to the triumvirate comes from Sextus Pompeius, sometimes known as Pompey, the son of Pompey the Great, who had fought an earlier stage of the civil war with Julius Caesar. The younger Pompey addresses the triumvirate: "To you all three / The senators alone of this great world, / Chief factors for the gods."[155] He thinks that the three have achieved the kind of dictatorship Julius Caesar was murdered for. This is what "hath made me rig my navy, at whose burden / The angered ocean foams."[156] Antony, though, is defiant, saying to Pompey: "Thou canst not fear us, Pompey, with thy sails. / We'll speak thee at sea. At land thou know'st / How much we do o'ercount thee."[157] Here, again, is the calculative applied to military might. "Speak" here implies an encounter or battle; and even though Antony's forces outnumber Pompey's by land, he is still prepared to meet him at sea.

Pompey strengthens his position when Lepidus allies with him, though the immediate impact is that Lepidus loses his soldiers and his share of the empire to Octavius. In the larger scheme of things the key impact is thus the increase in Octavius's strength, and indeed Shakespeare does not mention Lepidus's treachery, rather focusing on Octavius's power: "So the poor third is up, till death enlarge his confine"; "up" means locked or imprisoned, and "enlarge his confine" suggests this has set him free.[158] This then leaves the two men:

> Then, world, thou hast a pair of chaps, no more,
> And throw between them all the food thou hast,
> They'll grind the one the other.[159]

"Chaps," as well as being men, also means chops or jaws, with the symbol suggesting that they will "grind against each other and thus grind each other down."[160]

LAND AND SEA

Sextus Pompeius's challenge to the triumvirate is launched from the sea—the land masses being divided between the three. It is notable that his father had first achieved military greatness when charged with ridding the Mediterranean of pirates. We are told that the son "hath given the dare to Caesar and commands / The empire of the sea,"[161] and that "rich in his father's honour," he "makes his approaches to the ports of Rome."[162] While "port" is strictly speaking a gate, where the plowing of the sacred boundary had been stopped, carried (*portare*) in a break in the wall, here it is clearly a harbor.

Octavius is told that "Pompey is strong at sea,"[163] and he develops this theme in a speech rich with maritime imagery: "the ebbed man"; "This common body, Like to a vagabond flag upon the stream, / Goes to and back, lackeying the varying tide, / To rot itself with motion."[164] His aquatic reverie is interrupted by another messenger, who reports the challenge of "famous pirates."[165] This is another speech rich with maritime imagery, the key being the claim that "Many hot inroads / They make in Italy—the borders maritime / Lack blood to think on't."[166] The coastal regions are under threat and scared. Octavius suggests that Antony is not doing enough, and is distracted by entertainment. It is left to Lepidus to make preparations, saying he will report tomorrow on "Both what by sea and land I can be able / To front this present time."[167]

Later, Pompey says, "I shall do well / The people love me, and the sea is mine";[168] But he is reminded that Octavius and Lepidus "Are in the field. A mighty strength they carry."[169] Such a status is reinforced by an exchange between Antony and Octavius, the first asking, "What is his strength by land?" and the latter replying, "Great and increasing, but by sea / He is an absolute master."[170] The solution is one strikingly similar to the power of his father; Pompey says:

> You have made me offer
> Of Sicily, Sardinia; and I must
> Rid all the sea of pirates; then to send
> Measures of wheat to Rome. This 'greed upon,
> To part with unhacked edges, and bear back
> Our targes undinted.[171]

"Targes" here are shields. Later, when the pirate Menas meets the courtier Enobarbus, they exchange a sequence of comments about how each has done well in their respective elements:

> Menas: You and I have known, sir.
> Enobarbus: At sea, I think.
> Menas: We have, sir.
> Enobarbus: You have done well by water.
> Menas: And you by land.
> Enobarbus: I will praise any man that will praise me, though it cannot be denied what I have done by land.
> Menas: Nor what I have done by water.
> Enobarbus: Yes, something you can deny for your own safety: you have been a great thief by sea.
> Menas: And you by land.
> Enobarbus: There I deny my land service.[172]

As Wilders notes, the last is a pun, as "it could mean 'military as opposed to naval service' (*OED*)."[173] It also brings to mind the conversation between Alexander the Great and a pirate, that because the latter does his theft by sea and in a small boat he is so named, but the former does it on land and with a large army and is hailed as an emperor.[174]

Menas questions Pompey as to his intention: "Wilt thou be lord of the whole world?"[175] He continues, suggesting, "Whate'er the ocean pales or sky inclips / Is thine, if thou wilt ha't."[176] The spatial terms here are pronounced: "pales" means encloses or surrounds; "inclips" encircles. Menas pushes the point: "These three world-sharers, these competitors, / Are in thy vessel. Let me cut the cable."[177]

The land and sea contrast is marked through the play. Both Antony and Cleopatra seem content to let the sea absorb the land. We have already seen Antony's "let Rome in Tiber melt,"[178] but Cleopatra threatens, "melt Egypt into Nile" and "So half my Egypt were submerged and made / A cistern for scaled snakes"; and is content to "Sink Rome."[179] But this bravado is misplaced, because Antony and Octavius's final battle of Actium is at sea, leading to Antony's defeat, despite his being warned away from this. Antony, fatefully, is fixated on the idea: "we / Will fight with him by sea."[180] Canidus suggests that while Antony has challenged Octavius to meet him at Pharsalia, "Where Caesar fought with Pompey. But these offers, / Which serve not for his vantage, he shakes off, / And so should you."[181] There follows this dialogue:

Enobarbus. Your ships are not well manned,
Your mariners are muleteers, reapers, people
Engrossed by swift impress. In Caesar's fleet
Are those that often have 'gainst Pompey fought;
Their ships are yare; yours heavy. No disgrace
Shall fall you for refusing him at sea,
Being prepared for land.
Antony. By sea, by sea.
Enobarbus. Most worthy sir, you therein throw away
The absolute soldiership you have by land;
Distract your army, which doth most consist
Of war-marked footmen; leave unexecuted
Your own renowned knowledge; quite forgo
The way which promises assurance; and
Give up yourself merely to chance and hazard
From firm security.
Antony. I'll fight at sea.
Cleopatra. I have sixty sails, Caesar none better.

Antony. Our overplus of shipping will we burn,
And, with the rest full-manned, from the head of Actium
Beat th' approaching Caesar. But if we fail,
We then can do't at land.[182]

A muleteer is someone who drives mules, and reapers are harvesters: it is clear that Antony has conscripted land soldiers to serve at sea, improving his numbers by half measures. Octavius's ships are light and maneuverable, whereas Antony's are heavy. Antony is advised to avoid a battle at sea, but is fixed on his course. His lack of soldiers is such that he chooses to destroy the ships he cannot man, and heads to the harbor at the headland of Actium.[183] Even a common soldier warns against this course:

O noble Emperor, do not fight by sea;
Trust not to rotten planks. Do you misdoubt
This sword and these my wounds? Let th'Egyptians
And the Phoenicians go a-ducking; we
Have used to conquer standing on the earth
And fighting foot to foot.[184]

The soldier's message is clear: do not make us fall in the sea, because we are accustomed to fighting on land. Octavius is clear where he should fight: "Strike not by land; keep whole; provoke not battle / Till we have done at sea."[185] With the defeat, Scarus declares:

The greater cantle of the world is lost
With very ignorance. We have kissed away
Kingdoms and provinces.[186]

"Cantle" is a piece or slice, a segment of a circle or sphere.[187] Scarus then talks of Cleopatra's flight:

She once being loofed,
The noble ruin of her magic, Antony,
Claps on his sea-wing, and, like a doting mallard,
Leaving the fight in height, flies after her.
I never saw an action of such shame;
Experience, manhood, honour, ne'er before
Did violate so itself.[188]

"Loofed" means at a distance, heading away, an action that Cleopatra took, but Antony hoisted his sail, and followed her.[189] When they are reunited Cleopatra begs forgiveness: "Forgive my fearful sails! I little thought / You would have followed."[190] Antony replies to this: "Egypt, thou knewst too well / My heart was to thy rudder tied by th' strings / And thou shouldst tow me after."[191] Octavius also addresses Cleopatra as "Egypt," the personification of the country in its ruler.[192] After this defeat, Antony knows he must go to the "young man" Octavius to negotiate meekly, no longer able to sustain his position when "With half the bulk o'th'world played as I pleased."[193]

As it comes to the final battle, Antony declares forcefully, "Tomorrow, soldier, / By sea and land I'll fight."[194] The day comes:

> Antony. Their preparation is to-day by sea;
> We please them not by land.
> Scarus. For both, my lord.
> Antony. I would they'd fight i' th' fire or i' th' air;
> We'd fight there too. But this it is: our foot
> Upon the hills adjoining to the city
> Shall stay with us—order for sea is given;
> They have put forth the haven—
> Where their appointment we may best discover
> And look on their endeavour.[195]
> Caesar. But being charged, we will be still by land,
> Which, as I take't, we shall; for his best force
> Is forth to man his galleys. To the vales,
> And hold our best advantage.[196]

Given that the battle is already on land and sea, to add fire and air would complete the four elements. Octavius then defeats Antony. Cleopatra bemoans his loss:

> His face was as the heavens, and therein stuck
> A sun and moon which kept their course and lighted
> The little O, the earth. . . .
>
> His legs bestrid the ocean; his reared arm
> Crested the world. His voice was propertied
> As all the tuned spheres, and that to friends;
> But when he meant to quail and shake the orb,
> He was as rattling thunder. For his bounty,

There was no winter in't; an autumn 'twas
That grew the more by reaping. His delights
Were dolphin-like: they showed his back above
The element they lived in. In his livery
Walked crowns and crownets; realms and islands were
As plates dropped from his pocket.[197]

"O," the circular, is the idea of the three-part earth with the surrounding ocean, like the orb of the world (mentioned later). "Crowns and crownets" are kings and princes; "plates" are silver coins.[198] Cleopatra addresses Octavius as "Sole sir o'th'world"[199] before taking her own life; he is described by Thidias as "the universal landlord."[200]

While Octavius recognizes the loss of Antony—"The death of Antony / Is not a single doom; in the name lay / A moiety of the world"[201]—he also celebrates the victory to come: "The time of universal peace is near. / Prove this a prosp'rous day, the three-nooked world / Shall bear the olive freely." This "three-nooked world" could be the previously divided world of the triumvirate. But it also relates to the world as it was on "T-in-O" maps, with the three segments of Europe, Asia, and Africa. Isidore of Seville in late antiquity was one of the first writers to provide a map of the world in such a way.[202] These maps show the known world as a disk—the "O"—with the rivers Don and Nile running from left to right and the Mediterranean from the center to the bottom, forming a "T" shape within the "O." Asia occupies the upper segment; Africa the bottom right, and Europe the bottom left. Each has the name of one of the sons of Noah: Shem (Asia), Cham (Africa), and Japheth (Europe).[203] A "nook" could mean a sector or cantle of a circle. Surrounding the continents is the *Mare oceanum*, the oceanic sea. Outside the sea are the compass points, with *Oriens*, the east, to the top, hence "to orient." The key to the bottom right reads "Behold, how the sons of Noah divided the world after the flood."[204]

The geopolitical elements of the story are of course often cut in stage productions, not just because of the demands of staging them—armies, navies, and multiple locations—but also because of the play's length. Of course, cutting these scenes happens in other plays too, such as *Hamlet*. But the political and military scenes are crucial to the overall arc of the narrative, and to focus the action on the two main characters alone will miss much of what is important. As Wilders notes, "The impression that the play encompasses vast expanses of territory and that the conflict is one in which the politics of the world is at stake" does not require dramatic props and spectacle to achieve. "The impression is created more than sufficiently by

Shakespeare's language."[205] The reading here, though attentive to the wider world, its geophysical attributes and the geopolitics of the proponents, has based itself on the language.

From the well-known setting of *The Tempest* to the contest between Venice and the Ottoman Empire over Cyprus to the more historical settings of *Pericles* and *Antony and Cleopatra*, Shakespeare explores many colonial themes. These include practices of enclosure, land and territorial transformation and government, possession and dispossession. The plays discussed here are from a relatively few years of Shakespeare's career. The dates are disputed, of course, but *Othello* is the earliest of these, from 1603–4; *Antony and Cleopatra*, from 1606; *Pericles*, 1607; *The Tempest*, 1610–11. The relatively close proximity of dates may reflect colonial events and practices of Shakespeare's own time, though there are traces of such histories elsewhere—such as the shipwrecks that bankrupt Antonio in *The Merchant of Venice*. Shakespeare's use of past historical (*Antony and Cleopatra*), past semimythical (*Pericles*), near-contemporary (*Othello*), and fantastical/allegorical (*The Tempest*) settings helps to open up some of the issues around colonial territories. The themes of measure and division highlighted in *Antony and Cleopatra* are explored by Shakespeare in detail in many other plays. Chapter 7 turns its attention to those themes both in general terms and more specifically with regard to the relation between territory and technology. Many of the techniques of government, management of populations, and ordering of territories were developed, applied, or perfected in a colonial setting.

CHAPTER SEVEN

Measuring Territories: The Techniques of Rule

Shakespeare lived and worked at a time of profound technological transformation and development.[1] Adam Max Cohen has rightly suggested he "was well aware of the overlapping and interlocking technological revolutions that took place in England during his career."[2] Many plays include references to technical themes in general terms. As J. M. Nosworthy notes, while dominant imagery in *Cymbeline* concerns the natural world and classical references, "images connected with buying, selling, weight and exchange are surprisingly frequent and pervade the whole play."[3] The "rude mechanicals" of *A Midsummer Night's Dream* are skilled craftsmen,[4] and the term "mechanical" is also used to describe the carpenter, with his "leather apron" and "rule" in *Julius Caesar*.[5] "Mechanical" is often used by Shakespeare in a derogatory sense, such as "Base dunghill villain and mechanical" and "most mechanical and dirty hand."[6] On the other hand, there can be casual dismissals of those unschooled in mathematics. In *Love's Labour's Lost*, Armado says, "I am ill at reckoning; it fitteth the spirit of a tapster."[7] This same notion appears in *Troilus and Cressida*—of Troilus, Cressida says, "a tapster's arithmetic may soon bring / his particulars therein to a total";[8] and we are told Ajax "ruminates like an hostess that hath / no arithmetic but her brain to set down her reckoning."[9] This calculative logic is one that even a simpleton in a tavern can comprehend, perhaps counting on fingers, with hostesses and waiters not renowned for being good at addition.[10]

MEASURE FOR MEASURE

The relation of calculation to politics is, of course, a theme in *Measure for Measure*, from its title down.[11] While "measure" is naturally key, as Lezra has pointed out, the word "for," to express a relation of equivalence, is also crucial: "The play seeks to take the measure of the many uses of *for*—linguistic, aesthetic, juridical, and sexual—that arise when absence needs, as the Duke puts it, to be supplied."[12] In the play, the Duke removes himself from Vienna, handing control to his deputy Angelo:

> Of government the properties to unfold
> Would seem in me t'affect speech and discourse,
> Since I am put to know that your own science
> Exceeds, in that, the lists of all advice
> My strength can give you. Then no more remains
> But that, to your sufficiency, as your worth is able,
> And let them work. The nature of our people,
> Our city's institutions, and the terms
> For common justice, y'are as pregnant in
> As art and practice hath enriched any
> That we remember. There is our commission,
> From which we would not have you warp.[13]

This rather ponderous speech begins with the idea that he will outline the principles and qualities of ruling, but decides that this would be artificial, since Escalus appears to be wiser and more knowledgeable than him. "Institutions" here means "political or social customs," rather than the organizations of rule; "the common justice" is more likely to be administration generally rather than the narrower sense of the legal system.[14] Angelo is put in charge: "In our remove be thou at full ourself: / Mortality and mercy in Vienna / Live in thy tongue and heart."[15] Now, mortality and mercy are obviously key themes of the play, but given the Duke's praise of Escalus, and the latter's seniority, it is therefore remarkable that the Duke has elected to transfer power to Angelo:

> I have delivered to Lord Angelo—
> A man of stricture and firm abstinence—
> My absolute power and place here in Vienna.[16]

"Power and place" is an important phrase—his responsibility and situation, the two terms being commonly reinforced in political texts of this period. The people do not know what to make of this:

> The Duke is very strangely gone from hence;
> Bore many gentlemen—myself being one—

In hand, and hope of action: but we do learn,
By those that know the very nerves of state,
His giving out were of an infinite distance
From his true-meant design. Upon his place,
And with full line of his authority,
Governs Lord Angelo; a man whose blood
Is very snow-broth; one who never feels
The wanton stings and motions of the sense;
But doth rebate and blunt his natural edge
With profits of the mind, study and fast.[17]

The "nerves of state" are sinews, and as J. W. Lever indicates are therefore figurative "for the chief motivations of policy."[18] Again there is the reinforcement of Angelo in his "place," governing with given authority, and Angelo is described in terms of being cold-blooded, immune to sensory manipulation. The description of Angelo introduces another, crucial term. Bate notes that Angelo "is more than once described as 'precise,' a term applied to no other character in Shakespeare. It was a word that often went with 'puritan.' "[19] Bate's "more than once" is to be understood literally: Angelo is only ever described as precise twice. First, the Duke says that "Lord Angelo is precise,"[20] and later Claudio calls him "the precise Angelo."[21]

"Precise" is an interesting term, because it only appears in four plays by Shakespeare—once each in *Henry IV, Part 2*, *Love's Labour's Lost*, *The Merry Wives of Windsor*, and this play.[22] In *Love's Labour's Lost* it is used to describe poetry: "Taffeta phrases, silken terms precise."[23] The instances in the two other plays relate to traits of character, in both plays that of honor. In *The Merry Wives of Windsor* Falstaff uses it of himself: "Why, thou uncomfortable baseness, it is as much as I can do to keep the terms of my honour precise."[24] Here it has the sense of morally scrupulous, though the irony is clear. In *Henry IV, Part 2*, it comes in Lady Percy's lament for Hotspur—"O Never do his ghost the wrong / To hold your honour more precise and nice / With others than with him."[25] In this instance "precise" and "nice" are effectively synonyms, reinforcing the sense.

The term "precise" implies morally pure, spiritually strict, 'rigidly puritanical,'[26] though does it have the sense of numerical precision? In the other unquestionable use, in *Measure for Measure*, Claudio is described by Lucio as "ever precise in promise-keeping,"[27] glossed by N. W. Bawcutt as "punctilious, scrupulous."[28] The *OED* suggests that the term's meanings of "strict in the observance of rule, form, or usage," "strict in religious observance," "puritanical," and "exact, neither more nor less than" all date from the sixteenth century, and so could have been intended by Shakespeare in his use. "Precision" is a seventeenth-century word, not used by Shakespeare. Bate adds that Shakespeare's acquaintance John Florio defines the Italian phrase "gabba santi" as "a precise dissembling puritan, an hypocrite."[29]

There are other instances of "precise" in this play: Elbow talks of "precise villains,"[30] and Isabella suggests, "O, 'tis the cunning livery of hell / The damnedst body to invest and cover / In precise guards!"[31] "Guards" here are "ornamental borders or trimmings, fig. for a virtuous exterior."[32] Elbow's use, though, is not what it seems. It follows from his confusion between "benefactors" and "malefactors," and precedes his use of "profanation" when he probably means "profession."[33] Bawcutt suggests that "precise villains" is an error for "precious villains," in the specific sense of "arrant, egregious" (*OED* 4a), noting the use of "precious villain" with that sense in *Othello*.[34] Yet "precise villain" is a good description of Angelo, which may be deliberate on Shakespeare's part, creating "deeper parallels and interconnections between ostensibly separate dramatic actions and characters."[35]

Nor is Isabella's retort simple. Both in her reply, and in Claudio's "the precise Angelo," the Folio text actually reads "prenzie"—"The prenzie, *Angelo*" and "prenzie gardes"—which are usually emended to "precise" or "precious." "Prenzie" is of uncertain meaning, so different editors address this in different ways. Nosworthy follows the Second Folio's emendation to "princely Angelo" and "princely guards," suggesting that "prenzie" derives from an Italian word for "prince," *prenze*, and that "precise" is unlikely as an adjective before "guards." He stresses that the Folio punctuation seems to suggest the first use is a noun—"the prenzie, *Angelo*"—and the second an adjective—"in prenzie guards."[36] Bawcutt keeps "prenzie," and discusses this in some detail, calling it "the most puzzling crux in the play."[37] Brian Gibbons's Cambridge edition keeps "prenzie" and discusses the choice, suggesting both "princely" and "precise" are possible, "the former attractive for its irony," but "the latter orthographically more plausible." He wonders if "perhaps 'prenzie' is Shakespeare's coinage, fusing 'princely' and 'precise.'"[38] Bate and Rasmussen gloss "prenzie" as "unclear meaning; possibly an obscure heraldic term (the register of livery, invest, guards); perhaps a misprint for 'precise' (puritanical), 'priestly' or 'princely' (meant ironically)."[39] What we have then in *Measure for Measure* are five instances of the word, at most, compared to three in Shakespeare's other works. One is applied to Claudio, and one to Angelo. Another is Elbow's error, though Shakespeare may well be making a point about Angelo. The other two only appear if we accept that "prenzie" is a printer error for "precise."

Here there will not be a discussion of the complicated plots and subplots, with the exchanges between people, oaths, and bodies. The play is suffused with the language of calculation, from the description of Angelo as "Accountant to the law upon that pain"[40] through to the late lines that proclaim that "truth is truth / To the end of reck'ning"[41]—that is, in the final account, in the sense of both the final judgment and the furthest we can pursue calculation. In plotting for the resolution of some of the play's knotty problems, the Duke derives a plan that will see his "corrupt deputy scaled," that is, measured and assessed.[42] It is also notable that Angelo says that his marriage with Mariana was broken off "Partly for that her promised proportions / Came short of composition; but in chief / For that her reputation was disvalued / In levity."[43] The dowry he would have received for her was insufficient, in part, at least, because it was dependent on her brother's resources, which had been lost at sea:

> She should this Angelo have married: was affianced to her oath, and the nuptial appointed. Between which time of the contract and limit of the solemnity, her brother Frederick was wracked at sea, having in that perished vessel the dowry of his sister. But mark how heavily this befell to the poor gentlewoman. There she lost a noble and renowned brother, in his love toward her ever most kind and natural; with him, the portion and sinew of her fortune, her marriage dowry; with both, her combinate husband, this well-seeming Angelo.[44]

"Combinate" is bound by pledge, betrothed, but also implies a calculative rationality. The *OED* says it is "rare," dates it to 1583, and gives this passage from the play as its sole instance, glossing it as "betrothed, promised, settled by contract," following Johnson's *Dictionary*. It is the only use of the word by Shakespeare, though the Duke, when disguised as the friar, uses "combined" elsewhere to mean bound by a sacred oath.[45] While there are shipwrecks in many of Shakespeare's plays, most bring characters to a new place—in *The Tempest*, *Pericles*, and *Twelfth Night*. The shipwreck in *The Merchant of Venice*, however, brings financial ruin.[46]

THE MERCHANT OF VENICE

Ships, and their trade between the New World and the Old, are an absent presence through *The Merchant of Venice*.[47] They are conjured through the opening scene, when Salarino and Salanio are talking to Antonio—the merchant of the play's title. Salarino declares:

Your mind is tossing on the ocean,
There where your argosies with portly sail
Like signiors and rich burghers on the flood,
Or, as it were, the pageants of the sea
Do overpeer the petty traffickers
That curtsy to them, do them reverence
As they fly by them with their woven wings.[48]

Salanio adds that he does not have argosies—merchant cargo ships—at sea at the moment, and he says that if he did, then "I should be still / Plucking the grass to know where sits the wind, / Peering in maps for ports and piers and roads."[49] The divination of wind direction is, of course, crucial to navigation; whereas the "roads" he is trying to find in maps are routes, "sheltered stretches of water or coves near the shore where ships might anchor safely (*OED* 3)."[50] Even in this opening scene Salarino raises the risks inherent in such ventures, talking of "shallows and flats" and a ship "docked in sand,"[51] but Antonio says that he is not concerned: "My ventures are not in one bottom trusted, / Nor to one place; nor is my whole estate / Upon the fortune of this present year: / Therefore my merchandise makes me not sad."[52]

Antonio's fortune is made, and lost, through trade of goods; while the Jew Shylock's is made through moneylending. For John Drakakis, these are "different but related branches of mercantilist practice, the one engaged in overseas trade and the other in facilitating the flow of capital through contractual means."[53] The calculative is, then, a significant theme in *The Merchant of Venice*: the play is filled with language of measure, balance, exchange, counting, and weighing. Bassanio seeks the hand of Portia, but lacks the means. He asks Antonio to gift him the money. Antonio is willing, but claims, "all my fortunes are at sea; / Neither have I money, nor commodity; / To raise a present sum."[54] Instead Antonio obtains the money through a loan from Shylock. It is less often mentioned that Shylock himself cannot immediately raise the money:

I am debating of my present store,
And, by the near guess of my memory,
I cannot instantly raise up the gross
Of full three thousand ducats.[55]

Shylock realizes that he could borrow the money from Tubal, a fellow Jew. So Bassanio borrows from Antonio, who borrows from Shylock, who borrows from Tubal. Initially Shylock pretends to calculate the due interest: "Three thousand ducats, 'tis a good round sum. / Three months from twelve, then let me see the rate."[56] But he quickly comes up with another, quite different basis for the loan. The terms "interest" and "credit," which run through the play, shape both the calculation of debt and the assessment of desire and belief. The legal contract that is drawn up between Antonio and Shylock is precisely around an exchange of a financial sum for a forfeit:

> Go with me to a notary, seal me there
> Your single bond, and, in a merry sport,
> If you repay me not on such a day
> In such a place, such sum or sums as are
> Expressed in the condition, let the forfeit
> Be nominated for an equal pound
> Of your fair flesh, to be cut off and taken
> In what part of your body pleaseth me.[57]

There are two specific issues in this bond. First, that the place is unspecified, and second that "equal pound" is a "precise or just weight."[58] Corporeal imprecision is matched with calculative exactitude.

Bassanio tries to prevent Antonio taking a loan on these terms, but Antonio is confident that he will easily be able to repay the loan. "Within these two months, that's a month before / This bond expires—I do expect return / Of thrice three times the value of this bond."[59] It seems even quite early that Shylock anticipates that the money may not be repaid. As Salarino has already indicated, Antonio has risked much by having all his wealth invested in goods aboard ships. Shylock explains:

> My meaning in saying he is a good man is to have you understand me that he is sufficient. Yet his means are in supposition. He hath an argosy bound to Tripoli, another to the Indies; I understand moreover, upon the Rialto, he hath a third at Mexico, a fourth for England, and other ventures he hath squandered abroad. But ships are but boards, sailors but men; there be land rats and water rats, water thieves and land thieves—I mean pirates—and then there is the peril of waters, winds and rocks. The man is notwithstanding sufficient. Three thousand ducats: I think I may take his bond.[60]

This is a wide range of places—impossible for a Venetian merchant. Tripoli is likely a reference to the Levantine city, not the Libyan one.[61] As

M. M. Mahood has perceptively noted, Shylock "repeats words as if they were coins he was counting; his curt phrases are a sort of syntactical book-keeping; and his rare images have to be spelt out with heavy literalism."[62] While Shylock thinks it rash, "squandered" here has a sense of diversely scattered or distributed as much as thrown away. Antonio rather tempts fate by his clear indication to Bassanio: "Come on, in this there can be no dismay, / My ships come home a month before the day."[63] Indeed, not long after the bond is sealed, reports begin to come in of a shipwreck, as Salarino tells Salanio:

> I reasoned with a Frenchman yesterday
> Who told me, in the narrow seas that part
> The French and English, there miscarried
> A vessel of our country richly fraught.[64]

These rumors prove to be true, and the vessel to be Antonio's ship:

> Why yet it lives there unchecked that Antonio hath a ship of rich lading wracked on the narrow seas. The Goodwins, I think, they call the place: a very dangerous flat, and fatal, where the carcasses of many a tall ship lie buried, as they say, if my gossip Report be an honest woman of the world.[65]

On the Venice Rialto this story is told, but it is not contradicted: Antonio's ship is wrecked, and he cannot repay the sum. Tubal informs Shylock of a wreck coming from Tripoli, which delights the latter.[66] Antonio seems cursed with bad luck—there have been multiple wrecks and failures. Bassanio asks if this can be true:

> Hath all his ventures failed? What, not one hit,
> From Tripoli, from Mexico and England,
> From Lisbon, Barbary, and India,
> And not one vessel scape the dreadful touch
> Of merchant-marring rocks?[67]

Yet none have survived. Whether this is due to "peril of waters, winds and rocks" or the pirates that Shylock also mentions is unclear. Shylock is unmoved, even though many try to persuade him or even offer much greater sums in repayment. As Salerio remarks,

Twenty merchants,
The Duke himself and the magnificoes
Of greatest port have all persuaded with him,
But none can drive him from the envious plea
Of forfeiture, of justice and his bond.[68]

Even Shylock's daughter, Jessica, despairs of the situation:

When I was with him, I have heard him swear
To Tubal and to Chus, his countrymen,
That he would rather have Antonio's flesh
Than twenty times the value of the sum
That he did owe him; and I know, my lord,
If law, authority and power deny not,
It will go hard with poor Antonio.[69]

Antonio's letter confirms this:

Sweet Bassanio, my ships have all miscarried, my creditors grow cruel, my estate is very low, my bond to the Jew is forfeit, and since in paying it, it is impossible I should live, all debts are cleared between you and I if I might but see you at my death.[70]

Shylock is unmoved, repeating regularly, "I will have my bond."[71] Of course, Shylock's strict insistence on the terms of this agreement is his undoing. He will not accept a multiple of the capital later than the due date, because Antonio has failed to pay on the terms expressed. Any attempt at mitigation is unsuccessful, and he refuses even to provide a surgeon "to stop his wounds," since it is "not in the bond."[72] He insists on this to the Duke:

I have possessed your grace of what I purpose,
And by our holy Sabbath have I sworn
To have the due and forfeit of my bond.
If you deny it, let the danger light
Upon your charter and your city's freedom![73]

Freedom here seems to be constitutive of the city; a territorial determination of its politics.[74] It seems all is lost, especially when we discover that the part of the body that most "pleaseth" Shylock is his breast near his heart.

As Shapiro has noted, this is not what might be expected: "In the late sixteenth century the word *flesh* was consistently used, especially in the Bible, in place of *penis*."[75] Castration seems to be the initial bond; now Shylock opts for a wound that will surely kill Antonio: "The pound of flesh which I demand of him / Is dearly bought; 'tis mine, and I will have it."[76] But as Portia, disguised as the legal counselor Balthazar, explains, the terms must be followed precisely. As Jay Halio indicates, "Paradoxically, she frees Antonio from the tyranny of the bond by focusing more strictly than Shylock on the binding quality of its language":[77]

> This bond doth give thee here no jot of blood:
> The words expressly are 'a pound of flesh.'
> Take then thy bond: take thou thy pound of flesh.
> But in the cutting it, if thou dost shed
> One drop of Christian blood, thy lands and goods
> Are by the laws of Venice confiscate
> Unto the state of Venice. . . .
>
> Therefore, prepare thee to cut off the flesh.
> Shed thou no blood, nor cut thou less nor more
> But just a pound of flesh. If thou tak'st more
> Or less than a just pound, be it but so much
> As makes it light or heavy in the substance
> Or the division of the twentieth part
> Of one poor scruple: nay, if the scale do turn
> But in the estimation of a hair,
> Thou diest, and all thy goods are confiscate.[78]

Even though Shylock had brought his scales along with his knife, he realizes that he will inevitably fail to meet the exact terms. The "estimation of a hair" is either the scale moving a hair on the indicator or a single hair tipping the balance; a scruple is about a gram, and a twentieth would be a grain.[79] This is an impossibly precise definition of a pound. Now Shylock attempts to take the multiple of the loan offered, and then just the principal, reducing his demand by measured proportions, but he is told he has refused this in open court. Instead he is convicted of an attempt on the life of a citizen, leading to half his wealth going to the potential victim, and half to the state. His own life is in the hands of the Duke. The Duke pardons his life, but in another, forceful exchange, Shylock is converted to Christianity as part of his punishment. In the final scene the tables are turned again: Antonio receives a letter that will tell him, "three of your argosies / Are richly come to harbour suddenly."[80]

Other parts of the play equally trade on language of measure and exchange. Shylock's daughter Jessica, when she leaves him, disguised as a boy and stealing his jewels, admits, "I am much ashamed of my exchange."[81] Exchange can, here, be cross-dressing, her planned conversion to Christianity, the theft or deceit. Numerical multipliers are also frequent. These are not only in the attempts to get Shylock to accept much more than the three thousand ducats, but also in the attempts to win Portia. As she says to Bassanio,

> Though for myself alone
> I would not be ambitious in my wish
> To wish myself much better, yet, for you,
> I would be trebled twenty times myself,
> A thousand times more fair, ten thousand times more rich,
> That only to stand high in your account
> I might in virtues, beauties, livings, friends
> Exceed amount. But the full sum of me
> Is sum of something: who to term in gross,
> Is an unlessoned girl, unschooled, unpractised.[82]

NAVIGATION

If there are, clearly, accountable practices in some of Shakespeare's plays, how do these apply to place, space, and, even, territory?[83] This was a time when many practical spatial arts began to make use of the calculative developments of the period. Shakespeare does not say a great deal about navigation in his plays, but there are scattered references. Reference to the compass is the most common. The word "compass" was commonly used as a verb, meaning to include, surround, or achieve; and as a noun meaning scope, range, or bound. It is also used in *Coriolanus* in the sense of direction, when one of the citizens says: "I think if all our wits were to issue out of one skull, they would fly east, west, north, south, and their consent of one direct way should be at once to all the points o'th' compass."[84] They might agree to one direction, but then go off in all directions. In *The Two Noble Kinsmen* the jailer's daughter asks her father if he is "master of a ship," where his "compass" is, and then instructs him to "Set it to th' north / And now direct your course to th' wood." The sailing metaphor is reinforced by the "tackling," the rigging of the ship.[85] Hamlet tells Horatio in the graveyard scene, "We must speak by the card or equivocation will undo us"—a reference to a sailor's card or compass, the caution being like the modern "by the book."[86] Cohen glosses this as "to speak precisely, purposefully, choosing one's words to convey a specific meaning as carefully as a pilot uses the compass to choose a course for his ship."[87] This instrument also makes an appearance in *Macbeth*:

> I myself have all the other,
> And the very ports they blow,
> All the quarters that they know,
> I'th' shipman's card.[88]

"Quarters" are regions, directions, and the "card" a map or sea chart, perhaps marked with the points of the compass. "Card" here means much the same as the French *charte*.[89] The witch's invocation of this instrument comes after her reference to the "Master o'th' Tiger," a ship bound to Aleppo, and her sisters' offer of a wind to get her there.[90] It is followed by a reference to "a pilot's thumb / Wrecked as homeward he did come."[91] The specific reference is obscure, but the maritime navigation metaphor is clear; the pilot being a ship's steersman, and witches were supposed to produce and sell winds.[92] Indeed, Macbeth later blames them for turbulence: "you untie the winds and let them fight / Against the churches, though the yeasty waves / Confound and swallow navigation up."[93]

Of course, the play in which the maritime navigation sense is strongest is *The Tempest*. As Brayton indicates, "To begin a play on the deck of a ship, and one having a great deal of trouble with the wind, is to paint the stage as the space of navigation; it is also to invoke cartography, which was preoccupied with the depiction of wind as the driving force of navigation and, hence, the basis for geographic knowledge."[94] As discussed in chapter 6, Gonzalo proposes a trade of "a thousand furlongs of sea for an acre of barren ground."[95] Poole suggests that the play exhibits "a geodetic sensibility and epistemology typical of its time,"[96] as it portrays "a geometric way of understanding the lay of the land and the spatial surround. The recurrent references to units of measurements (furlongs, acres, fathom, leagues, cubits, inches) and elements of the survey indicate how naturalized such a vocabulary had become by the early seventeenth century."[97] The geographical language can appear in other instances too, such as the phrase "under the line," which is often taken to refer to the equator, though other senses are possible.[98] In *Sir Thomas More*, More's wife challenges his quiescence with these lines:

For keeping still in compass—a strange point
In time's new navigation—we have sailed
Beyond our course.[99]

The point seems to be that sailing in the direction of a simple compass point can lead to an error in navigation, because it fails to take account of the difference between magnetic and true north, and the curvature of the earth. On long sea voyages recalculations needed to be made at periodic intervals, which was a technique being perfected at the time of the play's writing. "Compass" here is both the navigational instrument and the idea of living within accepted limits. More's wife sees his constancy as a problem, and proposes adjustment in the light of the new situation. [100] The date of the composition of *Sir Thomas More* is disputed, but John Jowett has claimed it was ca. 1600. Edward Wright's *Certaine Errors in Navigation* was published in 1599, and it crucially showed how to reconcile magnetic compass direction with geographical direction.[101] This was through a combination of discussion, diagrams, and conversion tables. While not a scene from "Hand D," usually taken to be Shakespeare, this passage from *Sir Thomas More* gives a good indication of the intellectual context of fellow playwrights.

SURVEYING

One of the things that is interesting about Shakespeare's time is the increasing use of instruments and techniques developed for maritime navigation for land measurement.[102] This was certainly a key age in the development and use of land-surveying techniques: Leonarde Digges's *A Boke Named Tectonicon* was published in 1556 and went through multiple printings; John Norden's *The Surveyor's Dialogue* was first published in 1607 and was in its third edition by 1618.[103] The same was true in cartography. As well as small-scale projects such as the *Theatrum Orbis Terrarum*, there were also large-scale projects of cities and estates. William Cuningham's *The Cosmographical Glasse* was published in 1559 and contained a plan of the city of Norwich, Nordovicum, which was possibly the first plan of a British city.[104] It led to several others, including the *Civitates Orbis Terrarum*, published from 1572 onward, with hundreds of such plans, the first of which was London; and Norden's *Speculum Britanniae*, the first part of which treated London and Middlesex in 1593.[105]

Ben Jonson knew Edmund Gunter, the mathematician and geometer, and so even if Shakespeare did not meet Gunter he could have heard about him from Jonson.[106] Gunter invented the chain, quadrant, and scale that bear his name, all crucial for land surveying and navigation. While Cohen is correct to note that the plays "do not mention celestial navigational tools like the quadrant, the cross-staff, and the astrolabe even though these tools were present on English vessels,"[107] there is a reference to one such instrument in *The Merchant of Venice*. Shylock says, "By Jacob's staff, I swear, / I have no mind of feasting forth tonight."[108] Most editors gloss this as the staff of the biblical Jacob, and suggest that the oath sworn is "relevant and appropriate in that it refers to the progress of Jacob in Genesis, 32.10, from lowly beginnings to prosperity."[109] In the Bible Jacob says: "I had only my staff when I crossed this Jordan."[110] W. Moelwyn Merchant has suggested that "to Elizabethan ears there would be a further religious ambiguity in Shylock's possible reference to the staff of those who went on pilgrimage to the shrine of St James (Jacobus) of Compostella."[111] But a Jacob's staff is also a surveying instrument dating from the fourteenth century, a cross-staff, with a sliding wooden panel that allowed the relative height of objects or the altitude of stars to be determined.[112] It was later replaced by the much more accurate sextant, which dates from after Shakespeare's time. The term "Jacob's staff" was later reused in America simply to describe any staff supporting a compass. Both the staff of Jacob, in the biblical sense, and Jacob's staff, in the navigational one, are doubtless implied, with the second reinforced by the movement indicated by the idea of "feasting forth," attending a banquet, implying a journey, at night, when the stars could be used for navigation.

More commonly referenced is the "square." Falstaff complains, "If I travel but four foot by the square further afoot, I shall break my wind."[113] While the obvious sense is that if he has to go any further he will get out of breath, or fart, "four foot by the square" is a very precise distance, calculated by the square or set-square measuring tool.[114] This instrument is also mentioned in *Love's Labour's Lost*, with Berowne's question "Do not you know my lady's foot by the squier / And laugh upon the apple of her eye?"[115] Editors often keep the variant, now obsolete, spelling of square as "squier" to preserve the rhyme, and also because of the sexual pun: as well as "foot" implying the French *foutre*—just as it does in the scene between Katherine and Alice in *Henry V*—an "apple-squire" was a pimp.[116] Yet even so, the joke only works because of the straightforward meaning—to know the length of her foot, a proverbial expression for accurate measurement. These lines can be compared to Antony's comment "Read not my blemishes in the world's report. / I have not kept my square, but that to come / Shall all be done by th' rule."[117] In the past he has not kept to the appropriate measure, but will do so in the future. There are examples in other plays where "out of square" means disordered, such as the line in *The Two Noble Kinsmen* "and reduce what's now out of square in her into their former law and regiment."[118]

Cohen has suggested that estate surveying is regularly referenced in Shakespeare's plays,[119] of which the most significant passage comes in *Henry IV, Part 2*, when Lord Bardolph discusses the preparations for war:

When we mean to build,
We first survey the plot, then draw the model:
And when we see the figure of the house,
Then must we rate the cost of the erection
Which, if we find outweighs ability,
What do we do then but draw anew the model
In fewer offices, or at least desist
To build at all? Much more, in this great work,
Which is almost to pluck a kingdom down
And set another up, should we survey
The plot of situation and the model,
Consent upon a sure foundation,
Question surveyors, know our own estate,
How able such a work to undergo,
To weigh against his opposite; or else
We fortify in paper and in figures,
Using the names of men instead of men,
Like one that draws the model of an house
Beyond his power to build it, who, half-through,
Gives o'er, and leaves his part-created cost
A naked subject for the weeping clouds
And waste for churlish winter's tyranny.[120]

This is a fairly clear extension of the parable of the builder from the Gospel according to Luke, which says that planning and estimating before building are necessary, and then moves to compare this to a king preparing for war.[121] For Poole, this version "adds a trendy, emphatically geodetic vocabulary."[122] A "model" in Bardolph's use is an architectural plan, rather than a scale version; "figure" is a design as well as calculations; "rate" is to calculate; "offices" are rooms; "waste" is both wasteland and a waste of resources and effort.[123] The "plot of situation" implies both the physical location of the building and the plan of action, and of course the play itself.[124] The "surveyors" are both the construction supervisors and those who plan for war; the "estate" is landed and other property, as well as resources for war; reinforced by the figures on papers—again both the design and the calculation, either for the building or for the military preparation.[125]

The only surveyor to actually appear on stage is in *Henry VIII*. He works for the Duke of Buckingham, managing his estate. He is unnamed in the play, but historically was Buckingham's cousin, Charles Knyvet.[126] In the play he is an unsavory figure—accusing the Duke of treason, dismissed from his job because of complaints,[127] and potentially bought by others. The Duke himself says, "My surveyor is false. The o'er-great Cardinal / Hath showed him gold. My life is spanned already."[128] The lines are deliberately ambiguous: is the surveyor "false" because he lies or because he has betrayed the Duke? Is he paid by Cardinal Wolsey to tell the truth or to bear false witness?[129] What indeed is the surveyor? Is he simply the steward of an estate, one who oversees lands and property, or is his a more specialized technical profession? The technical meaning was certainly current at the time, and was an important part of estate management, given the complicated boundaries and legal possession of lands that supervision required. As Hainsworth notes of a slightly later period, estates "were not definitively surveyed, discrete blocks of land, . . . [but] hopelessly intertwined, their boundaries still reliant more on the memories of 'ancient men' than on surveyors' instruments. . . . It was often the lord's steward, rather than the lord's lawyer, who found himself first charged with the task of settling disputes about boundaries, feudal rights and feudal obligations, or of warding off encroachers, or of defending tenants against unjust taxes, levies or other impositions."[130] The Duke thinks his life is "spanned," that is, over, but there is an obvious sense of something being measured out.[131] The link between surveying and dishonesty is also found in the term "cheater" in *Henry IV, Part 2*.[132] "Cheater" was a word that "originally meant one who looked after the royal escheats (that is, land which reverted to the king), but in Shakespeare's day it was coming to mean such a person who operated dishonestly."[133]

CARTOGRAPHY

Shakespeare was clearly knowledgeable about advances in cartography, though many of his uses of "map" are largely metaphorical.[134] In *Titus Andronicus*, Titus describes Lavinia as "thou map of woe,"[135] an image or picture of despair. King Henry VI tries to save the Duke of Gloucester's life by saying he can see in his face "The map of honour, truth and loyalty."[136] This use appears in the Sonnets too: "Thus is his cheek the map of days outworn"; "And him as for a map doth nature store."[137] Lucrece's face is described as "that map which deep impression bears / Of hard misfortune, carved in it with tears."[138] This may relate to the printing process, using a copper plate, with her tears cutting into her face "like water that doth eat in steel."[139] This usage, of a map as an embodiment or representation, is not dissimilar to the idea of a face as a book. Lady Macbeth says to her husband, "Your face, my thane, is as a book, where men / May read strange matters."[140] More elaborately, in *Henry IV, Part 2*, Northumberland likens the face of a messenger to a book, and its title page, and then, quickly, to a coastline flooded by the sea. "Volume," here, works as both book and three-dimensional space:

> Yea, this man's brow, like to a title-leaf,
> Foretells the nature of a tragic volume:
> So looks the strand whereon the imperious flood
> Hath left a witnessed usurpation.[141]

Title-leaves, at this time, did more than just state a simple, short title, but gave an extended version with description of key plot points. The 1600 Quarto of this text, for instance, begins: "The second part of Henrie the fourth, continuing to his death, and coronation of Henrie the fift. With the humours of sir Iohn Falstaff, and swaggering Pistoll." Thus "a witnessed usurpation" on the "strand" after an "imperious flood" is "evidence of encroachment (that is, the lines or wrinkles left in sand by the retreating tide)."[142] In *Richard II*, in contrast, the Queen addresses the deposed king: "Ah thou, the model where old Troy did stand / Thou map of honour, thou King Richard's tomb."[143] Like the model, or ground plan, of Troy—that is, its ruins—the map is a "mere image (or outline) of former grandeur."[144] "Old Troy" is the classical place, in contrast to London, the new Troy. As Ure suggests, in many of his examples, Shakespeare uses "map" "figuratively for 'image,' 'embodiment' with emphasis on the *resemblances* between a map and that which it represents. . . . Here the stress seems to be on the way a map *differs* from the thing it represents."[145]

Of course, there are also instances of "map" and related words where the use is not metaphorical. Gillies claims seventeen uses of "map" and its variants in Shakespeare, only seven of which are "clearly geographic."[146] In *Henry V*, Fluellen tells Gower that he will see the similarities between Macedon and Monmouth "if you look in the maps of the world."[147] In *Cymbeline*, Cloten says, "I am near to th' place where they should meet, if Pisanio have mapped it truly."[148] As discussed earlier in the chapter, Salanio says that if his ships were at sea, he would be "Peering in maps for ports and piers and roads."[149] In *Twelfth Night* Maria says of the tricked Malvolio: "He does smile his face into more lines than is in the new map with the augmentation of the Indies."[150] While this is another instance of a face as a map, it clearly refers to an actual map. It has been surmised that this is a reference to Edward Wright's *Hydrographiae Descriptio* from 1600, a map that included the island Nova Zembla. Wright's map made use of the principles of his *Certaine Errors in Navigation*. The ill-fated exploration of the northeast passage is alluded to earlier in the scene.[151]

CALCULATION AND MILITARY TECHNOLOGIES IN *OTHELLO* AND *HENRY IV*

Additionally, there are instances where though a map is not mentioned, it is clearly intended. For instance, the Earl of Richmond says in *Richard III*: "Give me some ink and paper in my tent; / I'll draw the form and model of our battle";[152] and Gillies suggests: "The military application of cartography here identifies Richmond with two other Shakespearean soldier-technocrats: Ulysses and Michael Cassio."[153] Yet Ulysses's description is somewhat self-effacing: "They call this bed-work, mapp'ry, closet war."[154] Bevington has glossed this as "They call this armchair strategy, mere map-making, war planned in the study." He adds that "*Mapp'ry* is a Shakespearean coinage, used only this once by him."[155] Iago's description of Cassio is, similarly, a critical one:

> Forsooth, a great arithmetician,
> One Michael Cassio, a Florentine,
> A fellow almost damned in a fair wife
> That never set a squadron in the field
> Nor the division of a battle knows
> More than a spinster; unless the bookish theoric,
> Wherein the toged consuls can propose
> As masterly as he. Mere prattle without practise,
> Is all his soldiership—but he, sir, had th'election
> And I, of whom his eyes had seen the proof
> At Rhodes, at Cyprus and on other grounds,
> Christian and heathen—must be be-leed and calmed
> By debitor and creditor. This counter-caster
> He, in good time, must his lieutenant be,
> And I, God bless the mark, his Moorship's ancient![156]

Iago has been denied the position of Othello's lieutenant, and instead has the rank of "ancient," an ensign or standard-bearer. There are contrasts established here between Iago's own physical experience of battle and Cassio's more academic understanding of it. A "squadron" is a square of soldiers, possibly in a defensive formation, while a "division" is another arrangement of troops, but here posed for attack.[157] "Arithmetician" means that his knowledge is purely theoretical, reinforced by "bookish theoric." It is possible that "arithmetician" implies that he is trained in the more theoretical art of artillery rather than infantry—Galileo, remember, was a professor of military engineering, inventing a compass for the use of soldiers. The "toged consuls" are the civilian authorities, councillors wearing togas, with Venice following ancient Rome, where the senators would not wear military dress in the capital, though the Folio actually reads "tongued consuls," which would also work in this context as those who speak more than act. They "propose" and "prattle without practice," reinforcing this sense of discussion over experience. Iago has experience in battle with Rhodes and Cyprus, both to be revealed as key strategic sites in the battle for supremacy in the eastern Mediterranean with the Ottoman Empire, and more widely with the Crusades between "Christian and heathen." In contrast his power must be overruled by an accountant, "be-leed and calmed," that is cut off from the wind. Cassio as "debitor and creditor," that is, a bookkeeper, is reinforced by his description as a "counter-caster," a term of Shakespeare's own, but which clearly implies a reckoner or clerk, someone who uses tokens for counting.[158] Sanders suggests that Cassio's Florentine identity may be a reference to that city's reputation for banking and accountancy;[159] though of course Machiavelli was a Florentine, and so this may mean devious or manipulative.[160] Could Iago's critique of Cassio be an example of resistance to modern calculative technologies?

There are other passages where the technologies of war are discussed, beyond mere swords and shields. One passage comes when Hotspur's wife shows genuine concern for her husband's lack of sleep and preoccupation:

> In thy faint slumbers I by thee have watched
> And heard thee murmur tales of iron wars,
> Speak terms of manage to thy bounding steed,
> Cry 'Courage! To the field!' And thou hast talked
> Of sallies and retires, of trenches, tents,
> Of palisades, frontiers, parapets,
> Of basilisks, of cannon, culverin,

Of prisoners' ransom, and of soldiers slain,
And all the currents of a heady fight.[161]

There are many images in these lines. "Iron wars" are cruel; he gives commands to his horse. "Sallies and retires" are advances and retreats; "palisades" are fixed stakes, pales, possibly tipped with iron, used to reinforce a military position; "frontiers, parapets" are likewise ramparts, barriers, walls, and earthworks. "Basilisks . . . cannon, culverin" are, in descending size, types of cannon, the first named after a mythical fire-breathing snake; the last coming from the French for "adder." The ransom of captives, and the body count of the dead, are further forms of military accounting, and the "currents of a heady fight" are the battlefield maneuvers of a violent combat.[162] Kastan notes that "military terms had become an increasingly fashionable rhetoric," citing a passage from a poem by Sir John Davies that may have formed a model for Shakespeare.[163]

THE MAP DIVISION IN *HENRY IV, PART 1*

Of course, the most famous map in Shakespeare is found in the opening scene of *King Lear*, discussed in chapter 1. Another tripartite division of a kingdom, again using a map, occurs in *Henry IV, Part 1*. Cohen has said that this "anachronistic" scene, like that of *Lear*, is probably due to "the growing importance of Elizabethan-Jacobean estate maps."[164] Cartographic historian J. B. Harley explains:

> Maps were one of a number of instruments of control by landlords and governments; they were spatial emblems of power in society; they were artefacts in the creation of myth; and they influenced perceptions of place and space at a variety of geographical scales. Accordingly they can be considered to form the development of thought in early modern England.[165]

In the play the King is under threat from several directions. As noted in chapter 3, he voices his concerns in the very first lines of the play, suggesting that "short-winded accents of new broils" will prevent his promised pilgrimage to the Holy Land—"to be commenced in strands afar remote," "strands" being places, shores, or lands.[166] His wish for the end of bloodshed on the soil of England echoes the continual reference to this image throughout *Richard II*.[167] However, his attempt to get his country to work together—"Shall now in mutual well-beseeming ranks / March all one way and be no more opposed / Against acquaintance, kindred and allies"—already hints at what will undermine it.[168] Henry's claim to the English crown is contested by Lord Edmund Mortimer, the Earl of March, who is believed by many nobles to be Richard II's legitimate heir.[169] "Earl of March" means lord of the Welsh marches, the border region, which will become important. At the very outset of the play we hear that Mortimer had been defeated by the Welsh Owen Glendower, a battle that took place in 1402.[170] Now, Mortimer has allied with Glendower, who also mobilizes his forces against the King. In the north of England, the threat comes from Henry Percy, Earl of Northumberland, and his son of the same name, known as Hotspur.[171] There was an anticipation of the conflict to come in *Richard II*. Very late in that play Richard says to Northumberland:

> Northumberland, thou ladder wherewithal
> The mounting Bolingbroke ascends my throne,
> The time shall not be many hours of age
> More than it is ere foul sin, gathering head,
> Shall break into corruption. Thou shalt think,
> Though he divide the realm and give thee half
> It is too little, helping him to all.[172]

Northumberland has been the "ladder," the aid, by which Bolingbroke has seized the crown, but sooner rather than later Northumberland will regret this. Whatever promises Bolingbroke may have made him, presumably offering lands in the north of the country, Bolingbroke has gained more. Richard continues to say that Northumberland, already knowing how "to plant unrightful kings, wilt know again," that is, if he can replace Richard with Henry, he can potentially unseat Henry in his turn, "to pluck him headlong from the usurped throne."[173] In this play the rebels frequently refer to the King as Lancaster,[174] the title he legitimately inherited from his father, John of Gaunt, sometimes stressing that when he returned to England from exile he claimed that this was merely for the title Richard had denied him, not to seize the crown.[175] Their earlier support for Henry Bolingbroke had been on that basis; his exceeding it as the King legitimates their attempt at overthrow.

Hotspur is held up as the son that King Henry wishes he had, rather than the dissolute Hal.[176] In one exchange, the King likens Hal to Richard II, and Hotspur to himself, suggesting, "Now by my sceptre, and my soul to boot, / He hath more worthy interest to the state / Than thou, the shadow of succession."[177] "Interest," here, like "interest to this land" in *King John*,[178] means a title, which Hotspur deserves, though Hal has it, merely being the poor imitation of a worthy successor. Hotspur was, historically, older than King Henry IV, let alone his son, but Shakespeare makes him of the same generation as Hal for dramatic purpose. In this passage he suggests:

For, of no right, nor colour like to right,
He doth fill fields with harness in the realm,
Turns head against the lion's armed jaws,
And, being no more in debt to years than thou,
Leads ancient lords and reverend bishops on
To bloody battles and to bruising arms.[179]

Hotspur, through his actions, of leading armor in the field, even though he is no older than Hal, is eminently more suitable, even though he has no legal right, nor even the semblance of that. He "holds from all soldiers chief majority / And military title capital"—the reputation as the best military figure around, having proved his manhood.[180]

The threats are tied to each other by marriage: Mortimer's sister is married to Hotspur; Mortimer's wife is Glendower's daughter. As Kastan has noted, "Shakespeare's play conspicuously compresses and selects events of the reign, ignoring the conspiracy of the Abbot of Westminster as well as the troubled relations with France, and making the revolt of the Percys and their defeat at Shrewsbury the play's central historical spine."[181] Editors note that Shakespeare and his key source, Holinshed's *Chronicles*, confuse two Edmund Mortimers—one was married to Glendower's daughter, but was already dead by the time of the play's action; the other was his nephew, believed to be Richard II's heir. Though his status as heir is relayed in Holinshed and other chronicles of the time, more recent accounts are skeptical.[182] But the threats to King Henry are growing—reinforcements are coming from the Scots too; and Mortimer's brother, the Lord of Worcester, has defected, after having been dismissed from the court.[183]

On relaying to Hal that "Worcester is stolen away tonight" to join the rebellion, Falstaff adds: "Thy father's beard is turned white with the news. You may buy land now as cheap as stinking mackerel."[184] Land presumably is cheap because the risk of invasion means that it may soon be worth nothing; or because corruption is rife. Hal replies, trading on the obscure meaning of "maiden" as "fish," probably based on the sexual connotation of "mackerel": "Why then, it is like if there come a hot June and this civil buffeting hold, we shall buy maidenheads as they buy hobnails: by the hundreds."[185] In a time of civil war, the suggestion is virtue will be cheap. Continuing the sexual puns, but stressing commerce, Falstaff suggests that "it is like we shall have good trading that way."[186] In general, there is much economic language in the play, especially around ideas of counterfeit, commerce, and credit, and "there are more numismatic references" in this play than in any other by Shakespeare.[187] One economic reference to a dead man only taking pleasure in the rise of the price of oats is sometimes seen as a contemporary reference to the time of the play's composition.[188] But not much of this economic language concerns land, or territory, as it more clearly does in *Richard II*. Here, the focus is much more on the physical nature of land, and its cartographic representation.

The three leading rebels, Mortimer, Glendower, and Hotspur, are meeting to discuss how they will apportion the kingdom after the defeat of the King. The King is not yet beaten, and the division there is premature, but is presumably needed before they can agree to unite their forces. Mortimer suggests at the very outset of the scene that "These promises are fair, the parties sure, / And our induction full of prosperous hope."[189] Hotspur suggests that he has forgotten the map, but Glendower provides it.[190] As Bate notes, "A more politic customer would be most unlikely to have forgotten it—and would never admit as much if he had."[191] Hotspur, here and elsewhere, is a rather intemperate and rash figure, which makes the continual praise of Hotspur by the King rather strange. In this scene he is continually baiting Glendower in particular. For example, Hotspur mocks Glendower's boast about his defense of Wales:

> Glendower: Three times hath Henry Bolingbroke made head
> Against my power; thrice from the banks of Wye
> And sandy-bottomed Severn have I sent him
> Bootless home and weather-beaten back.
> Hotspur: Home without boots, and in foul weather too!
> How scapes he agues, in the devil's name.[192]

The Wye and the Severn are two rivers dividing Wales from England, both emptying into the Bristol Channel. The sources of both rivers are within a couple of miles of each other, on Plynlimon in the Cambrian Mountains. From that point, the Wye runs predominantly southeast, east, and then south, while the Severn runs east, northeast, east, south, and then southwest. The Severn is continually to the east of the Wye, and the rivers mark the edges of contested land between Wales and England, sometimes known as the Welsh marches. The present-day border runs along the Wye for part of its length. Hotspur's response wonders how the King had escaped a fever, given the weather and his lack of boots—a literal reading of "bootless," which, for Glendower, simply means defeated.[193] Perhaps because of this prelude, what follows is often played as a comedy, but it need not be.[194] There is more than enough humor in this play—it has Falstaff, in particular—to not need to find it elsewhere. What if we took the scenes seriously?

Returning to the key issue, Glendower begins:

Come, here is the map. Shall we divide our right
According to our threefold order ta'en? [195]

The "right" implies that the territory of the kingdom is being claimed; the "threefold order," a triple entente or agreement, implies that the basic principle has already been decided, and that now they need specify its geographical division. Mortimer explains:

The Archdeacon hath divided it
Into three limits very equally:
England, from Trent and Severn hitherto,
By south and east is to my part assigned;
All westward, Wales beyond the Severn shore
And all the fertile land within that bound,
To Owen Glendower;—and, dear coz, to you
The remnant northward lying off from Trent.
And our indentures tripartite are drawn,
Which, being sealed interchangeably—
A business that this night may execute—[196]

The Archbishop is mentioned nowhere else in the play, but is the Archdeacon of Bangor in Holinshed's *Chronicles*. In Holinshed, the scene takes place in the house of the archdeacon.[197] He has divided the kingdom into three equal parts or portions—"regions defined by boundaries," or what we would now call "territories," and indeed the *OED* (3) gives this as the instance of the use of "limit" in that sense: "The tract or region defined by a boundary." A similar use is found in *Edward III*, "Shall the large limit of fair Bretagne / By me be overthrown";[198] in Sonnet 44, "I would be brought / From limits far remote, where thou dost stay";[199] and in *Venus and Adonis*, "within this limit."[200] Onions glosses this term as "tract, region."[201] In relation to its use in *Henry IV, Part 1*, Holinshed's *Chronicles* simply says: "all England from Seuerne and Trent, south and eastward, was assigned to the earle of March: all Wales, & the lands beyond Seuerne westward, were appointed to Owen Glendouer: and all the remnant from Trent northward, to the Lord Persie."[202]

The dividing lines are not drawn onto the map, but rivers have been chosen. The river Severn, already mentioned by Glendower, will divide off England from Wales; and the river Trent will form the boundary between Mortimer's southern England and Hotspur's north.[203] Mortimer claims the English crown, but knows its limits.[204] This decision seems to have already been agreed, because "indentures tripartite"—a three-part contract—have already been sealed, and each will take a copy with the other's mark. Mortimer therefore continues to discuss battle plans, first with Hotspur and then with Glendower. With Hotspur he discusses how the two of them, along with Worcester, will meet with Hotspur's "father and the Scottish power" at Shrewsbury.[205] Shrewsbury, where the battle will take place, is on the northern bank of the river Severn. To Glendower, whom he addresses as father (as father-in-law), Mortimer notes that the forces are not yet ready, but that they do not need Welsh support for two weeks, though Glendower suggests that he may be able to join them sooner.[206] You can imagine the men parting at this point, with the agreement made and the plans laid.

But Hotspur is not ready to settle just yet:

> Methinks my moiety, north from Burton here,
> In quantity equals not one of yours.
> See how this river comes me cranking in
> And cuts me from the best of all my land
> A huge half-moon, a monstrous scantle, out.
> I'll have the current in this place dammed up,
> And here the smug and silver Trent shall run
> In a new channel fair and evenly.
> It shall not wind with such a deep indent
> To rob me of so rich a bottom here.[207]

This is a departure from Holinshed, and is a rich passage with several key phrases. It is clear that Hotspur thinks his share or portion ("moiety") is unfair. He points to the town of Burton-on-Trent, where the river Trent turns northeastward. "Cranking" implies direction, but Bruce Avery has pointed out that it has not only the meaning winding or turning

> but also an association with military defeat and weakness. Its Old English root verb, according to the *OED*, meant "to fall in battle," and "to shrink, give way, become weak or ill." Eventually, it evolved to mean

> shifty and devious, as well as developing the senses of twisting and turning for which the *OED* cites this passage as exemplary. Given what we know of Hotspur and his conspirators, it's not pushing too hard on this word to see all of these meanings suspended in it at once, so his use of it hints at his fear that the map and its river might represent the ebbing of his power.[208]

The "half-moon" is the shape of the area that, Hotspur thinks, is on the wrong bank of the river, an area he describes as a "monstrous scantle" in the Quarto, or "monstrous Cantle" in the Folio texts of the play. A scantle is a small piece, while a cantle is a segment, usually of a circle or sphere. For Peter Davison, the Folio's reading of "cantle" is one of "two emendations . . . that every editor is grateful for."[209] Presumably a "monstrous" small piece seems contradictory, and a "monstrous" segment more in keeping with the complaint. The suggestion is that, in printing, the final "s" sound of "monstrous" was mistakenly attached to the beginning of "cantle."[210] An alternative reading suggests that the Folio editors thought "scantle" a mistake for this reason, and, to their mind, corrected it to "cantle." If "scantle" is correct, the contradiction is presumably to show Hotspur is arguing on a trifle, part as bluff.[211] In order to remedy this geographical inequity, Hotspur proposes damming the river and changing its direction, so that the smooth ("smug") flow of the river will lead to a more favorable distribution. The "indent" or indentation it currently makes shall not deny him the "bottom," a river valley, which is currently on Mortimer's side of the line.

Glendower is incredulous: "Not wind? It shall; it must. You see it doth." For him, seemingly, geography, or at least the representation of it on the map, determines the division. Mortimer tries to reason with Hotspur:

> Mortimer: Yea, but mark how he bears his course and runs me up
> With like advantage on the other side,
> Gelding the opposed continent as much
> As on the other side it takes from you.[212]

Mortimer's point is that the direction of the river means that while it may put some land on his side at Hotspur's expense, it soon works the other way, "gelding the opposed continent," that is, cutting from the other bank. Mortimer's use of "runs *me* up" indicates that the disadvantage caused by a later turn is his. (This use of the pronoun is an instance of the ethical dative, or dative of indirect interest, like Hotspur's "comes me cranking in."[213])

Worcester, who has not yet spoken in this scene, contributes a suggestion, supported by Hotspur, but disputed by Glendower:

> Worcester: Yea, but a little charge will trench him here,
> And on this north side win this cape of land,
> And then he runs straight and even.
> Hotspur: I'll have it so; a little charge will do it.
> Glendower: I'll not have it altered.[214]

"A little charge" means a small expenditure, through which the river can be trenched, that is, diverted into a different channel, and the "cape" or spur of land can be redistributed to Hotspur's north side. What is interesting is that it is Glendower who argues against Hotspur, even though it would change the boundary between Mortimer and Hotspur.[215] Perhaps it is the general principle—either of redistribution or of challenging nature—which is the matter for him. Strikingly, if Glendower's earlier boast of defeating the King when he crossed the Severn before is to be believed, he seems to have largely been given the land he already effectively controls, although perhaps his gain is the land between the Severn and the Wye. They argue, and Hotspur loses sight of the issue, before Glendower relents: "Come, you shall have Trent turned." Hotspur now affects indifference:

> I do not care. I'll give thrice so much land
> To any well-deserved friend;
> But, in the way of bargain, mark ye me,
> I'll cavil on the ninth part of a hair.
> Are the indentures drawn? Shall we be gone?[216]

Hotspur seems to be at pains to say that he is generous, but in a bargain, will fight or argue ("cavil") over the tiniest portion. He hopes the agreement ("indentures") has been completed, but this will take time. Glendower says that he can be gone by nightfall, which will not be a problem as "the moon shines fair," but he will "haste the writer,"[217] that is, hurry the person who will amend the agreement and put it into a final, fair copy. The scene continues, with an exchange between Mortimer and his wife, and an admonishment of Hotspur for his treatment of Glendower. The agreement between the men is mentioned in two further passages, once when they are still waiting for the "book" to be drawn; and then at the end of the scene when "indentures be drawn," and it only needs to be sealed.[218]

This is a fascinating scene, and one, I am suggesting, that should be taken very seriously. As Kastan notes, "The rebels have differing commitments and values, ironically demonstrated both in the embarrassing division of the nation played out upon the map in 3.1 and in the fact that they are unable to come together to fight at Shrewsbury."[219] Both those moments—the disagreement on the map, and the inability to coordinate—are highlighted in this scene; two dramatic representations of the split in the rebel forces. Reason enough to take it as significant. It is one of the ways, as Greenblatt has noted, that Shakespeare undercuts the pretensions of political figures: "His kings repeatedly discover the constraints within which they must function if they hope to survive. His generals draw lines on maps and issue peremptory commands, only to find that the reality on the ground defies their designs."[220] But what does the cartographical dispute entail? Two interesting readings come from Sullivan and Avery. Sullivan notes:

> What is most important about this scene, however, is that Hotspur imagines not altering the *map* so as to increase the value or nature of his property, but rather he plans to alter the *land* itself. . . . Of course the irony is that if Hotspur does redirect the river, then the cartographic representation of the land must be modified.[221]

Avery develops this point of the relation between the map and the land:

> When Glendower answers "Come, you shall have Trent turned," the effect of Hotspur's disruption becomes clear. The map is now unstable, arbitrary, and malleable not just for Hotspur but for everyone else in the room, and finally for the Globe audience, as well. . . .
>
> It seems from here no great leap to argue Hotspur and his cronies see the territory as a representation of the map. Rather than simply rewriting the map—say, drawing a new boundary across the disputed territory—the conspirators want to rewrite the *territory*, so the subsequent drawn map will reflect the "natural" boundary between Glendower and Hotspur. Never mind that the new channel will be artificial, and forget that a map evoked the desire for the change in the first place. The conspirators want to preserve the image of the map as a mimetic, not prescriptive, representation, because such an image is necessary to maintain its rhetorical force.[222]

Why might this be? There is no other obvious river that might form a boundary, and the choice of the Severn and the Trent might be thought to be emblematic of how boundaries of the time were drawn. Contrary to Fluellen's claim in *Henry V* that the Wye at Monmouth and the river at Macedon—presumably the Vardar/Axios—"'tis all one, 'tis alike as my fingers is to my fingers, and there is salmons in both,"[223] rivers are not all alike. Historically, Hotspur may have had a point, because the Trent—one of the relatively few English rivers to run north—did once run more easterly, emptying into the Wash, rather than in Hotspur and Shakespeare's time, as now, into the Humber estuary.

Avery, however, thinks that the land dispute is between Glendower and Hotspur, which is a position he shares with Gillies.[224] This does not make geographical sense. While Glendower and Hotspur disagree, the dispute over land to be apportioned is between Hotspur and Mortimer. If the river cranks, turns, it does so at Burton from a largely southeastern course to turn northeast, and then north.[225] To suggest Hotspur's objection is to the way it runs before it reaches Burton would not be to divert its flow, but to seek a different source. Hotspur's complaint is that the current division denies him some of Nottinghamshire and all of Lincolnshire, a county that on the map falls on Mortimer's side of the line. The reason he and Glendower disagree is not because Glendower will lose out. Rather, it continues their carping from earlier in the scene. More fundamentally it is because Glendower objects to the idea of challenging the natural course of the river. It is Mortimer who suggests that the advantage to him is canceled out by an advantage to Hotspur as the river turns again—"mark how he bears his course and runs me up / With like advantage to the other side." This is further downstream, as the river continues to flow on its course, the next turning point presumably when it bears east to empty into the Humber, meaning that the East Riding of Yorkshire is to the north of the river, and thus part of Hotspur's share.[226]

The map presents the men with a representation of the land as it is, and rather than chose a different line, they do indeed agree to change the territory. The map would then need to be amended, of course. But Hotspur is proposing geo-engineering as a technique for the production of territory, a territorial strategy.[227] That this new channel will be "artificial" is beside the point, because the dynamics of a river flow render any such division unable to be fixed long-term in any case. Many river boundaries today are reinforced by channels or controls to get the geophysical to conform to the geopolitical. Even if a river is believed to be a "natural boundary," the choice of which river or other landscape marker is of course arbitrary: the Rhine and not the Rhône; this mountaintop linked to that one, rather than another; this island group based on proximity but not that one.

Of course, none of this comes to pass in the play. Hotspur is killed in battle at Shrewsbury, losing his life in single combat with Prince Hal. This is a marker of Hal's transition from wayward son to military hero. Yet the geographical returns even in death. Hal stands over Hotspur's dying body and completes his final sentence:

> Hotspur: . . . O, I could prophesy,
> But that the earthy and cold hand of death
> Lies on my tongue. No, Percy, thou art dust
> And food for—[*He dies.*]
> Prince: For worms, brave Percy. Fare thee well, great heart.
> Ill-weaved ambition, how much art thou shrunk!
> When that this body did contain a spirit
> A kingdom for it was too small a bound,
> But now two paces of the vilest earth
> Is room enough. This earth that bears thee dead
> Bears not alive so stout a gentleman.
> If thou wert sensible of courtesy,
> I should not make so dear a show of zeal:
> But let my favours hide thy mangled face;
> And, even in thy behalf, I'll thank myself
> For doing these fair rites of tenderness.
> Adieu, and take thy praise with thee to heaven!
> Thy ignominy sleep with thee in the grave,
> But not remembered in thy epitaph.[228]

While alive, Hotspur's ambition was such that it could not be contained in the kingdom—"too small a bound"—and of course, he wanted to change the bounds of the land he sought to claim by the King's overthrow. But this is now shrunken, like a poorly made garment, as now his dead body can be contained in a mere "two paces of the vilest earth." He had told Glendower he would "cavil on the ninth part of a hair,"[229] the tiniest portion of land. "Ill-weaved ambition," of course, means badly purposed, as well as poorly made. Just as Alexander the Great and "Imperious Caesar" were in Hamlet's graveside reflections (see chapter 2), he has been reduced from territorial ambition to corporeal limits.

The battle at Shrewsbury is the defeat of Hotspur, but not of all the rebels against the King. Fortified by this victory, and the change of Hal's stature through his valiant defeat of Hotspur, the King makes plans for the rest. He will split his forces in two, one half to attack Hotspur's father; the other to deal with Glendower and Mortimer:

> This then remains, that we divide our power.
> You, son John, and my cousin Westmorland,
> Towards York shall bend you with your dearest speed
> To meet Northumberland and the prelate Scrope,
> Who, as we hear, are busily in arms.
> Myself and you, son Harry, will towards Wales
> To fight with Glendower and the Earl of March.
> Rebellion in this land shall lose its sway
> Meeting the cheek of such another day;
> And since this business so fair is done,
> Let us not leave till all our own be won.[230]

What comes through in these references to navigation, surveying, cartography, and geo-engineering is not fully consistent. In the late sixteenth to early seventeenth century, the time when Shakespeare was writing, these techniques were being developed, tested, and applied. As Poole has perceptively noted, this does not mean that there was a wholesale replacement of earlier divine geographies with geometric space. For her the two coexist in complicated ways, and the standard story is reductive:

> The recurrent pattern of telling this history . . . posits a transition that is at best far too abrupt, and at worst historically inaccurate. I think we should be discussing space in much the same way that Jonathan Gil Harris has written about time. Incorporating the terminology of Michel Serres, Harris considers "the time of Shakespeare" as being polychronic and multitemporal; this time is understood as layered, or pleated, rather than linear and sequential. The space of Shakespeare's time period was similarly folded, similarly polydimensional and multi spatial, as the rigors of geometry and the suppleness of the supernatural coexisted in a range of relationships.[231]

The reference to Harris is helpful. His analysis of the shaping of early modern subjects and early modern objects puts temporality to the fore, but this is not just read synchronically, but also diachronically and polychronically.

Thus objects are read not just synchronically in relation to "moment of production or use" but for other times that can be seen within them, not merely in terms of a past and future that can be lined up chronologically, but in terms of other logics of "supersession, explosion, conjunction."[232] Shakespeare's ambivalent attitude to calculation may be explained in terms of this tension. He explores themes around measure, calculation, and accountancy in general terms in some of his plays, of which the key examples here were *Measure for Measure* and *The Merchant of Venice*. But the application of technological developments also figures in his thinking about the question of territory, and related questions of land and rule. Not all calculative practices are territorial; but territory is increasingly calculated. The adoption of maritime navigation techniques for use in land surveying, and the development of cartographic practices, make possible the assessment and control of much greater extents of land. Land becomes an object of calculative measurement, especially when territorial projects are at stake.

CHAPTER EIGHT

Corporeal Territories: The Political Bodies of *Coriolanus*

This chapter examines Shakespeare's late tragedy *Coriolanus* in terms of the various political bodies that structure the narrative and its language.[1] The relation between territorial ambition and the bodies of women was a theme in chapter 5 with Katherine and the Countess of Salisbury. While male fantasies may extend to the conquest of both, not all political bodies are female. *Coriolanus* is a play about the political body of the polity itself, its inside and outside, the aggressive wars to keep it safe externally, and its internal health and well-being. The play also raises a range of questions about what it is to contribute to a political community, and who should rule. But literally and figuratively these themes are often performed through the physical bodies of its characters, with language invoking wounds, contagion, animals, and a variety of body parts. This book has shown how territory cannot be reduced to mere political and geographical questions, and that a range of economic, strategic, legal, and technical issues are tangled up in the concept and practice. Yet, just as chapter 7's reading of *Henry IV, Part 1* showed that the geophysical is important; this chapter insists on the affective, emotional, physical, and biophysical aspects of territory, of bodies in places and places embodied.

Coriolanus has been described as Shakespeare's "most political play,"[2] and as "hugely, indeed grotesquely, political."[3] The play has often been mined for its political leanings, particularly because it gives voice both to the common people and their rulers. It has polarized opinion in part because it presents different perspectives with such passion, leading William Hazlitt to suggest in the early nineteenth century that if it is studied it obviates the need to read Burke or Paine or listen to debates in the House of Commons.[4] As Patterson notes, it "featured in the nineteenth century by Charlotte Brontë in *Shirley* as a model for industrial ethics, and by George Eliot in *Felix Holt* as a lesson in Victorian electoral politics."[5] Both T. S. Eliot and Bertolt Brecht, with very different political motives, attempted to adapt the play in the twentieth century.[6]

While set in ancient Rome, and based upon the *Life* of the title character written by Plutarch, the resonances with Shakespeare's own time have often been remarked upon, especially as regards the corn riots and resultant popular uprisings in the English Midlands around the time he wrote the play, in 1607. The Midland Revolt was about the enclosure of land into sheep pasture, and the problems this created for agricultural labor and the production of grain. The winter of 1606–7 was hard, and a poor harvest was expected in 1607.[7] Robin Headlam Wells suggests that the allusion to the corn riots in the opening scene would have therefore had a strong contemporary resonance for Shakespeare's audience.[8] Yet Shakespeare took the corn riots from Plutarch's account of Coriolanus's life, which means it was not a gratuitous addition but part of the story itself.[9] However, as Philip Brockbank suggests, "he could not have been unaware of their closeness to his own time,"[10] and it is of course plausible that the parallels inspired Shakespeare to take up this story as the basis for his play. Brockbank notes that this is a recurrent theme in Shakespeare, from the Jack Cade uprising in *Henry VI, Part 2*, his involvement in *Sir Thomas More*, and the popular voice in *Julius Caesar*. *Coriolanus* is, for Brockbank, the dramatic consummation of this theme, stretching from one of Shakespeare's earliest plays to his last tragedy.[11] Frank Kermode argues that while this contemporary political crisis was certainly the immediate context, the political aspects of the play transcend this:

> More abstractly: it is a study in the relationships between citizens within a body politic; the relationship of crowds to leaders and leaders to led, of rich to power. The polis has its troubles: dearth, external enemies, enmity between classes. The patricians have a ruthless but narrow and selfish code of honour. The people are represented by tribunes who are in their own way equally ruthless, scheming politicians. The monarchic phase of Roman history has recently ended, the kings replaced by an oligarchy tending to be oppressive, committed to warfare as the ultimate proof of valour and worth, and largely indifferent to social obligation.[12]

Interestingly, despite the elaborate examination of the body politic in this play, his last, collaborative English history, *Henry VIII*, does not examine this theme, despite its use of body imagery.[13] Like the earlier *Troilus and Cressida*, *Coriolanus* is a play about eating, bodies, animals, and disease.[14]

Only that play has more animal imagery than this one.[15] Additionally, *Coriolanus* is nakedly about force and violence: as Patterson notes, the word "power" is used thirty-eight times in this play, compared to eighteen in *Richard II*, the closest competitor.[16]

Caius Martius is a victorious Roman general, yet he is intensely disliked by the people as a whole because he is seen as one of the reasons behind their hunger while the grain stores are full. An encounter between the people and Martius goes badly, with Martius scornful of their claims. Martius leads a victorious army, defeating the Volscians at Corioli and fighting the Volscian commander Tullus Aufidius, who is dragged away by his soldiers. Martius returns to Rome as a conquering hero, taking the honorific *cognomen* (more properly an *agnomen*) of Coriolanus. He is persuaded by his mother to run for consul, and he is elected with the backing of the people. However, two tribunes, Junius Brutus and Sicinius Velutus, conspire against him, leading to a popular revolt. Coriolanus condemns the people, which means Brutus and Sicinius have him convicted of treason. He is condemned to execution by being thrown off a cliff, but is banished as an alternative. Coriolanus goes to the Volscian city of Antium, initially to allow his enemy Aufidius to kill him to spite Rome. Instead they together plot the city's downfall. On hearing the two men are leading an army against the city, Rome attempts to persuade Coriolanus against this course. While the soldier Cominius and the patrician Meninius fail, a delegation including Coriolanus's son, wife, and mother, Volumnia, succeeds. A pact is sealed with Rome, but Coriolanus is denounced as a traitor by the Volscians and killed by Aufidius and others.

The political questions in the play are thus not merely secondary aspects, but the very heart of the drama. Kermode has suggested that its plot is "probably the most fiercely and ingeniously planned and expressed of all the tragedies."[17] As Agnes Heller notes, "Coriolanus is in my mind such a thoroughly political play that the politically uninteresting scenes (such as the introductory part of 2.1) also seem dramatically redundant."[18] The aim here is to avoid the narrow, uninteresting political reading of the play that Stanley Cavell suggests "is apt to become fairly predictable once you know whose side the reader is taking, that of the patricians or that of the plebeians."[19] Among these many themes, the focus will be on bodies.[20]

First, and most obviously, is the question of the body politic, which is recounted in a fable told to hungry citizens. The chapter moves from this to discuss actual physical bodies and wounds, and then turns to the recurrent language of infection and infected bodies. Bodies and pathology are related to the question of nonhuman animals, frequently mentioned in approbation or condemnation, particularly in the context of the clash between the elite and the multitude; the singular and plural; and the understanding of the city as being both the people and the place. There are a number of examples of animals as the hunter and hunted, consumer and consumed. Finally, the chapter addresses the theme of banishment and the conquest of territories, in terms of bodily aspects in other political senses. In *Coriolanus*, territory as the body of the state is only one aspect of its corporeal nature.

The theme of bodies has, naturally, been discussed before. As Bristol notes,

> The image of the body is, of course, a familiar topic in the critical discussion of *Coriolanus*. The play is saturated with concrete situations in which the fate and condition of bodies is of paramount importance. Both literal and symbolic implications of the analogy between the private individual body and the body politic are elaborated in nearly every scene.[21]

Some of what I say here will, then, be familiar to students of the play. The discussion of bodies in a literal sense, particularly the fable of the belly in the opening scene, and the ideas of infection and contagion that run throughout the play, are relatively well known and examined. But even here I think there are issues less commonly remarked upon, and I hope in terms of wounds, animals, and territories the reading goes beyond this.

BELLIES

The initial complaint of the people is hunger in the face of dearth. They feel that the state has sufficient resources, which it does not release to them. The patrician Menenius Agrippa tries to suggest that it is the gods that have created this situation, and that rather than carry arms against the Roman state, the people should get on their knees to pray. He tries to explain the situation to the people by means of a corporeal parable that reverses this idea of grumbling bellies.

> Menenius. There was a time, when all the body's members
> Rebelled against the belly; thus accused it:
> That only like a gulf it did remain
> I'th' midst o'th' body, idle and unactive,
> Still cupboarding the viand, never bearing
> Like labour with the rest; where th'other instruments
> Did see, and hear, devise, instruct, walk, feel
> And, mutually participate, did minister

Unto the appetite and affection common
Of the whole body. The belly answered—
2 Citizen. Well, sir, what answer made the belly?
Menenius. Sir, I shall tell you. With a kind of smile,
Which ne'er came from the lungs, but even thus—
For, look you, I may make the belly smile
As well as speak—it tauntingly replied
To th' discontented members, the mutinous parts
That envied his receipt; even so most fitly,
As you malign our senators for that
They are not such as you.
2 Citizen. Your belly's answer—what?
The kingly crowned head, the vigilant eye,
The counsellor heart, the arm our soldier,
Our steed the leg, the tongue our trumpeter,
With other muniments and petty helps
Is this our fabric.[22]

The sheer number of bodily parts is overwhelming. Menenius sets the other members of the body against the belly. It cupboards the viand, storing the food, but not distributing it to the other parts. Basic instincts—seeing, hearing, devising, instructing, walking, feeling, participating—are not served by the belly; the appetites and affections are not provided for. Menenius describes the complainants as "discontented members, the mutinous parts," and notes that the belly is the root of this, not the lungs. The second citizen replies by listing the parts of the entire body politic: the head, the eye, the heart, the arm, the leg, the tongue. The body politic is a recurrent theme in political theory. Perhaps the most important medieval example is found in John of Salisbury's twelfth-century *Policraticus*,[23] but the most striking early seventeenth-century example is Edward Forset's *Comparative Discourse of the Bodies Natural and Politique*.[24] Forset's work was published in 1606, shortly after his involvement in the prosecution of Guy Fawkes and his fellow conspirators. This play is Shakespeare's most thorough treatment of the idea.[25] Elsewhere in the play, many of the other body parts are the physical organs of speech—tongue, mouth, throat, lips.[26]

One of the things that is unusual is that Shakespeare does not allow the fable of the belly (which has its root in Aesop before Plutarch) to simply stand as a metaphor of politics. In Plutarch the people are told the fable outside of the city walls, where they are complaining about usury. The people are actually hungry in *Coriolanus*.[27] Usury is mentioned only briefly in

Shakespeare; in Plutarch the grain riots come later in the narrative. The exchange on the fable continues for some time, with the first citizen describing it as "the cormorant belly,"[28] trading on the idea that this is a voracious bird. Menenius counters that "your most grave belly was deliberate, / Not rash like his accusers" and that it acts as "the storehouse and the shop / Of the whole body."[29] What is stored there is distributed appropriately:

> I send it through the rivers of your blood
> Even to the court, the heart, to th' seat o'th' brain,
> And, through the cranks and offices of man,
> The strongest nerves and small inferior veins
> From me receive that natural competency
> Whereby they live.[30]

It is clear what he intends, with the people getting the flour and only the bran being taken, though the crowd seem confused, so he makes it explicit:

> The senators of Rome are this good belly,
> And you the mutinous members. For examine
> Their counsels and their cares, digest things rightly,
> Touching the weal o'th' common, you shall find
> No public benefit which you receive
> But it proceeds or comes from them to you,
> And no way from yourselves. What do you think?
> You, the great toe of this assembly?[31]

The use of body parts to discuss the polity continues at other times in the play: "the navel of the state" and "the fundamental part of state";[32] "lords and heads o'th'state."[33] Even the last, the well-known idea of a "head" of state, follows from this basic idea. And the bodily sustenance can be to disadvantage too, with Coriolanus claiming that whoever gave our stored corn freely "nourished disobedience, fed / the ruin of the state."[34] As R. B. Parker has shown, a range of language in the play relates to the grain and its growth and storage, even before it reaches the hungry.[35]

WOUNDS

As Cavell among others has argued, ideas of hunger and cannibalism run through the narrative.[36] Yet the play is not concerned simply with the workings of the body, and the distribution of nourishment throughout it. As Zvi

Jagendorf notes, "The physical is inescapable in this most unerotic of plays; everywhere we encounter legs, arms, tongues, scabs, scratches, wounds, mouths, teeth, voices, bellies, and toes, together with such actions as eating, vomiting, starving, beating, scratching, wrestling, piercing, and undressing."[37] This is perhaps one of its strongest contrasts to *Troilus and Cressida*—in that play there is a sexuality to the bodies and their appetites. The only erotic connection here, it seems, is between Coriolanus and his Volscian foe, Aufidius. Combat is as close to physical proximity as he seems to get, despite a wife and son. Equally, the bodies of this play tend to be male, with the rare exception of Volumnia's talk of her womb—her physical connection to her son.[38]

Much more significant in a corporeal register are the wounds borne by soldiers on behalf of the city.[39] Martius notes, "I have some wounds upon me, and they smart / To hear themselves remembered."[40] Menenius and Volumnia wish him to come home from Corioli with wounds upon him. It was nearly much more—his soldiers see him surrounded and expect "To th'pot, I'll warrant him"'—that is, he will be cut up as meat for the cooking vessel.[41] But he survives against the odds, and Menenius is pleased to receive a letter from Martius:

> A letter for me? It gives me an estate of seven years' health, in which time I will make a lip at the physician. The most sovereign prescription in Galen is but empiricutic, and, to this preservative, of no better report than a horse-drench. Is he not wounded? He was wont to come home wounded.[42]

Galen is of course a medical writer from antiquity who came after the action of the play, but the point is well made. ("Empiricutic" appears to be a combination of "empiric" and "pharmaceutic," perhaps with a critical sense of quackery, based on folklore rather than learning.[43]) While Coriolanus's wife, Virgilia, is concerned, his mother, Volumnia, learns news that she is pleased with: "O, he is wounded; I thank the gods for't"; and Menenius concurs: "So do I too, if it be not too much: brings 'a victory in his pocket? The wounds become him."[44] They then have the following exchange:

> Menenius: . . . Where is he wounded? [*to the Tribunes*] God save your good worships! Martius is coming home: he has more cause to be proud. [*to Volumnia*] Where is he wounded?
>
> Volumnia: I'th' shoulder and i'th' left arm there will be large cicatrices to show the people, when he shall stand for his place. He received in the repulse of Tarquin seven hurts i'th' body.
>
> Menenius: One i'th' neck, and two i'th' thigh—there's nine that I know.

Volumnia: He had, before this last expedition, twenty-five wounds
upon him.
Menenius: Now it's twenty-seven: every gash was an enemy's grave.[45]

This passage enumerates the wounds, a calculus of pain from the battlefields that can be expended in the political marketplace.[46] Despite the shaky arithmetic, the message is clear. But breaking with custom, the newly named Coriolanus is unwilling to show his wounds to the crowd. Cynthia Marshall has argued that this is his attempt to remain "author of himself," though as such he is unwilling to let others read the script, revealing little of his motives.[47] Brutus is critical of this decision:

I heard him swear,
Were he to stand for consul, never would he
Appear i'th' market-place, nor on him put
The napless vesture of humility;
Nor showing (as the manner is) his wounds
To th' people, beg their stinking breaths.[48]

Coriolanus confirms this reluctance himself: "Your honours' pardon, / I had rather have my wounds to heal again / Than hear say how I got them."[49] And again:

I do beseech you,
Let me o'erleap that custom, for I cannot
Put on the gown, stand naked and entreat them,
For my wounds' sake, to give their suffrage. Please you
That I may pass this doing.[50]

Siculus and Brutus see this failure to display the heroic body as an affront, as do the citizens:

> We have power in ourselves to do it, but it is a power that we have no power to do; for if he show us his wounds and tell us his deeds, we are to put our tongues into those wounds and speak for them. So, if he tell us his noble deeds, we must also tell him our noble acceptance of them. Ingratitude is monstrous, and for the multitude to be ingrateful were to make a monster of the multitude; of the which we, being members, should bring ourselves to be monstrous members.[51]

Aside from the rather crude sexual reference in the final line, the crowd are uncertain how to react. Coriolanus is reluctant to simply stand and tell the crowd of the wound he received, and outlines how he imagines it is desired he do:

What must I say?
'I pray, sir'? Plague upon't, I cannot bring
My tongue to such a pace. 'Look, sir, my wounds!
I got them in my country's service, when
Some certain of your brethren roared and ran
From th' noise of our own drums.'[52]

This is the contradiction: he does not wish his wounds to speak of his actions.[53] Crucially he cannot train his tongue to a strict rhythm, as a trained horse. He later concedes, "I have wounds to show you, which shall be yours in private," to one citizen,[54] but refuses to confirm another's claim that "You have received many wounds for your country" by replying, "I will not seal your knowledge with showing them,"[55] though he later boasts of "wound two dozen odd; battles thrice six."[56] Again there is the calculus, but he does not wish them visible, only known. The citizens later interpret this as disdain: "He used us scornfully: he should have showed us / His marks of merit, wounds received for's country."[57] Instead of being part of a collective body, Coriolanus wishes to remain private, separate, his body to himself. He was earlier described as "a carbuncle entire."[58] This idea is returned to in the third act, where Menenius tells the crowd:

The warlike service he has done, consider. Think
Upon the wounds his body bears, which show
Like graves i'th' holy churchyard.
Coriolanus: Scratches with briers,
Scars to move laughter only.[59]

INFECTIONS

As well as the actual damage to a physical body, the play uses the recurrent theme of corruption, pollution, and disease of the political body.[60] Something is rotten in the state, but the trope is also used to refer to different elements or characters. Following his earlier description of the crowd as making themselves scabs by "rubbing the poor itch of your opinion,"[61] Martius condemns the soldiers who are on the verge of defeat at Corioli in strident terms, invoking animals and disease:

All the contagion of the south light on you,
You shames of Rome! You herd of—boils and plagues
Plaster you o'er, that you may be abhorred
Farther than seen, and one infect another
Against the wind a mile! You souls of geese
That bear the shapes of men, how have you run
From slaves that apes would beat! Pluto and hell!
All hurt behind, backs red, and faces pale
With flight and agued fear.[62]

Later, after being denied by the crowd, despite his sacrifices, he returns to taunt them:

As for my country I have shed my blood,
Not fearing outward force, so shall my lungs
Coin words till their decay against those measles
Which we disdain should tetter us, yet sought
The very way to catch them.[63]

Measles could be the modern disease, but it could also be leprosy—the words have confused etymologies, while tetters are skin eruptions.[64] There is further specialized veterinary language—Volumnia mentions a "red pestilence," which may be typhus, and a citizen uses "murrain," a cattle plague, as a curse.[65] Later in the same scene, Coriolanus suggests that there is a need "to jump a body with a dangerous physic / That's sure of death without it—at once pluck out / The multitudinous tongue: let them not lick / The sweet which is their poison."[66] Yet how he describes the people also is applied to him. He has already been labeled a "poison";[67] now it is his turn to be described as a contagion:

Sicinius. He's a disease that must be cut away.
Menenius. O, he's a limb that has but a disease:
Mortal, to cut it off: to cure it, easy.
What has he done to Rome that's worthy death?
Killing our enemies, the blood he hath lost—
Which I dare vouch is more than that he hath
By many an ounce—he dropped it for his country;

And what is left, to lose it by his country
Were to us all that do't and suffer it
A brand to th'end o'th' world.
Sicinius. This is clean cam.
Brutus. Merely awry. When he did love his country,
It honoured him.
Sicinius. The service of the foot,
Being once gangrened, is not then respected
For what before it was.
Brutus. We'll hear no more.
[*to the Citizens*] Pursue him to his house and pluck him thence,
Lest his infection, being of catching nature,
Spread further.[68]

Sicinius and Brutus argue that his past deeds need to be forgotten, just as a gangrenous foot is no use for its previous purpose, and that in order to prevent further infection he needs to be cut away. Menenius suggests that the loss of his blood in the service of Rome outweighs his current situation. Yet both use the same mode of language in their opposed views. Coriolanus's exile has unpleasant implications. His mother, Volumnia, condemns Sicinus: "Hadst thou foxship / To banish him that struck more blows for Rome / Than thou hast spoken words?"[69]

ANIMALS

Throughout the play the people as a whole and particular characters are described as animals.[70] There are many beastly bodies. Coriolanus describes the people as a herd;[71] and Menenius talks of "being the herdsmen of the beastly plebeians."[72] The people are described as "multiplying spawn" by Menenius,[73] and Coriolanus accuses Siculus of being the "Triton of the minnows."[74] There are other insults used of the crowd, such as Martius's evocative "Go get you home, you fragments!"[75] a word that makes clear the collective nature of the crowd, sometimes described as the "rabble"—often in stage directions as well as dialogue, which may indicate Shakespeare's own views.[76]

But there is a veritable bestiary beyond this. Curs, lions and hares, foxes and geese, all appear in Martius's first speech,[77] shortly after a contrast was drawn by Menenius between "Rome and her rats."[78] Coriolanus mocks the crowd's hunger and their use of proverbs such "dogs must eat,"[79] and links this to their unwillingness to fight with the lines "The Volsces have much corn; take these rats hither / To gnaw their garners."[80] Later exchanges bring in a "fawning greyhound in the leash,"[81] a cat and mouse,[82] mules, camels, dogs and sheep,[83] moths,[84] conies,[85] a serpent,[86] an "old goat,"[87] and a "poor hen."[88] Even the term "rascal," as well as meaning a person of low status or a mob, can also be a reference to an ill-bred dog.[89] The herd, "souls of geese," and "slaves that apes would beat" all appear in the condemnation of his soldiers.[90] He later claims he will "never / Be such a gosling as to obey instinct."[91] However, the use of animals is sometimes rather lazy, with them simply listed and the auditor (or reader) having to do the work of embellishing the image. As Maxwell notes, a comparison with Macbeth is telling. In that play we find the more evocative claim that "A falcon towering in her pride of place / Was by mousing owl hawked at and killed."[92] The terms here are quite specific: "towering" is an upward flight; "pride of place" is the peak of a swoop on prey, while the lower-flying owl, hunting for mice, takes the falcon on the wing, the sense of "hawked at."[93] Things are turned around.

One of the most striking things in *Coriolanus* is how often the animals function as pairs, one the hunter and the other the hunted, one consumer and one consumed. The animals are frequently their bodies, meat.[94] When Menenius says that the people do not love Martius, Sicinius declares that "nature teaches beasts to know their friends,"[95] which generates an exchange where the crowd is compared to a wolf looking to devour a lamb, but Brutus notes that Martius sounds like a bear, to which Menenius replies he's a bear who lives like a lamb.[96] The implication appears to be that in politics he is a lamb, ready to be devoured, when his true strength is in battle. This is confirmed by Coriolanus later referring to a "wolvish toge," civilian dress covering his martial instincts.[97] He had earlier welcomed war as a "means to vent / Our musty superfluity."[98] What these contrasts mean is that while many of the terms are pejorative, there are also animals that evoke strength or nobility ("noble" is itself recurrent throughout the play). There are several instances where humans are approvingly compared to an animal. Martius is described as a bear,[99] and he declares Aufidius "a lion / That I am proud to hunt."[100] That said, in their first encounter, Aufidius vows, "If I fly, Martius, hollo me like a hare,"[101] with "hollo" meaning "chase with shouts," and the hare standing as a figure of timidity.[102] Later in the play, as he is sent into exile, Coriolanus complains that "the beast / With many heads butts me away,"[103] and compares his fate "to a lonely dragon that his fen / Makes feared and talked of more than seen."[104] The multitude expels the individual. The dragon is echoed in Menenius's later description: "There is differency between a grub and a butterfly; yet your butterfly was a grub. This Martius is grown from man to dragon; he has wings, he's more than a creeping thing."[105] Menenius twice describes Coriolanus as a tiger, speaking of him having "tiger-footed rage,"[106] and saying there is "no more mercy in him than there is milk in a male tiger."[107]

Aufidius declares Coriolanus is ruled by animal instincts: "I think he'll be to Rome / As is the osprey to the fish, who takes it / By sovereignty of nature."[108] This relation of the osprey to the fish, a noble animal to prey, is found in other places in the play. It comes through in the idea that the people might have upset the natural balance of things, a reversal of power Coriolanus suggests would allow "the crows to peck the eagles,"[109] with the eagle a somewhat anachronistic imperial emblem. He later describes Rome as the "city of kites and crows," the kite being at Shakespeare's time a common sight scavenging in London, and mentions "daws," that is, jackdaws.[110] There are other times, however, where the power of the strong is clear, with the suggestion that the Volscian assault on Rome will be of soldiers following Coriolanus "with no less confidence / Than boys pursuing summer butterflies / Or butchers killing flies."[111] Additionally, a Volscian servant reports the early defeat of Aufidius: Coriolanus "scotched him and notched him like a carbonado," meat prepared for grilling.[112]

The animal imagery also is important in considering the relation of the people and the patricians to the city. In the opening scene Martius suggests that "the rabble should have first unroofed the city,"[113] a claim that makes more sense in a later context:

> 1 Senator. To unbuild the city, and to lay all flat.
> Sicinius. What is the city but the people?
> Plebeians. True, the people are the city. . . .
>
> Cominius. That is the way to lay the city flat,
> To bring the roof to the foundation,
> And bury all which yet distinctly ranges
> In heaps and piles of ruin.[114]

In distinction, Coriolanus sees them as in Rome but not Roman:

> I would they were barbarians, as they are
> Though in Rome littered; not Romans, as they are not,
> Though calved i'th' porch o'th' Capitol![115]

He wishes they were barbarians, so he could dispense with them as in a war—"On fair ground I could beat forty of them."[116] There is an interesting structure to these lines, with the wish followed by the "as they are"; and the negative description reinforced by the "as they are not," with the two instances of "though" used to qualify still further. Additionally there is a use of terms that refer to their mere bodies. As Paul Cantor notes, "Coriolanus' use of terms for animal procreation, 'littered' and 'calved,' reveals his point: the plebeians no more deserve to be called Romans than do the tame beasts who happen to be born every year within the city's walls."[117] He also threatens to dispatch them with his sword, and "make a quarry / With thousands of these quartered slaves"—a pile of dead bodies, usually the deer from a hunt.[118]

But the contrasts go deeper than this. The people believe they are the city, whereas the patricians clearly see a physicality to the city itself that is in danger of being destroyed.[119] The people care more about their own health, but Coriolanus, the Senator, and Cominius talk of the destruction of the buildings, literally an "unbuilding"; the city laid flat; a "falling fabric."[120] Sicinius reenters, looking for Coriolanus in order to pass sentence:

> Where is this viper,
> That would depopulate the city, and
> Be every man himself?[121]

One of the citizens sees the enacting of the sentence as setting both the politics and the bodies in balance again. Thrown from the rock, "he shall well know / The noble tribunes are the people's mouths / And we their hands."[122]

When he goes into exile and seeks out Aufidius, Coriolanus addresses the city itself:

> A goodly city is this Antium. City,
> 'Tis I that made thy widows. Many an heir
> Of these fair edifices fore my wars
> Have I heard groan and drop. Then know me not,
> Lest that thy wives with spits, and boys with stones,
> In puny battle slay me.[123]

In the final scene, discovering his Coriolanus has been persuaded against the assault on Rome, Aufidius slurs him with the accusation of being a "boy of tears," to which Coriolanus replies with the terrible lines that lead to his death:

> Cut me to pieces, Volsces men and lads,
> Stain all your edges on me. 'Boy,' false hound!

If you have writ your annals true, 'tis there
That, like an eagle in a dovecote, I
Fluttered your Volscians in Corioles.
Alone I did it. 'Boy'![124]

The description of himself as the eagle, the return to the scene of his greatest victory at Corioli, and the use of the animal descriptions of "hound" and fluttering those in the "dovecote" return to a number of key themes of the play. Coriolanus thus ends as he began, setting himself up alone against a crowd, with the people as a whole crying "tear him to pieces" as a reply to his own "cut me to pieces" before many blows rain down.[125] Recall the condemnation of the crowd as "fragments" and a "herd" in the first scene. The individual people only make sense as part of a whole; Coriolanus can himself be reduced to pieces. For Jagendorf, the manner of his death "represents the number over singularity, of the limbs over the belly, of the spread of power over its concentration."[126]

TERRITORIES

These corporeal bodies—bodies as sites, sites as bodies—relate in a number of ways to the question of territory in a narrower, more specific sense. The key here is the expulsion of Coriolanus from the civil body, and the attack he then leads against it. When the tribunes rule him guilty of treason Sicinius notes that there are three options—"For death, for fine or banishment"—and asks the Aedile to instruct the crowd to follow his lead.[127] Coriolanus is unrepentant:

Let them pronounce the steep Tarpeian death,
Vagabond exile, flaying, pent to linger
But with a grain a day, I would not buy
Their mercy at the price of one fair word,
Nor check my courage for what they can give,
To have't with saying 'Good morrow.'[128]

A few lines later Sicinius picks up the theme:

. . . in the name o'th' people,
And in the power of us the tribunes, we,
E'en from this instant, banish him our city,
In peril of precipitation

From off the rock Tarpeian, never more
To enter our Rome gates. I'th' people's name,
I say it shall be so.[129]

The people reply enthusiastically, "It shall be so, it shall be so! Let him away! / He's banished, and it shall be so."[130] There is then the important exchange and reversal:

Brutus. There's no more to be said, but he is banished,
As enemy to the people and his country.
It shall be so.
All Citizens. It shall be so, it shall be so!
Coriolanus. You common cry of curs, whose breath I hate
As reek o'th' rotten fens, whose loves I prize
As the dead carcasses of unburied men
That do corrupt my air: I banish you! . . .

. . . Despising
For you the city, thus I turn my back.
There is a world elsewhere![131]

The "curs" here, as in the first scene, invokes the dogs baiting the bear.[132] The Aedile claims, "the people's enemy is gone," to which the crowd reply, "our enemy is banished, he is gone!"[133]

Greenblatt has suggested that "Rome cannot stomach Coriolanus, however important he has been to the city as a virtually unstoppable war machine, and it has driven him out. He walks through the city gates into a mythical space, as he imagines it, and assumes an identity that no longer conforms to human shape [the lonely dragon]. His participation in the human community is finished."[134] Greenblatt makes the point that Shakespeare could have ended things here, but chose not to.

Coriolanus does not remain in exile and heads for the Volscian city of Antium. Antium is only about thirty miles from Rome—Corioli was even closer, about halfway between the cities. This is the early days of Rome, when the city had many neighbors and rivals in the Italian peninsula. Corioli is situated between Rome and Antium, a contested point on the map that changes hands between the sides, from its conquest by Rome to its recapture by Antium, both enabled by the military power of Coriolanus. Peter Holland has described Corioli as the play's "crucial border point, . . . the marking of the fluidity of the liminality defined by ownership in the condition of permanent conflict the play records."[135]

In the compressed dramatic action of the play, Coriolanus's journey in exile is short, and his arrival in Antium, in disguise, follows quickly after his departure. Stage productions tend to shield his identity with a hood. But Ralph Fiennes's recent film version captures this rather differently.[136] The film shows the exile through a lengthy journey whose passage of time is tracked by the transformation in Fiennes's appearance, where shaven head and face become ever lengthening hair and a thick beard. By the time Coriolanus arrives in Antium, the transformation is such that any artifice is unnecessary. But the geographical complications are shown in this version. The play is set in early republican Rome, not long after the uprising against and expulsion of the Tarquin kings. The play is written in early seventeenth-century England. The film is set in a near-contemporary pseudo-Balkans, shot in Serbia and Montenegro, but the largely urban scenes of decay and despair could have been anywhere. The film, making effective use of newsreel and TV, shows that the Volscians are in close proximity to the "place calling itself Rome."[137] The initial war footage, of the siege of the Volscian city of Corioli, talks of a "border dispute." That proximate location, or contested front between the sides, seems largely absent when he is making his way to Antium. The means used to mark the transition at other points in the film, where a motorway is punctuated by roadblocks, with a kind of no-man's-land between them, was more effective. But if this is so, and the two neighbors share a narrow, effectively modern border, a boundary, where does Coriolanus go when he moves into exile? Why does it take him so long to move between these places?

And yet, in early republican Rome, it is indeed the case that there would have been areas outside of Rome that were not yet part of its neighbors. Cities were one thing, but beyond their cultivated lands there would still be wild areas, and these could be contested or, in contrast, unclaimed. There were places that were not yet spaces; lands that were not yet cultivated, not yet territory. In the early seventeenth century, Shakespeare could effectively play this spatial politics. Banishment was still a potential punishment and, as previous chapters have explored, features importantly in his history plays. The transportation of convicts to the New World or slavery was merely a modern example of an age-old practice. In the later seventeenth century John Locke would discuss the "Indian who knows no Inclosure, and is still a Tenant in common" and yet still laid claim to private property and thus a nascent form of civil society,[138] with Locke declaring that "in the beginning, all the World was America."[139] Not all places within Shakespeare's England were yet enclosed, much less if Wales, Scotland, and Ireland were included.

But in the late twentieth and early twenty-first century, and especially in the Balkan setting that is otherwise so effective in Fiennes's adaptation, the idea of a place *outside* territory is harder to grasp. Where *is* Coriolanus as he moves through that sequence of locations, sleeping rough and his hair growing ever longer? He could be in isolated locations. He is undoubtedly making his way through war-ravaged landscapes, contested places in the present or recent past. But given the modes of modern warfare and territorial settlements, he is either still within the place calling itself Rome" or behind enemy lines. It is hard to conceive of a no-man's-land of such extent that the time could have passed in such a way. He is effectively either in one territory or another. It is hard to imagine him outside of territory but for early Rome, or even in Shakespeare's England, it is not so difficult. This theme of what lies outside territory will be explored in the concluding chapter.

When Coriolanus arrives in Antium to talk to Aufidius he speaks of how the people were allowed by nobles to be cheered into exile, with the "voice of slaves to be / Whooped out of Rome."[140] "Whooped" is a hunting term, though it can also express "exultation, definance, intimidation" (*OED*).[141] Aufidius clearly sympathizes, recalling later how "Being banished for't, he came unto my hearth, / Presented to my knife his throat. I took him, / Made him joint-servant with me."[142] Back in Rome, Menenius condemns the crowd—"the clusters"[143]—for being those "that made the air unwholesome when you cast / Your stinking greasy caps in hooting / At Coriolanus' exile."[144] The citizens quickly claim that they said to banish him reluctantly: "we willingly consented to his banishment, yet it was against our will."[145] Later Menenius suggests that those who sent him into exile should beg forgiveness: "Go you that banished him; / A mile before his tent fall down, and knee / The way into his mercy."[146] In his attempt to mediate, Menenius is criticized by the Antium guard:

> Can you, when you have pushed out your gates the very defender of them, and in a violent popular ignorance given your enemy your shield, think to front his revenges with the easy groans of old women, the virginal palms of your daughters, or with the palsied intercession of such a decayed dotant as you seem to be? Can you think to blow out the intended fire your city is ready to flame in with such weak breath as this? No, you are deceived. Therefore back to Rome and prepare for your execution. You are condemned; our general has sworn you out of reprieve and pardon.[147]

Toward the end of the play, a Roman senator expresses a wish to "Unshout the noise that banished Martius, / Repeal him with the welcome of his mother."[148]

The play is also one of the few in the Shakespearean corpus to use the actual word "territories." First Aufidius declares:

> Worthy Martius,
> Had we no other quarrel else to Rome, but that
> Thou art thence banished, we would muster all
> From twelve to seventy and, pouring war
> Into the bowels of ungrateful Rome,
> Like a bold flood o'erbear't. O, come, go in,
> And take our friendly senators by th' hands,
> Who now are here, taking their leaves of me
> Who am prepared against your territories,
> Though not for Rome itself.[149]

This territorial ambition is then reported by an Aedile:

> There is a slave, whom we have put in prison,
> Reports the Volsces with several powers
> Are entered in the Roman territories,
> And with the deepest malice of the war
> Destroy what lies before 'em.[150]

It is also told by a messenger to Menenius:

> You are sent for to the senate.
> A fearful army, led by Caius Martius,
> Associated with Aufidius, rages
> Upon our territories, and have already
> O'erborne their way, consumed with fire and took
> What lay before them.[151]

All of these passages are in a single act, when Coriolanus is mustering Volscian forces to attack Rome. It is Rome that has territories, and seemingly not its neighbors; Rome that has lands at risk. Rome, though, is distinct from its territories. This only hints at the geographical complexities of Rome's rule, especially in the later empire—there were lands currently part of its *imperium*, and those that could be at some point in the future. Rome did not see the political geographical status of its neighbors in the same terms.[152] Aufidius talks of the territories to attack; the functionaries of the threat at home. And yet these places are also related to bodies, and not simply Aufidius's description of "the bowels of ungrateful Rome."[153] Coriolanus is content to "Let the Volsces / Plough Rome and harrow Italy,"[154] that is, reduce it to a colony or *territorium*. An important moment in the founding of a new colony was the plowing of the sacred boundary of the city walls, the *pomerium*. We hear about this from Cicero, Varro, and Tacitus, for example, and in the early Middle Ages Isidore of Seville even derives the etymology of the term from this practice: "a *territorium* is so called as if it were a *tauritorium*, that is 'broken by a plow' [*tritum aratro*] and by a team of oxen (cf. *taurus*, "bull")—for the ancients used to designate the borders of their possessions and *territoria* by cutting a furrow."[155] In this sense, even the "geopolitics" of the play is figured as a body politics, if not always as a biopolitics.

Having been expelled, and taking refuge in Antium, Coriolanus frequently uses this language. He invokes "those maims / Of shame seen through thy country," that is, injuries done to the body politic, perhaps territorial losses akin to the loss of a body part.[156] He talks to Aufidius of his "cankered country with the spleen / Of all the under-fiends."[157] Aufidius continues the polluted body trope in the passage quoted above, suggesting that they "muster all / From twelve to seventy, and pouring war / Into the bowels of ungrateful Rome, / Like a bold flood o'erbear't."[158] War seems like an enema or emetic, purging the Roman body. Returning to this theme of cannibalism, the most powerful figure in the play, the most sufficient, the one who bore Coriolanus in her body, and prevents his turning against the body of Rome, is Volumnia. And it is she who utters the horrific line "Anger's my meat: I sup upon myself / And so shall starve with feeding."[159] Finally she pleads with Coriolanus not to make "the mother, wife and child to see / The son, the husband and the father, tearing / His country's bowels out."[160] She then shifts the body to her own, suggesting that "thou shalt no sooner / March to assault thy country than to tread— / Trust to't, thou shalt not—on thy mother's womb, / That brought thee to this world."[161] Coriolanus has become Rome's other, its outside; his mother tells him not to tread on her womb, which gave him birth, but also on Rome itself, the site that made him possible.

Territory is, for obvious reasons, generally associated with geopolitics. At its most fundamental it is way of exercising political control over space. But as this book has shown, territory encompasses a range of complicated and contested processes. The strategic, economic, and legal aspects of territory also govern conduct within spaces, conduct of bodies and relations between them. The regulation of bodies is a crucial aspect of colonialization, and as well as technologies shaping land surveying and cartographic practices, they also partner the development of population statistics and censuses. I have argued elsewhere that we need to interrogate biopolitics and geopolitics together, rather than privileging one over the other, just as we should see population and territory as the dual objects of modern government as it is forged in the late Middle Ages and early modern period.[162] Yet territory does not just sit alongside governance of bodies, either of individuals or of the collective social body of population. Territory is often conceptualized *as a body*, a corporeality that gives flesh to more abstract determinations. The most striking in Shakespeare, perhaps, is the dual possession of Katherine's body and the body of France by King Henry V. But *Coriolanus* shows us that the body of the polity, imagined through the physique of its heroes, with its health and vitality, is a key determinant of political rule. The parts of the body, whether metaphorically through the fable of the belly, or the components of the polity, are likened to animals, contaminated or cured, wounded and displayed. Expulsion and invasion are physical transformations of the body politic. Territory has a biopolitics as well as a geopolitics; the corporeal is a crucial aspect of territory.

CHAPTER NINE

Outside Territory: The Forest in *Titus Andronicus* and *As You Like It*

If territory has a history, with the concept and practice emerging at a particular historical and geographical conjuncture, then it follows that before this there were political-spatial orderings that were not territory. We can therefore think of examples of configurations of the relation between power and place that were not territorial, that is, outside territory *historically*. Initially, "territory" *was* the outside. *Territorium*—an extremely rare word in classical Latin—was the area surrounding a place, perhaps a town or a colony. The suffix *-orium* means area around—a notion we maintain in words like "sanatorium," "auditorium," and "crematorium" The territory was the area around a settlement, the surrounding agricultural lands. This is the way it is used in Cicero, in Varro—who claims it is the area trodden on most—and in Seneca. (Of the other uses of the term in a classical Latin author, Pliny the Elder employs it in the neutral sense of an area. Other instances can be found in legal and technical works.) In the later Roman Empire, Ammianus Marcellinus makes the point that while they avoided the towns, the Germanic tribes frequented their territories. So, there is a story to be told about how territory moved from being the thing that was outside, outside the city walls and external to the *urbis*, to becoming the thing within which the city was located, within which the law was exercised.[1]

Shakespeare's *Coriolanus* is in this sense anachronistic: it is reading a later way of making sense of political-spatial orderings back into prerepublican Rome. There are political spaces, then, that were other than territory. These could be polities such as the Greek *polis*, which had surrounding lands, *khora*, but was not inside them in a complete way; the Roman *urbis* or the empire, which while divided into *territoria* did not see the areas outside as such; and the medieval church or kingdom. Did the Native Americans or

African tribes have their own *territory*, in the specific sense, before Europeans reordered the spaces of their lands? Or did they have a different way of experiencing, ordering, and understanding the relation between people, power, and place that was transformed into territories and through "territory"? Each of those different configurations would need to be understood through the words, concepts, and practices that would have made sense to those who lived in, fought over, worked in, and wrote about them. Equally there were areas that were outside territory in the sense of spaces between such designated and designed sites. The idea that outside any one territory is always another territory is an even more recent invention.

If the first text of the Western tradition, the *Iliad*, is about a struggle over a city and its lands, or what we might today call territory, Homer's *Odyssey*, in which Odysseus journeys home after the Trojan War, is about what it means to be outside it. In Sophocles's tragedy *Oedipus at Colonus*, one of Oedipus's most plaintive pleas in exile is when he asks, "Will they even shroud my body in Theban soil?"[2] In the play about one of his daughters, and the fate of his sons, *Antigone*, the whole story revolves around the question of burial and site, and the dislocation of familial and political relations.[3] In the Anglo-Saxon poem *Beowulf* there are isolated human settlements such as the village of Heorot where the mead hall stands. But outside, especially after dark, there are spaces of great danger, what we might today call the wild. Grendel, for instance, is described as a *mearcstapa*, a march-stepper or border-walker, a figure who prowls the margins, the edges, the limits, the liminal.[4] The *mere*—the pool of deep water—where his mother dwells is similarly beyond the reach of ordinary man, though not the hero Beowulf himself. Similarly dangerous places, outside territory or dominant ordering of political space, punctuate the Norse myths of the *Edda*.[5]

So it is with Shakespeare. Perhaps the most famous of such places is the heath in *King Lear*, where Lear in his madness, Kent in his exile, Edgar in his disguise as Poor Tom, and Gloucester in his blindness all end up.[6] All are outside, outsiders in some way; all are outside territory, the political space that was divided by the King between his daughters in the opening scene: "Since now we shall divest us both of Rule, / Interest of territory, cares of state."[7] This heath is famous, and the parts of the play on it are regularly described as the "heath scenes," even though the word "heath" never appears in the text of the play itself, nor in the Quarto or Folio stage directions.[8] There is also the "blasted heath" of *Macbeth*, where Macbeth and Banquo encounter the witches.[9] Later in that play Birnam Wood moves toward Dunsinane Hill, heralding the English invasion to unseat Macbeth. In *Troilus and Cressida*, Aeneas and Troilus discuss the war to come, "without the walls of Troy":[10]

> Aeneas: Hark, what goods sport is out of town today!
> Troilus: Better at home, if 'would I might' were 'may.'
> But to the sport abroad. Are you bound thither?[11]

"Out of town" means outside the city walls, on the battlefield, a notion that is referred to as the idea of "abroad."[12]

Alternatively, there is the confusion produced when figures from outside arrive, such as the shipwrecked twins in *Twelfth Night*. Here Viola and Sebastian believe each other dead. Viola disguises herself as a man, Cesario, but falls for the Duke Orsino, who sees Viola, as Cesario, as a confidant and messenger. Indeed he sends her to woo Lady Olivia on his behalf, but Viola creates the wrong kind of impression on Lady Olivia who thinks her a man and desires her (as him) for herself. Only the arrival of Sebastian, after much confusion, allows Olivia to have Sebastian, and Orsino to have Cesario, now revealed as Viola. Alternatively, *Measure for Measure* hinges on both the outside and the new arrival inside, when Duke Vicentio's seeming departure from Vienna and subsequent disguise allows him to analyze the relations within that place in a way he never could have done had he remained, or had he really left (see chapter 7).

Several other plays have banishment as a theme; this has already been discussed in relation to *Coriolanus*, *Richard II*, and *The Tempest*. Shakespeare's late collaboration with John Fletcher, *The Two Noble Kinsmen*, also treats this theme.[13] Yet in these plays there is not so much examination of the places where the exiled characters go, unless they go to other polities or to sites they recreate as such. Coriolanus ends up in Antium, to ally with the Volscians and then lead the attack on Rome. Bolingbroke of course returns before the end of his exile, and we know little of what happened to Mowbray, other than that he served in the Crusades and died in Venice. Prospero creates a new dukedom on the island, though this requires the dispossession of Caliban. *The Two Noble Kinsmen* moves between the cities of Athens and Thebes, in which Arcite and Palamon are variously banished, imprisoned, and dream of returning to. The play also features the forest outside of Athens to which Palamon escapes, where the jailer's daughter goes mad. Yet in that play the specific characteristics of the forest are underdeveloped—it functions more as a place between others. Other plays put much more emphasis on such liminal or external spaces. In this final chapter, I want to examine the use of forests or woods in Shakespeare more generally, as spaces outside of ordered politics.[14]

FROM *THE TWO GENTLEMEN OF VERONA* TO *TIMON OF ATHENS*

Shakespeare's early comedy *The Two Gentlemen of Verona* anticipates this theme, with the Duke's exile of Valentine from Milan:

> But if though linger in my territories
> Longer than swiftest expedition
> Will give thee time to leave our royal court,
> By heaven, my wrath shall far exceed the love
> I ever bore my daughter or thyself.[15]

The speech is similar to the King's expulsion of the Duke of Suffolk in *Henry VI, Part 2* and has parallels with speeches in later plays.[16] Valentine's response is equally one that Shakespeare would reuse in tone many times—that banishment is worse than death, because of the person he is now exiled from.[17] There are echoes in speeches by Romeo and the Duke of Suffolk.[18] A different attitude to banishment, as we have seen, comes in the defiant speech of Coriolanus.

While most of *The Two Gentlemen of Verona* is set in either Verona or Milan, there is also a third location—a forest outside of Milan, on the way to Mantua.[19] This is described as a "wilderness,"[20] and a "mountain foot,"[21] but is only three leagues from the abbey in Milan.[22] There are outlaws in this forest who ally themselves with Valentine.[23] As Bate describes it, the forest "is a place where the polished veneer of civil society is stripped away, allowing people to act impulsively on their desires."[24] Proteus's attempted rape of Silvia takes place in the forest, just as the rape and mutilation of Lavinia does in *Titus Andronicus*. Yet it is also a space in which, just as in *As You Like It*, different thoughts occur. As Valentine says in *The Two Gentlemen of Verona*,

> How use doth breed a habit in a man!
> This shadowy desert, unfrequented woods,
> I better brook than flourishing peopled towns;
> Here can I sit alone, unseen of any,
> And to the nightingale's complaining notes

Tune my distresses and record my woes,
O thou that dost inhabit in my breast,
Leave not the mansion so long tenantless,
Lest, growing ruinous, the building fall
And leave no memory of what it was.[25]

A desert here is a deserted, uninhabited place, a location Valentine says he can cope with better than inhabited urban areas. Shakespeare does little with this idea in this early play, but it is a theme to which he will return several times.

One play that treats the question of land is *Arden of Faversham*, an anonymously published and probably collaborative play in which Shakespeare may have had a hand in at least one scene.[26] Arden inherits lands belonging to the Abbey of Faversham, granted by the Duke of Somerset.[27] His wife and various other characters all plan to kill him, for which seizing the land is a major motive, finally succeeding but being caught and executed. The play uses the recurrent phrase "plot of ground" to weave together the lands under dispute, the place where Arden's body is left, and the interrelated elements of the play's structure.[28] Another anonymous play, *Mucedorus*, again sometimes attributed in part to Shakespeare, has a pastoral theme, with the Prince of Valencia disguising himself as a shepherd, killing a bear that was pursuing Amadine, Princess of Aragon. The play moves from woods to Aragon and back again.[29]

Two major treatments of what lies outside are found in *Titus Andronicus* and *As You Like It*, which will be the focus of analysis here. But important roles are also played by the wilderness, often a forest, in other plays.[30] In *A Midsummer Night's Dream* there is a clear contrast between the ordered, structured space of Athens and the dreamworld of the woods. In the former, women are forced to do men's bidding; in the latter they are allowed license to choose their own partners. Lysander suggests a plan to Hermia:

I have a widow aunt, a dowager,
Of great revenue, and she hath no child.
From Athens is her house removed seven leagues,
And she respects me as her only son.
There, gentle Hermia, may I marry thee,
And to that place the sharp Athenian law
Cannot pursue us. If thou lov'st me then,
Steal forth thy father's house tomorrow night;

And in the wood, a league without the town— . . .

There will I stay for thee.[31]

Of course, Lysander and Hermia never reach the aunt's house, and when they meet they are interrupted first by Demetrius and Helena, and then the fairies. This play is structured around this division, with two opening scenes in Athens, the forest and dream scenes in the middle, and then two closing scenes back in Athens. The forest scenes proper are framed by the "rude mechanicals," who rehearse their play in a forest clearing.[32] *As You Like It* also has a rural interlude and a homecoming, as does *Cymbeline* and *The Winter's Tale*. Yet other plays do not stage a return to the city and its territorial order.

A play that sets up a more serious contrast between city and outside is *Timon of Athens*.[33] Timon's wealth and generosity had been tied to his status as a landowner. Nicholas Walton has noted that the play establishes its places only in the vaguest terms, but the comparison is nonetheless stark.[34] The play was likely cowritten with Thomas Middleton, and in one of the early passages from Middleton, Timon declares:

Methinks I could deal kingdoms to my friends,
And ne'er be weary. Alcibiades,
Thou art a soldier, therefore seldom rich—
It comes in charity to thee, for all thy living
Is 'mongst the dead and all the lands thou hast
Lie in a pitched field.[35]

To "deal" is to dispense or distribute, while a "pitched field" is "a planned battle set out on an agreed terrain (with a punning reference back to *lands* . . .)."[36] Alcibiades's reply plays on "pitch" as staking out and as bitumen: "Ay, defiled land, my lord."[37]

Yet Timon's assets are squandered, and when he is considering how to deal with the creditors who are asking for money back, he finds that no one will come to his aid. He turns from immediate property and money to real estate: "Let all my land be sold."[38] But his servant Flavius quickly disabuses him of this notion:

'Tis all engaged, some forfeited and gone,
And what remains will hardly stop the mouth

Of present dues; the future comes apace.
What shall defend the interim, and at length
How goes our reckoning?[39]

Some of his land is mortgaged or pawned, some owed to creditors, and some already lost. The remainder is not enough to stave off immediate concerns, let alone future ones. The sense of ravenous creditors is clear with the mouth devouring all. What then can be done in the immediate term and further ahead to deal with this "reckoning," an account of resources?

Timon is incredulous: "To Lacedaemon did my land extend"[40]—Lacedaemon being an alternative word for Sparta. Interestingly, the Spartans' attitude to land was rather different than simple property. As Cicero and Plutarch both report, they claimed all the land that they could touch with a spear: the extension of their military power.[41] Flavius says that Timon would give the world away as quickly as he could say the word "world." He adds that there is nothing wrong with his management of Timon's accounts, and that "if you suspect my husbandry of falsehood," auditors can put his practices to the test, to which he swears an oath.[42]

In the second half of the play—in passages probably by Shakespeare—Timon is outside the city, moving into the wilderness. The scene has the stage direction "Enter Timon in the woods." Unusually, this is Shakespeare's own statement of a location, which is significant, in part because it is rare, and in part because it refers back to a previous play's depiction of the woods outside Athens.[43] But here, this is far from the idealized vision of what lies outside found in *A Midsummer Night's Dream*. Unlike many of Shakespeare's other tragic heroes, Timon does not return from this exile—even Lear has his moment of reconciliation with Cordelia.[44] Timon's wilderness is a desolate landscape, though Timon recognizes that the city itself has just as many problems:

Let me look back upon thee. O thou wall
That girdles in those wolves, dive in the earth
And fence not Athens![45]

Timon thus looks back at the city, and its wall, which encloses the voracious citizens he has turned his back upon. He suggests that the wall should fall to the ground, collapse, or sink down.[46] "Fence" here has a sense of both to border and to defend the city. This sense of the collapse between the wild outside and the turmoil within continues in the play, with Apemantus noting that "The commonwealth of Athens is become a forest of beasts."[47] A very similar phrase is found in the much earlier *Titus Andronicus*: "Rome is but a wilderness of tigers."[48]

THE FOREST IN *TITUS ANDRONICUS*

The forest plays a crucial role in *Titus Andronicus*.[49] The play is one of Shakespeare's earliest, and has often been dismissed—either as juvenile excess or not by Shakespeare's hand.[50] While many editors suggest a collaboration with George Peele, Eugene Waith has claimed that the play was written by Shakespeare alone, and probably performed 1590–92, and that there was a revision before the performance of 1594, including one additional scene (Act III, scene ii). That latter scene only appears in the First Folio.[51] *Titus Andronicus* is not based on a historical source, unlike Shakespeare's other Roman plays. There are two other sources of the story, a prose description and a ballad. Editors of the text generally suggest that the prose version is the earliest, and therefore Shakespeare's key source (Waith, Sonia Massai and Jacques Berthoud, Alan Hughes), or that the play comes first (Bate). Most editors see the ballad as subsequent, though Hughes thinks that the play was not one of its sources.[52] Even if the poetry of the play is not always up to Shakespeare's later standards, it is a striking and very bloody spectacle, with a complicated structure and intricate plot.[53]

Titus has returned from "weary wars against the barbarous Goths,"[54] with their captured queen, Tamora, three of her sons, and the Moor Aaron. But victory has come at considerable cost. As Titus says in his opening words, "Hail, Rome, victorious in thy mourning weeds!"[55] Several of Titus's sons are interred in the burial vault. Despite Tamora's pleas for clemency, he executes her eldest son, Alarbus, in revenge for the death of several of his own. He thereby creates an enmity with Tamora and her other sons, Demetrius and Chiron. At the same time, brothers Saturninus and Bassianus are feuding over who should be Rome's next emperor, following the death of their father. The people's view is that Titus should be emperor, to "help set a head on headless Rome."[56] Yet Titus declines—"Give me a staff of honour for mine age, / But not a scepter to control the world"[57]—and instead endorses Saturninus as the elder son. Saturninus is therefore chosen, but compounds the feud with his brother by demanding Titus's daughter Lavinia, already betrothed to Bassianus, as his wife. Titus supports this, but Lavinia escapes and is aided by her brother, Mutius, who is killed by Titus for defying him. Saturninus choses Tamora as his bride and "empress of Rome" instead.[58]

Much of the subsequent action takes place in the forest, a wild site, in contrast to urban Rome. The forest seems to relate to the homeland of the

Goths Tamora and her sons, but distinctions between a supposedly civilized Rome and the wild outside become increasingly blurred.[59] In a hunt on the second day of the play's action many key elements of the plot are played out. Titus had invited the Emperor to hunt "the panther and the hart . . . / With horn and hound" the previous evening,[60] and the next day proclaims: "The hunt is up, the morn is bright and grey, / The fields are fragrant and the woods are green."[61] As Bate has indicated, "Hunting for sport is 'civilized' society's way of getting back in touch with the wild."[62] Initially the forest appears as it does in other Shakespeare plays, as a space of love and wonder. Slipping away from the hunt, Tamora meets up with her secret lover, Aaron. She speaks of the forest in positive, pastoral tones:

> The birds chant melody on every bush,
> The snakes lies rolled in the cheerful sun,
> The green leaves quiver with the cooling wind
> And make a chequered shadow on the ground.[63]

She continues, suggesting that the forest canopy makes a "sweet shade" for her and Aaron to sit. But the forest quickly becomes a much darker, transgressive place. Aaron says that while Venus may "govern" her desires, "Saturn is dominator" of his.[64] He outlines the plan for the death of Bassianus and the violation of Lavinia. The two Roman lovers enter, and realize that Tamora and Aaron are together. When Tamora's sons appear, she turns the accusation back on Bassianus and Lavinia, suggesting that they have enticed her to this site. Tamora herself now describes the location in much darker, more negative terms:

> A barren detested vale you see it is;
> The trees, though summer, yet forlorn and lean,
> O'ercome with moss and baleful mistletoe;
> Here never shines the sun, here nothing breeds
> Unless the nightly owl or fatal raven.[65]

The contrast is striking, and her two descriptions make use of the same imagery, but put to different uses.[66] Following Tamora's injunction, Demetrius and Chiron kill Bassianus, throwing his body into a pit. Demetrius and Chiron then abduct and rape Lavinia, cutting out her tongue and chopping off her hands so that she cannot say or write what has happened and who is responsible.

Demetrius and Chiron had previously described Lavinia as their desire, and had likened her to "a doe," "borne . . . cleanly by the keeper's nose," that is, poached and stolen away without the gamekeeper noticing.[67] Edward Berry suggests that "Aaron, Chiron and Demetrius can thus be seen as nightmare versions of Elizabethan poachers," and claims that Shakespeare's experience might have been more poaching than hunting, given his own social setting.[68] Aaron suggests that the forest served this purpose well: "the woods are ruthless, dreadful, deaf and dull."[69] He urges:

> My lords, a solemn hunting is in hand;
> There will the lovely Roman ladies troop.
> The forest walks are wide and spacious,
> And many unfrequented plots there are
> Fitted by kind for rape and villainy.
> Single you thither then this dainty doe,
> And strike her home by force. [70]

The description of Lavinia as a female deer only reinforces the hunting sense that pervades these scenes, literally and figuratively. Later Demetrius recalls the telling phrase: "Chiron, we hunt not, with horse nor hound, / But hope to pluck a dainty doe to ground";[71] as does Titus's brother Marcus when he discovers her later: "found her, straying in the park, / Seeking to hide herself, as doth the deer / That hath received some unrecuring wound."[72] Titus's later invocations of her as "my dear" and "dear Lavinia, dearer than my soul" further stress the identification.[73] Berry has rightly described it as "Shakespeare's most obviously tragic treatment of the hunt in relation to male sexual aggression . . . in which the rape of Lavinia by Tamora's sons is portrayed as a grisly parody of a hunt, even to the symbolic dismembering of the victim."[74] In addition, the way that Lavinia's violated body is described uses arboreal imagery: "what stern ungentle hands / Hath lopped and hewed and made thy body bare / Of her two branches, those sweet ornaments / Whose circling shadows kings have sought to sleep in."[75] "Chop," "lop," "hew," and "cut," with their forestry and agricultural connotations, are frequently used words, and actions, in the play.[76]

Titus's other sons, Martius and Quintus, are lured into the pit and discover the dead body. The "loathsome pit" is described as a "subtle hole . . . Whose mouth is covered with rude-growing briers, / Upon whose leaves are drops of new-shed blood / As fresh as morning dew distilled on flowers."[77] Though the pit is a hunting trap, the link between the hole and the rape is, of course, far from subtle, and continues in further description: "fatal place," "this unhallowed and bloodstained hole," "this detested, dark, blood-drinking pit," "the ragged entrails of this pit," "this fell devouring receptacle, / As hateful as Cocytus' misty mouth."[78] Aaron, by way of a forged letter and a buried stash of gold, frames Martius and Quintus for Bassianus's death, and the Emperor sentences them to death.

The forest therefore functions as a space outside of the city, a place of desire, both of adultery and of rape, murder, muting, and mutilation. As Alan Sommers suggests, "The essential conflict in *Titus Andronicus* is the struggle between Rome, and all that this signifies in the European tradition to which we, and Shakespeare, belong, and the barbarism of primitive, original nature."[79] Yet it is clear that the two sides of this become increasingly blurred. With the trial of Quintus and Martius, and the banishment of his remaining son, Lucius, for trying to rescue them, Titus exclaims that "Rome is but a wilderness of tigers."[80] The wild animal, and the space outside, the wilderness, now constitute the city that was previously seen as the bulwark against them. Moments later, Marcus brings the mutilated Lavinia to him, suggesting that "this was thy daughter."[81] Titus exclaims that this is all too much:

> Speak, Lavinia, what accursed hand
> Hath made thee handless in thy father's sight?
> What fool hath added water to the sea?
> Or brought a faggot to bright-burning Troy?
> My grief was at the height before thou cam'st,
> And now like Nilus it disdaineth bounds.[82]

The idea of the Nile washing over existing boundaries is important, though the river is seasonal, and there is no sense Titus's grief will ebb. Thus any boundary between Roman civilization and Gothic barbarism, which might have been supposed on the basis of earlier parts of the plays, is rendered much more complicated. Titus realizes he is alone: two sons are due to be executed, the remaining son is banished, and his daughter has been violated:

> For now I stand as one upon a rock,
> Environed with a wilderness of sea.
> Who marks the waxing tide grow wave by waves,
> Expecting ever when some envious surge
> Will in his brinish bowels swallow him.[83]

Aaron tells him that the sons will be spared if Titus, Marcus, or Lucius will cut off their hand as punishment. Titus has his hand cut off, and sent to the Emperor, but in return is presented with his sons' heads and his own hand. He tells Lucius to leave Rome: "Thou art an exile and thou must not stay; / Hie to the Goths and raise an army there."[84] Lucius's banishment is equally intriguing: expelled from Rome itself, he turns to those who constitute its barbarian outside to try to redeem Rome from itself. Lucius replies that

> If Lucius live, he will requite your wrongs
> And make proud Saturnine and his empress
> Beg at the gates like Tarquin and his queen.
> Now will I to the Goths and raise a power,
> To be revenged on Rome and Saturnine.[85]

Lavinia indicates what has happened to her by finding the story of Philomela in a copy of Ovid's *Metamorphoses*.[86] Titus realizes that the same violation of Lavinia has taken place as what the poem recounts of Philomela: "surprised . . . Ravished and wronged . . . Forced in the ruthless, vast and gloomy woods?"[87] He adds that there was such a place where they hunted, and regrets ever going to that place, "by nature made for murders and for rapes."[88] Lavinia then drags a stick through the earth to name the crime and her attackers: "Stuprum—Chiron—Demetrius."[89]

In the meantime Tamora had given birth to Aaron's child, and Aaron has to kill the nurse to protect it. Aaron, for Sommers, is an elemental figure, "a creature of wild nature."[90] In the infamous final scene, Titus has his revenge by killing Demetrius and Chiron, baking their blood, bones, and flesh into a pie, which he feeds to Tamora. Tamora thus ingests the flesh she had previously carried in her body—another crossing of boundaries that is entirely unnatural. Titus kills his daughter to end her suffering, kills Tamora, and is killed by Saturninus. Lucius kills Saturninus in vengeance, and is proclaimed Emperor. As all this indicates, this play is one of Shakespeare's most shocking. At the end Tamora's body is thrown outside the city for "beast and birds to prey," just after she has been described by Lucius as a "ravenous tiger";[91] and Aaron—himself earlier called exactly the same by Lucius[92]—is buried chest-deep within the earth to starve and thirst to death.[93] The two tigers are thus cast out into the wilderness again. Titus and Lavinia are interred in the family tomb. The burial of the dead, which began the play, also ends it; Rome inters its own bodies within the site, an attempt to put things back in place.

The blurring of Rome and the outside is a continual theme. At the beginning of the play it is the "Roman rites" of sacrifice and Titus's denial of Tamora's passionate plea for mercy that sparks the revenge plot.[94] At the end, it is an army of Goths that Lucius leads to Rome, as Marcus had hoped: "Join with the Goths and with revengeful war / Take wreak on Rome and for this ingratitude, / And vengeance on the traitor Saturnine."[95] Other invading figures in Shakespeare's plays lead foreign forces against their homeland—Malcolm uses English forces to overthrow Macbeth; Cordelia leads an army from France; the Earl of Richmond is supported by Bretons against Richard III. It is striking how often this theme of invasion appears in this way, with the outside forces allowing the place to return to its appropriate identity. Shakespeare would return to the idea of a Roman leading external forces against Rome in *Coriolanus*. Indeed, this play makes the relation explicit:

> Arm, arm my lords! Rome never had more cause:
> The Goths have gathered head, and with a power
> Of high-resolved men bent to the spoil
> They hither march amain under conduct
> Of Lucius, son to old Andronicus,
> Who threats in course of this revenge to do
> As much as ever Coriolanus did.[96]

"Bent to the spoil" suggests a clear and resolute purpose; "amain" is full force or speed. Saturninus realizes that this may be his end:

> Is warlike Lucius general of the Goths?
> These tidings nip me and I hang the head
> As flowers with frost or grass beat down with storms.
> Ay, now begins our sorrows to approach.
> 'Tis he the common people love so much;
> Myself hath often heard them say,
> When I have walked like a private man,
> That Lucius' banishment was wrongfully,
> And they have wished that Lucius was their emperor.[97]

The Goths are historically unspecific, suggesting various wars with Germanic tribes from Julius Caesar onward, but more especially the overrun of the Roman Empire in the fifth century CE.[98] Bate summarizes the shift in register: "As the play begins, Titus has spent ten years expanding the empire, as if during its heyday; when it ends, an army of Goths is in the city, as if during its decline."[99] This is important for the general political context of the play, for the Roman attitude to *territorium* was transformed over time, moving from land outside the walls of a city to parts of the empire, and, in time, to the area over which jurisdiction was exercised. The latter, developing out of a medieval reworking of Roman law, became central to the development of the territory-sovereignty relation in the seventeenth century. Shakespeare's own time saw that being developed especially in the Holy Roman Empire, with political theorists such as Knichen and Althusius.[100]

THE FOREST OF ARDEN IN *AS YOU LIKE IT*

As You Like It, one of Shakespeare's last comedies, before the "problem plays" of *Measure for Measure* and *All's Well That Ends Well*, returns to an idea explored in his first: exile and outlaws in a forest. This is the Forest of Arden, where a deposed Duke and his courtiers live in exile. Duke Senior (Ferdinand) has been overthrown by his brother Duke Frederick. In the early scenes some exposition explains the situation:

> There's no news at the court, sir, but the old news: that is, the old Duke is banished by his younger brother the new Duke, and three or four loving lords have put themselves into voluntary exile with him, whose lands and revenues enrich the new Duke; therefore he gives them good leave to wander.[101]
>
> They say he [the old Duke] is already in the Forest of Arden and a many merry men with him, and there they live like the old Robin Hood of England. They say many young men flock to him every day and fleet the time carelessly as they did in the golden world.[102]

The lords that have left the court have forfeited their lands, and find themselves in a place outside of modern time, harking back to a previous golden age. Duke Senior's daughter Rosalind initially stays behind, because of her close friendship with Duke Frederick's daughter, Celia. But Rosalind too is banished:

> You cousin.
> Within these ten days if that thou be'st found
> So near our public court as twenty miles,
> Thou diest for it.[103]

Rosalind and Celia both disguise themselves as men and flee the confines of the court, also pursuing love. Along with the clown Touchstone they head for the forest. The play thus establishes some apparent contrasts. The court is dominated by constraint, whereas the forest is a space of freedom. This distinction between the court and the forest is paralleled in the liberation of Rosalind and Celia when they take on their disguises. As Celia says when they leave, "Now we go in content / To liberty and not to banishment."[104] Juliet Dusinberre notes that "content" is a key word linked to pleasure and liking, particularly pointing to "settled low content" and an exchange between the shepherd Corin and Touchstone.[105]

In the forest Duke Senior says to the assembled lords:

> Now my co-mates and brothers in exile,
> Hath not old custom made this life more sweet
> Than that of painted pomp? Are not these woods
> More free from peril than the envious court?
> Here we feel not the penalty of Adam,
> The seasons' difference—as the icy fang
> And churlish chiding of the winter's wind. . . .
>
> And this is our life, exempt from public haunt,
> Finds tongues in trees, books in the running brooks,
> Sermons in stones, and good in everything.[106]

The "penalty of Adam" is usually work, but here it is the change of seasons, from a perpetual spring to a long winter. Nonetheless, for the Duke this does not damage his perception of liberation. The crowd is absent, and natural features replace the conversation, learning, and religion of society.

The Duke continues to indicate what must be done:

> Come, shall we go and kill us venison?
> And yet it irks me the poor dappled fools,
> Being native burghers of this desert city,
> Should in their own confines with forked heads
> Have their rough haunches gored.[107]

Crucially this is not, now, hunting for sport, but for food. Hunger is a common refrain through the play, especially in Act II.[108] While "venison" at this time meant meat in general, rather than specifically deer, it is clear that the latter are intended with the description.[109] The deer, which appear in the motley of fools (but are themselves fools, understood as unfortunates, rather than idiots), are the burghers of this city, deserted by people but not without citizens.[110] Confines are "territory, limits," while "forked heads" are the types of arrows used.[111] We are then told how the melancholy Jaques mourns the death of the deer, and it is reported that he "swears you do more usurp / Than doth your brother that hath banished you."[112]

> Thus most invectively he pierceth through
> The body of country, city, court,
> Yea, and of this our life, swearing that we
> Are mere usurpers, tyrants and what's worse
> To fright the animals and to kill them up
> In their assigned and native dwelling-place.[113]

Duke Senior's overthrow of the deer in their domain, a usurpation of their rights, thus relates to his own overthrow by his brother.[114]

In the forest Rosalind and Celia encounter Jaques, and while they are kept apart from the Duke and his courtiers, Jaques acts as a transitional figure between the groups. It is clear that Jaques is not an exiled courtier, but a man who has sold his goods and lands and chosen the forest as his home. As Bate rightly notes, if he was part of the court, why does he not recognize Touchstone?[115] Rosalind says to Jaques: "A traveller! By my faith, you have great reason to be sad. I fear you have sold your own lands to see other men's."[116]

There are parallels with some of Shakespeare's other plays. These obviously include the well-worn trope of women disguised as men (and, of course in Shakespeare's time, it would have been boys playing these women), and then falling in love with men for whom they cannot declare, and becoming objects of affection for misguided women. In addition, as Dusinberre suggests, Valentine's leadership of the outlaw band in *Two Gentlemen of Verona* acts as a kind of preview of the situation of the deposed Duke and his courtiers in the Forest of Arden:[117]

> The time of the play's setting is unspecified, but it is likely to be a near present to Shakespeare's own. The idea of time is gently mocked: When asked "what is't o'clock?"; Orlando replies: "You should ask me what time o'day. There's no clock in the forest."[118]

While its curious mix of flora and fauna suggests a place outside of any time or geography, and the play seems confused as to a French or English setting, in an important political sense the Forest of Arden functions as a kind of site outside of enclosure.[119] In this graduated sense of an outside it is perhaps more subtle than some of Shakespeare's earliest plays on the city/country contrast.[120] Enclosure was, of course, a key issue at Shakespeare's own time. Yet it would be a mistake to think that Arden is completely devoid of this. Rosalind pays for food and shelter, and buys "the cottage, pasture and the flock" from a landlord, so the wood is not entirely outside modern political-economic ordering.[121] The cottage is later described as "the cottage and the bounds," so it has come with surrounding lands.[122] There is something of an ambiguity here—some of these are places outside ordering; others entirely within it. It is therefore clear that the Forest of Arden comprises two different elements of the outside here. One is the marginal space of agriculture, shaped by enclosure; the second is the deeper forest of the outlaws and hunting. Rosalind as Ganymede and Celia as Aliena remain in the first; Jaques and Orlando move between the two. Rosalind and Celia do not meet the Duke and his courtiers until the play's finale.[123]

It would therefore be a mistake to think that the contrast of the play is straightforward. Arden is not a rural idyll; life there is hard work, and its weather can be unforgiving.[124] A. Stuart Daley in particular challenges the idea that Shakespeare simply praises the country over the city and court.[125] The play also ends on an ambiguous note because the new society created in the forest is replaced by the return of the old, as the exiles, with the exception of Jaques, return to the place they left.[126] Dusinberre suggests this shows a clear contrast with Touchstone: "At the end of the play the fool returns to court, where he can make a living out of his folly. The satirist remains in the Forest, weeping tears into the brook where he sees a reflection: is it of a wise man or a fool?"[127]

The tension in the play concerns the word "forest," which appears twenty-three times in this play, but "no more than three times in any other play."[128] What is meant by "forest"? When the women first arrive and Rosalind says, "well, this is the forest of Arden,"[129] they are on a sheep farm. Daley argues that at the time Shakespeare was writing, the primary meaning of "forest" was still "a mostly untilled tract used mainly for grazing, propagating and preserving game, and, usually, growing some timber (hence a diversified landscape)." It was only later that "in popular usage [it] came to suggest a large extent of fairly densely wooded country."[130] John Manwood's late sixteenth-century definition is useful here:

> A Forest is a certain Territory of wooddy grounds and fruitful pastures, priviledged for wild beasts and fowls of Forest, Chase and Warren, to rest and abide in, in the safe protection of the King, for his princely delight and pleasure, which Territory of ground, so priviledged, is meered and

> bounded with unremovable marks, meers and boundaries, either known by matter of record, or else by prescription.[131]

Manwood goes on to discuss the territory in relation to the Latin *territorium* and the French *territoire*, suggesting that the meaning is the same and specifies "all the fields and country lying within the bounds of a City, which doth extend far without the walls of the City round about, by certain meers and boundaries, without any other inclosure belonging to the same."[132] The forest in the general sense is more closely specified, as Forest in a more specific sense, Chase and Warren. These places relate to different animals that are found and hunted there: hart, hind, hare, boar, wolf (Forest); buck, doe, fox, martin, roe (Chase); hare, coney, pheasant, partridge (Warren).[133] Buck and doe are fallow deer; hart (or stag) and hind are red deer. Rackham expands on this theme:

> We owe the word *forest* to the mysterious Merovingian Franks, who lived in north-east France after the fall of the Roman Empire. Its original meaning is unknown. The medievals thought it signified a region outside (Latin *foris*) the ordinary laws. The learned brothers Grimm, of *Grimm's Fairy Tales*, conjectured a place of fir (Old High German *forha*) trees. Both guesses are still related by modern etymologists, but there is really nothing to be said for either: scholars clutch at any straw sooner than admit ignorance.
>
> The word may originally have meant land covered with trees, but it soon came to mean a region subjected by the king to special laws concerned with preserving game. This latter idea, and the word, became widespread in Europe long before they reached England. Throughout our Middle Ages *a Forest was a place of deer, not necessarily a place of trees*.[134]

Shakespeare's "forest" is therefore potentially quite different from how we imagine it. Daley stresses the point that the forest in the play is often a harsh, inhospitable landscape: "The dialogue in the country scenes devotes abundant imagery to the perils and hardships experienced there." Daley draws on Caroline Spurgeon's work on language of sickness, disease, and medicine, indicating that the use of such vocabulary in *As You Like It* is "equaled or exceeded in number only by those in *2 Henry IV*, *Hamlet*, *Troilus and Cressida*, and *Coriolanus*."[135] Daley also stresses that, in addition, "apart from forest or wood, Arden is six times called a 'desert.'"[136]

At a crucial part of the play, Oliver is sent by Duke Frederick to find his brother:

Find out thy brother whereso'er he is;
Seek him with candle. Bring him dead or living
Within this twelvemonth, or turn thou no more
To seek a living in our territory.
Thy lands, and all things that thou dost call thine
Worth seizure, do we seize into our hands,
Till thou canst quit thee by thy brother's mouth
Of what we think against thee.[137]

Searching "with candle" implies looking in all the corners, like the Bible verse of the woman who has lost a piece of silver, though it may also be a reference to a vigil over a dead body, or even the rite of excommunication.[138] The Duke threatens banishment and confiscation of lands and property. When Oliver declares he never loved his brother, the Duke is hypocritical:

More villain thou! Well, push him out of doors,
And let my officers of such a nature
Make an extent upon his house and lands.
Do this expediently, and turn him going.[139]

As Oliver travels into the forest, one other crucial word for understanding the geography of this time comes when Oliver asks the disguised women:

Pray you, if you know
Where in the purlieus of this forest stands
A sheepcote fenced about with olive-trees.[140]

This is the only instance of the word "purlieu" in Shakespeare, a term referring to a border area on the edge or outskirts of a forest. A purlieu is an edge or fringe, a border region or frontier rather than a boundary line.[141] Indeed, Rosalind herself says that she and Celia are living "in the skirts of the forest, like fringe upon a petticoat."[142] Richard Marienstras claims that "it certainly refers to a territory intermediary between the political world and the 'sanctuary' of the forest,"[143] and, drawing on Manwood, expands: a purlieu was "a territory adjoining the forest and indicated by recognized and declared markers. From a historical point of view, these were forest zones that had been disafforested—that is to say that between the beginning of the thirteenth and that of the fourteenth centuries they lost their status as forestland."[144] But the problem here is that the term "territory" is used to describe this transitional space. "Territory" used to mean the outside, and came to be the container. Here we have gradations: the political, colonized space—the new sense of territory; a transitional region surrounding it—the old sense of *territorium*; and what lies beyond, outside of that.

Berry has explored the play in detail in relation to the theme of hunting. While there is much said about hunting, Duke Senior and his group are "technically poaching, since they live as outlaws, taking deer from land that is under the authority of Duke Frederick," even though he is the usurper.[145] Berry argues that for the deposed Duke and his followers, "their social status" has been "untouched by exile; they have not become shepherds but have merely exchanged one form of hunting for another. In exile they hunt out of necessity; in power they are privileged to hunt for sport."[146] Intriguingly, given his liminal position in the play, it is Jaques who weeps over the death of a hunted deer. Berry argues that Jaques's attitude perhaps implies a critique of hunting for sport, and sympathy for wild nature, even if hunting for food is a necessary burden.[147]

Robert Pogue Harrison has claimed that

> the Shakespearean comedies that take place in the forests—*A Midsummer Night's Dream* and *As You Like It*, for example—follow . . . comic patterns . . . : disguise, reversals, and a general confusion of the laws, categories, and principles of identity that govern ordinary reality. In this respect there is nothing new in Shakespeare's forests with regard to the underlying logic of comedy (which in no way means that his comedies bring nothing new to the genre). The same cannot be said, however, of his dramas about civic barbarism. Those dramas are unique for the way they bring the shadow of natural law to bear on the religious, moral, and social crises that were shaking the traditional foundations of society.[148]

Harrison's reading is largely of *Macbeth*, but the point surely extends to *Titus Andronicus*.

Beyond Shakespeare, the theme of enclosure is explored perhaps most interestingly in Edward Bond's play *Bingo: Scenes of Money and Death*.[149] In that play, Shakespeare himself is a landowner, trying to work out if he should accept William Combe's offer of protection for his own lands at the expense of local farmers.[150] Bond himself writes of Shakespeare that while he "created Lear, who is the most radical of all social critics,"[151] his own "behaviour as a property-owner made him closer to Goneril than Lear. He supported and benefitted from the Goneril-society—with its prisons, workhouses, whipping, starvation, mutilation, pulpit-hysteria and all the rest of it."[152] These ambiguities are often spatially distributed as well as conditioned.

In a number of his plays, then, Shakespeare explores the relations between politically ordered spaces and differently constituted ones. Sometimes this is a contrast between a city and the country, such as in many of the plays discussed in this chapter, or in the contrast of the court of King Cymbeline and the wilds of Wales. It can also be found in *The Winter's Tale* and the opposition between what appears to be urban and wintery Sicilia and alternately isolated and bucolic Bohemia. In *The Winter's Tale*, of course, Shakespeare has been famously criticized for giving Bohemia a coastline.[153] Is this a simple geographical error on Shakespeare's part, as Ben Jonson famously thought,[154] or is something more at stake? As John Pitcher puts it, "Even the shoreline isn't real. Bohemia was a real place on the map, but as Shakespeare and his contemporaries knew full well, it was land-locked, hundreds of miles from the sea."[155] Well-known contemporary maps from Mercator (1569) and Abraham Ortelius's *Theatrum Orbis Terrarum* (1570) would have clearly shown this: on the latter Bohemia is strikingly encircled by a mountain range, as it is in Jodocus Hondius's 1609 reprint of Mercator's plates.[156]

The geographical anomaly in *The Winter's Tale* has been widely discussed, but the wider point is surely to allow Sicilia to have an outside.[157] King Polixenes wants his newborn daughter removed from the kingdom, and instructs his servant Antigonus to do this task:

> We enjoin thee,
> As thou art liegeman to us, that thou carry
> This female bastard hence, and that thou bear it
> To some remote and desert place, quite out
> Of our dominions; and that there thou leave it,
> Without more mercy, to it own protection
> And favour of the climate.[158]

Bohemia, while clearly a different territory than Sicilia, is described in terms of its isolation—"the deserts of Bohemia,"[159] and "places remote enough are in Bohemia."[160] "Desert" and "deserts" here are again deserted places, wilderness, uninhabited shores, rather than literal deserts.

Yet how many of these places are genuinely outside of territory? In premodern settings, such as *Coriolanus*, *Timon of Athens*, or *Titus Androni-*

cus, it is entirely possible to imagine places that were not yet spaces, wildernesses that were not yet incorporated into political, geographical, and technical orderings. Lear's wilderness and Macbeth's "blasted heath" seem to serve similar purposes. In *As You Like It* the Forest of Arden is a more ambiguous space. While exile to the forest seems to be removal from the political structures of the court, the agricultural practices and the traces of enclosure demonstrate that few places remained outside of territorial orderings. This is in terms of both enclosure—the first Act of Enclosure dating from 1604, as chapter 1 noted—and colonization. Both at home and abroad lands are incorporated, boundaries are drawn, territories are established. The idea of being outside territory becomes a historical memory.

Coda: Beyond Pale Territories

Shakespeare's use of the words "territory" and "territories" is, as this book has discussed, rare. The plural "territories" appears in some of his earliest plays, from *Two Gentlemen of Verona* to the *Henry VI* plays, in midcareer histories such as *Richard II*, and then again in the late *Coriolanus*. The singular "territory" can be found only in the midcareer *As You Like It* and Folio text of *King Lear*. It does not appear that the *word* undergoes any significant shift in his usage over time, though as chapter 3 noted, there is an intriguing use of a definite rather than possessive article in one use in *King John*. Nor is it clear that Shakespeare's conceptualization of territory changes between early and late plays. The chapters in this book have used a range of his plays to examine different aspects of the question of territory as concept and practice, and while these have tended to be mid- to late career plays, this selection has more to do with the status of those plays as exemplars, rather than an argument about the maturity of his thought on this question. Yet the cumulative nature of Shakespeare's work adds to the complexity of his and our understanding of territory, as the plays indicate different aspects of the question.

Shakespeare's conceptualization of territory deepens over time, but in addition one of the key places he explores as a territory certainly undergoes a significant transition. Shakespeare's career spans the latter part of Elizabeth I's reign and the early years of James I's. The national projects of this period feature in important ways in his plays.[1] As Shapiro has noted, the words England and English appear regularly in Elizabethan plays, and only rarely afterward—largely in *Henry VIII*. The word British never occurs in Elizabethan plays, and Britain only twice, whereas both are much more common in Jacobean plays.[2] British and Britain at the time did not signify quite the same as today, and until the rule of James, Briton largely referred to the Welsh or Bretons as descendants of the ancient race.[3]

There are themes, especially in the history plays, where the national sense of England, and its relation to continental neighbors and the Roman Catholic Church, are explored. Examples include *King John*, *Richard II*, and *Henry V*, as discussed in chapters 3, 4 and 5. The notion of Britain comes through as a major theme in *King Lear* and in *Cymbeline*, both of which look at an earlier Britain through dramatization of semimythical figures. In *King Lear*, as explored in chapter 1, this is a pre-Roman Britain, which is divided—a collapse back into a disunited kingdom. Yet in no play more than *Cymbeline* is the question of Britain more apparent. Valerie Wayne has counted that the play "uses forms of 'Britain,' 'Briton,' or 'British' 52 times, far more than any other text of Shakespeare, and it never refers to 'England' or the 'English.'"[4] While there are themes relating to the identity of Britain itself, a key theme concerns the relation of Britain to Rome, and more generally to the outside world. The immediate issue is the question of whether Britain should pay Rome tribute following an invasion fifty years before by Julius Caesar. Rome, by this time, under Augustus Caesar, has conquered most of Western Europe. The exiled Posthumus, now in Rome, declares:

> I do believe—
> Statist though I am none, nor like to be—
> That this will prove a war; and you shall hear
> The legions now in Gallia sooner landed
> In our not-fearing Britain than have tidings
> Of any penny tribute paid.[5]

"Statist" here means statesman, and while Posthumus claims that this is not his vocation, he realizes the potential threat. Britain is fearless and will not pay either one penny or one penny rate of tax on a larger sum, if the term should read 'penny-tribute' as a compound.[6]

Nonetheless, the Roman forces in Gaul may well look to invade again. While the wiser counsel of Posthumus sees the threat of Rome, back in Britain the Queen and her son, Cloten, are urging resistance without perhaps much thought. Cloten first declares that "Britain's a world / By itself";[7] and then the Queen follows with advice for the King:

> Remember sir, my liege,
> The kings your ancestors, together with
> The natural bravery of your isle, which stands
> As Neptune's park, ribbed and paled in
> With oaks unscalable and roaring waters,

With sands that will not bear your enemies' boats,
But suck them up to th'topmast. A kind of conquest
Caesar made here, but made not here his brag
Of 'came and saw and overcame.' With shame—
The first that ever touched him—he was carried
From off our coast, twice beaten, and his shipping,
Poor ignorant baubles, on our terrible seas
Like eggshells moved upon their surges, cracked
As easily 'gainst our rocks.[8]

While not reaching the heights of John of Gaunt's description in *Richard II*, this is still a rich passage invoking national pride in Britain's geography and identity. Neptune has the island of Britain as his estate, which is enclosed ("ribbed") and reserved ("paled") as a park for his use and enjoyment. "Paled" means fenced in with pales. This island is surrounded by oaks—although sometimes emended to "rocks" or "banks"[9]—seas, and quicksand. "Oaks"—the Folio has "oakes"—makes sense because of the image of the deer-park and the pales.[10] These together provide its "natural bravery," its excellence and status, a defiant political stand against the Roman Empire, but enabled in part by natural, physical geography.[11] Caesar only made a "kind of conquest," not close to the successes he achieved elsewhere. Whereas in *Richard II* the sea is a "moat defensive," while the island's "rocky shore beats back the envious siege / Of wat'ry Neptune",[12] here it is more aggressive—destroying the ships, tossed like toys or eggshells, either on the rocks or by sucking them below the surface.[13]

This understanding of Britain is returned to, though with more critical distance, by the King's daughter Innogen (sometimes called Imogen). She is now in Wales, and realizes that returning to the court presents danger, so considers leaving its shores:

Hath Britain all the sun that shines? Day, night,
Are they not but Britain? I'th'world's volume
Our Britain seems as of it but not in't,
In a great pool a swan's nest.[14]

Britain is a part of the world, in all its volume, which is both a book, and the volume of the ocean, but apart from it, a nest in the pool.[15] Unlike the Queen's image of complete isolation, Britain is remote but certainly part of the world. But the image cuts both ways: while safely defended by the sea, hard to invade, and insulated from contagion, Britain is also cut off from the rest of the world, isolated and detached.

There are parallels in another part of the plot, concerning the wager between Posthumus and the Roman Iachimo concerning Innogen's fidelity. Iachimo's deception mirrors the larger political intrigues of the play.[16] As Glenn Clark suggests, "Iachimo threatens to cross both the geographic borders of Britain and the physical borders of Imogen's body."[17] Iachimo even associates himself with Tarquin,[18] who had viewed Lucrece's body in geographical terms in Shakespeare's early poem. There, sleep is seen as "the map of death," and her breasts are described as being "like ivory globes circled with blue, / A pair of maiden worlds unconquered."[19] Yet it is not just Iachimo who sees Innogen as a potential conquest; the idea of her as the property of Posthumus or her father is clear. As Peter Stallybrass describes the situation of women at this time generally, "Economically, she is the fenced-in enclosure of the landlord, her father, or husband."[20]

Even politically, Britain is not the only area resisting Rome. Cymbeline taunts Caius Lucius:

> I am perfect
> That the Pannonians and Dalmatians for
> Their liberties are now in arms, a precedent
> Which not to read would show the Britons cold;
> So Caesar shall not find them.[21]

The Pannonians inhabited lands today part of Hungary, south of the Danube; the Dalmatians the coastal region of modern Croatia, on the Adriatic Sea. Pannonia and Dalmatia were adjoining provinces of the Roman Empire. Nosworthy notes that "the revolt referred to appears to have taken place during the reign of Cymbeline's father, Tenantius," but adds that "it would be both idle and pedantic to probe too deeply among Holinshed's dates."[22] Back in Rome, the senators and tribunes discuss this same revolt:

> This is the tenor of the Emperor's writ:
> That since the common men are now in action
> 'Gainst the Pannonians and Dalmatians,
> And that the legions now in Gallia are
> Full weak to undertake our wars against
> The fall'n-off Britons, that we do incite
> The gentry to this business.[23]

As well as suppressing the revolt there, forces are needed to send to Britain, from the province of Gaul. These legions are mentioned several times in the play, forming both the original invasion by Julius Caesar, and the one led by Caius Lucius.[24]

Yet while Britain is resisting one empire, at the time the play was written it was already beginning an empire of its own. Wayne describes the play as looking, "Janus-like, in two temporal directions at once . . . combining a backward glance at Britain's past with a forward look at England's colonizing future."[25] As Willy Maley notes, "The very legislation that freed England from Roman authority, the Act of Restraint of Appeals (1533), did so by declaring England to be an empire in its own right."[26] Roman here is, of course, the Roman Catholic Church. Clark finds an echo in the suggestion in the play that "Th'imperious seas breeds monsters."[27] But what an empire was at this time is open to debate. Maley quotes Brian Levack, who suggests that "during the sixteenth century the word empire usually designated a sovereign territorial state which was completely independent of the pope and all foreign princes."[28] This may be correct, in relational terms, but is misleading conceptually—neither "territory," "sovereign," nor "state" is quite right at this point. Nonetheless, Maley is right to see the Roman model: "A state forged in Wales and subsequently fitted with parts in Scotland before being exported to Ireland had an obvious blueprint. Britain was made in Rome."[29]

As Maley continues, building on McEachern's work, the irony is that this is a nation being created as an empire.[30] This empire building, here, is through the contiguous colonization of its closest neighbor, Wales. Yet Wales is not yet fully tamed, a kind of wild only partly under British control. Sullivan highlights this point, suggesting that Wales is both separate and incorporated. In the play, "Britain really refers to England, . . . the court of Cymbeline is seen as an English one," yet this is not true in all registers: "On the one hand, Wales is a country distinct from a Britain suggestive of England; on the other it is where 'British' battles against Roman troops take place. Thus, within the logic of the play Wales is simultaneously other than, and conceptually annexed by, Britain."[31]

In Shakespeare's time, King James had made his son the Prince of Wales, in 1610, possibly the date of the play's composition.[32] But the wider project of state-, indeed nation-, formation concerned England and Scotland. Initially, as chapter 3 indicated, this was one king of two kingdoms, even though James had ambitions for more when he invoked Britain. James liked to see himself as the second Brutus, the legendary founder of Britain, the descendant of Aeneas of Troy. James was his successor because he was the first since then to unite the island of Britain.[33] Brutus's wife, notably, was named Innogen; Posthumus Leonatus's name may signify the 1608 "Case of the Post Nati," which confirmed the status of those Scots born after 1603 as citizens of Britain.[34] It has even been suggested that the late claim by the Soothsayer that "from a stately cedar shall be lopped branches, which, being dead many years, shall after revive, be jointed to the old stock, and freshly grow" can be read politically as well as personally.[35] The Soothsayer clarifies that "the lofty cedar, royal Cymbeline, / Personates thee," and the branches his sons,[36] but as Maley has noted, there is another homonym here:

> A "stately cedar" might suggest a state that ceded, in the sense of ceding territory, in this case a state that ceded branches—limbs or members—that are now being grafted back onto the main body of the state.[37]

He continues, developing this theme:

> *Cymbeline* was written at a time when a new British imperial monarch with two sons, one the Duke of Albany, the other Prince of Wales, had effected a union between two warring kingdoms—Scotland and England—and made possible the conquest of a third, Ireland. Where Lear had divided the ancient kingdom of Britain with disastrous consequences, Cymbeline preserves its integrity while keeping the peace with Rome. It is hard not to hear a contemporary resonance in the closing speeches of Shakespeare's king of Britain.[38]

But James also saw parallels with two other figures in British history, mythical or historical, King Arthur and the Earl of Richmond. Richmond, the future King Henry VII, lands at Milford Haven in Shakespeare's *Richard III* on his way to defeat Richard at Bosworth Field.[39] From the house of Lancaster, son of Edmund Tudor, he married Elizabeth of York and so put an end to the Wars of the Roses. Milford Haven thus is a kind of "sacred site" for Tudor politics.[40] Milford Haven is of course an important location in *Cymbeline*, the most frequently named place and where several important episodes occur. But, as Sullivan notes, there is more to Milford Haven than the future Henry VII; it was a landing place for foreign invaders as well as domestic rescuers from tyranny. In the play the invasion is by the Roman army; in Shakespeare's time the threat was from Spain.[41] Innogen describes it as the "blessed Milford," and asks "how Wales was made so happy as / T'inherit such a haven."[42]

Wales thus figures in an important way in the play. Cymbeline asks that the Roman lord Lucius be escorted from the kingdom, but only so far:

"Leave not the worthy Lucius, good my lords, / Till he have crossed the Severn."[43] The river Severn, as was shown in chapter 7, has functioned as something of a natural border between England and Wales. As Pitcher suggests,

> Lucius will have an escort until he crosses the Severn into Wales, a country which is still part of Cymbeline's territory, but which, because of its terrain, is beyond immediate control and a hiding-place for outlaws (among them, Belarius and the princes).[44]

Yet this is significant too, because all the characters we encounter in Wales are actually from the court or from Rome—the seeming Welsh are really the King's sons and his former courtier. The major political plot is crucial here, for Cymbeline's refusal to pay a tribute as a result of Rome's earlier incursion does lead to a new invasion, but one improbably repelled in Wales by a tiny army of Cymbeline's sons and their adoptive, kidnapping father. Belarius, we later learn, was actually banished from Cymbeline's court because of an unjustified accusation that he "was confederate with the Romans." Following his banishment, he says that for "twenty years / This rock and these demesnes have been my world."[45] "Demesnes," "domains" are lands or estates. Belarius notes that he stole Cymbeline's children, "thinking to bar thee of succession as / Thou reft'st me of my lands."[46] Geography figures in the repelled invasion, too, in the conjunction of "A narrow lane, an old man, and two boys";[47] a "strait lane" or "strait pass" described as "ditched, and walled with turf."[48] Belarius had declared to the Romans: "Stand, stand! We have th'advantage of the ground, / The land is guarded. Nothing routs us but / The villainy of our fears."[49] Paradoxically, though Britain defeats Rome, in reconciliation Cymbeline does agree to pay the tribute, and Rome agrees to leave Britain alone.[50] As Pitcher suggests, "Britain in *Cymbeline* is a provincial monarchy on the northern edge of Augustus Caesar's superstate, the Roman Empire."[51] Supporting this peace, the Soothsayer declares:

> For the Roman eagle,
> From south to west on wing soaring aloft,
> Lessened herself, and in the beams o'th'sun
> So vanished; which foreshowed our princely eagle,
> Th'imperial Caesar, should again unite
> His favour with the radiant Cymbeline,
> Which shines here in the west.[52]

Invoking the key imperial imagery of the sun and the eagle, these are tellingly used to describe both Rome and Britain. Cymbeline replies to say that "a Roman and a British ensign wave / Friendly together" as they march through Lud's town—the old name for London, named after Cymbeline's grandfather—and continues:

> And in the temple of great Jupiter
> Our peace we'll ratify, seal it with feasts.
> Set on there. Never was a war did cease,
> Ere bloody hands were washed, with such a peace.[53]

The year 1610, if indeed the date of the play's composition, fits this as well—the "only year, of this period, in which all the European states were at peace."[54] That said, the assassination of France's King Henri IV just a few days before the investiture of the Prince of Wales indicated the religious wars to come.[55] The play thus ends ambiguously: celebrating union, and peace, but not quite the union and peace of James's time. Here it is not England and Scotland, but Britain and Rome.[56]

The Queen's speech in *Cymbeline* is one of the frequent places in which Shakespeare uses the term "pale."[57] A sampling from his work would include the fairies telling Puck of their journey "Over hill, over dale / Thorough bush, thorough brier, / Over park, over pale" in *A Midsummer Night's Dream*;[58] and Adriana complaining of her husband Antipholus in *The Comedy of Errors*: "But, too-unruly deer, he breaks the pale / And feeds from home; poor I am but his stale."[59] A 'stale' is a decoy, a stalking horse, and the point is she has become a fool because of his actions. Previous chapters have highlighted *Hamlet*'s "pales and forts of reason" and how Macbeth imagines himself afraid and confined in the double sense of "pale."[60] In *Troilus and Cressida*, Ulysses speaks to Ajax of his wisdom: "Which, like a bourn, a pale, a shore, confines / Thy spacious and dilated parts."[61]

In the early poem *Venus and Adonis*, Venus tries to distract Adonis from hunting and to share her love. Adonis resists, and returns to hunt, where he dies in an accident. One passage is especially striking:

> 'Fondling,' she saith, 'since I have hemmed thee here
> Within the circuit of this ivory pale,
> I'll be a park, and thou shalt be my deer;

Feed where thou wilt, on mountain or in dale
Graze on my lips, and if those hills be dry
Stray lower, where the pleasant fountains lie.
Within this limit is relief enough,
Sweet bottom-grass and high delightful plain,
Round rising hillocks, brakes obscure and rough,
To shelter thee from tempest and from rain:
Then be my deer, since I am such a park;
No dog shall rouse thee, though a thousand bark.'[62]

This is a rich set of spatial terms and boundary practices. To hem is to enclose, restrain, or bound, in a surrounding perimeter of an "ivory pale." As well as referring to a picket fence, these are her white arms around him. Her body is thus imagined as a park and Adonis as the deer that roam and feed within, a set of sexual and hunting metaphors follow. The "limit" is bounded or enclosed territory, a tract or region; "relief" the act of seeking food; a "bottom" a valley; and "brakes" a bush or thicket. The hunting terms return with the dog that pursues the deer, "rouse" being a hunting term of drawing a beast from its lair.[63] Rackham provides a useful clarification of these terms. The word "park" comes from the Anglo-Saxon *pearroc*, which "was any piece of land within a fence. Later it meant a place for keeping deer."[64] To enclose the deer "they were confined by a *park pale*, a special palisade of cleft-oak stakes."[65]

Toward the end of *The Winter's Tale* the thief Autolycus sings a song with the line "for the red blood reigns in the winter's pale."[66] Some editors have emended "reigns" to "runs," suggesting that red blood runs pale in the winter. But this is unnecessary, as is the change from "in" to "o'er." The line is a masterpiece of concision; one that admits multiple readings. The most obvious, perhaps, is that the red blood of spring emerges from the depths of winter, and that the darkest days are behind us. Or the "pale" is a cheek, a pallor, but with the blood visible within. The red blood dominates or conquers, the red blood perhaps rains down, if the homonym is allowed. Of this line, Pitcher suggests two likely readings: "(1) spring returns because, however long winter is, male seed has the sovereign power of blood in it; (2) the capacity to give and return to life survives harsh challenges."[67] But the pale is, of course, as other plays show, also an area of land, a reserve, domain, or park, and within this, the red blood, perhaps anger or rage, reigns. The king of Sicilia rules, is sovereign, within his realm, in a particular way. Alternatively, on this geographical reading, the line can be understood as a shift: "The red blood reigns in the former domain of winter," as much as "the red blood reigns in place of the pallor of winter."[68]

The term "pale" is used in the histories too—in the garden scene of *Richard II*, and as a description of England in *King John*,[69] as explored in chapters 3 and 4, and in this passage of *Henry VI, Part 1*:

> How are we parked and bounded in a pale—
> A little herd of England's timorous deer,
> Mazed with a yelping kennel of French curs.
> If we be English deer, be then in blood:
> Not rascal-like to fall down with a pinch,
> But rather, moody-mad and desperate stags,
> Turn on the bloody hounds with heads of steel
> And make the cowards stand aloof at bay.
> Sell every man his life as dear as mine
> And they shall find dear deer of us, my friends.
> God and Saint George, Talbot and England's right,
> Prosper our colours in this dangerous fight.[70]

The term "pale" comes, as was shown in chapter 2, from hunting practices, an enclosed part of a forest, set aside for control, and taking its name from the stakes or pickets, the *palus*, that marked them out. The term took on a more general, and political, sense: the Pale of Ireland, mentioned in *The First Part of the Contention* and in Marlowe's *Edward II*,[71] or the Pale of Calais. These demonstrate the importance of territorial control and the extent of jurisdiction. "Pale" is therefore an archaic term for a jurisdiction.

Yet to see territory simply *as* a pale, as a bounded space set aside for a particular purpose, defended from the outside and enclosing the inside, would be reductive. This is one sense, certainly, but not the only one. As this book has demonstrated, Shakespeare shows us how territory encompasses a much richer and wider set of spatial practices. Different plays have been used here to interrogate these multiple aspects: the economic through a reading of *Richard II*, the legal in *Henry V*, and the strategic in *Hamlet* and *Macbeth*. The *Henry VI* plays, *Richard III*, and, especially, *King Lear* look at questions of the division of territories and the political instability that can follow the unraveling of territorial settlements. These aspects relate to themes I have explored elsewhere in contemporary political and historical-conceptual work—in the two books *Terror and Territory* and *The Birth of Territory*.

While every instance of the *words* "territory" and "territories" in Shakespeare has been discussed here, the analysis goes far beyond this. There is a linguistic issue, explored here through a reading of *King John*, but Shakespeare more significantly helps us to understand the concept and practice of territory. Territories and territory are often described with different language, sometimes explicit and sometimes implicit. Yet a profound territorial imagination runs through Shakespeare's plays, as does a concern with territorial practices. As I have tried to show, many of his plays concern exile, banishment, land politics, spatial division, contestation, conquest, and succession. Shakespeare's time of writing is revealing—this was the time that the modern concept of territory was being developed in political thought, and states were actively pursuing territorial accumulation, consolidation, and assessment. With an eye to the historical past—especially republican and imperial Rome and medieval England—and an eye to his own present, Shakespeare helps to dramatize this crucial moment of transition.

These themes are particularly pronounced in the major tragedies, especially *King Lear*, *Hamlet*, and *Macbeth*, discussed in chapters 1 and 2. The English history plays show the contestation of the spaces under the rule of kings, not only in terms of who should rule, but also what the limits of their rule should be, in both the sense of what is permissible and where it applies. In chapters 3, 4, and 5 the discussion focused on *King John*, *Richard II*, *Henry V*, and *Edward III*, but the theme runs through all of Shakespeare's historical dramatizations. In these plays the question of who owns, who rules, and who controls indicates both a range of political and geographical questions about territory, and how we should understand territory in a broader way. Politics and geography need to be understood in relation to, at least, economic, legal, and strategic questions.

Yet while economic, legal, and strategic concerns around the politics of land are long-standing historically, two key developments of Shakespeare's time were their playing out in new spaces and the development of new technologies. The new spaces are explored in relation to the colonial conquest of territories in chapter 6. Measuring and calculative technologies are a focus of many of Shakespeare's plays, but the specific focus here has been its application to navigation, cartography, military operations, and surveying. These themes are explored in chapter 7 through a number of plays, with a particularly detailed reading of the dispute over a map division in *Henry IV, Part 1*. That play forces an examination of the physical landscape over which territory is imposed.

The discussion of the colonial context and of this scene in *Henry IV, Part 1* is an example of how interrogating Shakespeare's plays with these issues in mind has required me to deal with themes I had insufficiently addressed in previous work. In *The Tempest*, but also in a number of plays set in the eastern Mediterranean, Shakespeare explores colonial projects, both in his own time and in antiquity, in detail. *Henry IV, Part 1* turns our attention to the physical landscape and its interaction with human agency. In *Coriolanus* the corporeal is a significant theme of bodies in places and places embodied, the biopolitics of territory, which is a focus of the discussion in chapter 8. The colonial, corporeal, and geophysical thus deepen and extend the work I have done elsewhere, even as other plays provide illustrations of previous arguments. Finally, in chapter 9, through an examination of several plays, especially *Titus Andronicus* and *As You Like It*, the book examined what it might mean to be outside territory. This is, again, an instance of the particular time of Shakespeare's writing. While such places were fast disappearing, he is still able to imagine places that were not yet spaces, not yet transformed by the political technologies that were being shaped as he was writing.

In Shakespeare's discussions of these political geographical issues, territory becomes an ever wider set of concerns. There are many more aspects to territory than the narrow, bounded definition that is often taken to be definitive and ahistorical. *Shakespearean Territories* has shown just how far such a theme operates in his writings. Territory must be thought beyond the pale.

REFERENCES TO SHAKESPEARE'S PLAYS

Unless otherwise noted, references to Shakespeare's plays are to the editions in the Arden Shakespeare Third Series, except for *All's Well That Ends Well*, *King John*, and *Measure for Measure*, where I have used the editions in the Arden Second Series, and *Edward III*, the Cambridge edition. Where editors have not, especially in Second Series texts, I have emended the contracted ending "'d" to "ed." Full references to texts used are provided on first occurrence in the notes.

NOTES

INTRODUCTION

1. Generally, see Jonathan Dollimore and Alan Sinfield, eds., *Political Shakespeare: Essays in Cultural Materialism,* 2nd ed. (Manchester: Manchester University Press, 1994); and David Armitage, Conal Condren, and Andrew Fitzmaurice, eds., *Shakespeare and Early Modern Political Thought* (Cambridge: Cambridge University Press, 2009). On the histories, see Peter Saccio, *Shakespeare's English Kings: History, Chronicle, and Drama,* 2nd ed. (1977; Oxford: Oxford University Press, 2000); Alexander Leggatt, *Shakespeare's Political Drama: The History Plays and the Roman Plays* (London: Routledge, 1988); John Julius Norwich, *Shakespeare's Kings,* (London: Penguin, 1999); Hugh Grady, *Shakespeare, Machiavelli, and Montaigne: Power and Subjectivity from Richard II to Hamlet* (Oxford: Oxford University Press, 2002); Jean E. Howard, "Dramatic Traditions and Shakespeare's Political Thought," in David Armitage, ed., *British Political Thought in History, Literature, and Theory, 1500–1800* (Cambridge: Cambridge University Press, 2006), 129–44; and David Womersley, *Divinity and the State* (Oxford: Oxford University Press, 2010), pt. 3. A more politically and theoretically reactionary account can be found in Howard Erskine-Hill, *Poetry and the Realm of Politics: Shakespeare to Dryden* (Oxford: Clarendon Press, 1996), pt. 1.

2. Andrew Gurr, "Shakespeare's Localities," in Agostino Lombardo, ed., *Shakespeare a Verona e nel Veneto* (Verona: Accademia di agricoltura scienze e lettere, 1987), 55–66.

3. *The Comedy of Errors,* ed. Kent Cartwright (London: Arden Shakespeare, 2017), III.ii.114–44.

4. Charles Whitworth, "Introduction," in *The Comedy of Errors,* ed. Charles Whitworth (Oxford: Oxford University Press, 2002), 52. On this generally, see Richard Levin, "Anatomical Geography in 'The Tempest,' IV, 1, 235–38," *Notes and Queries,* n.s., 11 (1964): 142–46.

5. Whitworth, note to *The Comedy of Errors,* ed. Whitworth, 132. For a fuller discussion, see Paul G. Chamberlain, "The Shakespearean Globe: Geometry, Optics, Spectacle," *Environment and Planning D: Society and Space* 19 (2001): 317–33; Lesley B. Cormack, "Glob(al) Visions: Globes and Their Publics in Early Modern Europe," in Bronwen Wilson

and Paul Yachnin, eds., *Making Publics in Early Modern Europe: People, Things, Forms of Knowledge* (London: Routledge, 2010), 139–56.

6. See A. J. Hoenselaars, "Mapping Shakespeare's Europe," in A. J. Hoenselaars, ed., *Reclamations of Shakespeare* (Amsterdam: Rodopi, 1994), 223–48, esp. 240–41, and figs. 4 and 5.

7. See Ros King, "Introduction," in *The Comedy of Errors*, ed. T. S. Dorsch (Cambridge: Cambridge University Press, 2004), 38–40; Gavin Hollis, *The Absence of America: The London Stage, 1576–1642* (Oxford: Oxford University Press, 2015).

8. John Gillies, *Shakespeare and the Geography of Difference* (Cambridge: Cambridge University Press, 1994); Garrett A. Sullivan Jr., *The Drama of Landscape: Land, Property, and Social Relations on the Early Modern Stage* (Stanford: Stanford University Press, 1998); Steve Mentz, *At the Bottom of Shakespeare's Ocean* (London: Bloomsbury, 2009); Kristen Poole, *Supernatural Environments in Shakespeare's England: Spaces of Demonism, Divinity, and Drama* (Cambridge: Cambridge University Press, 2011).

9. See, for example, Jane Kingsley-Smith, *Shakespeare's Drama of Exile* (London: Palgrave Macmillan, 2003); Joan Fitzpatrick, *Shakespeare, Spenser, and the Contours of Britain: Reshaping the Island Archipelago* (Hatfield: University of Hertfordshire Press, 2004); Lisa Hopkins, *Shakespeare on the Edge: Border-Crossing in the Tragedies and the "Henriad"* (Aldershot: Ashgate, 2005); Mary Floyd-Wilson and Garrett A. Sullivan Jr., eds., *Environment and Embodiment in Early Modern England* (London: Palgrave Macmillan, 2007); Monica Matei-Chesnoiu, *Geoparsing Early Modern English Drama* (London: Palgrave Macmillan, 2015). For shorter discussions, see Walter Cohen, "The Undiscovered Country: Shakespeare and Mercantile Geography," in Jean E. Howard and Scott Cutler Shershow, eds., *Marxist Shakespeares* (London: Routledge, 2001), 128–58; Dan Brayton, "Angling in the Lake of Darkness: Possession, Dispossession, and the Politics of Discovery in *King Lear*," *English Literary History* 70, no. 2 (2003): 399–426; Jerry Brotton, "Tragedy and Geography," in Richard Dutton and Jean E. Howard, eds., *A Companion to Shakespeare's Works* (Oxford: Blackwell, 2003), 1:219–40; Garrett Sullivan Jr., "Shakespeare's Comic Geographies," in Dutton and Howard, *A Companion to Shakespeare's Works*, 182–99; François Laroque, "Shakespeare's Imaginary Geography," in Andrew Hadfield and Paul Hammond, eds., *Shakespeare and Renaissance Europe* (London: Arden Shakespeare, 2005), 193–219; Chee-Seng Lim, "Crossing the Dotted Line: Shakespeare and Geography," in Christa Janson, Lena Cowen Orlin, and Stanley Wells, eds., *Shakespeare without Boundaries: Essays in Honor of Dieter Mehl* (Newark: University of Delaware Press, 2011), 253–63; and Sarah Dustagheer, "Shakespeare and the 'Spatial Turn,'" *Literature Compass* 10, no. 7 (2013): 570–81.

10. See, among many others, Richard Helgerson, "The Land Speaks: Cartography, Chorography, and Subversion in Renaissance England," in Stephen Greenblatt, ed., *Representing the English Renaissance* (Berkeley: University of California Press, 1983), 327–61 (reprinted in Helgerson, *Forms of Nationhood: The Elizabethan Writing of England* [Chicago: University of Chicago, 1992], chap. 3); Frank Lestringant, *L'atelier du cosmographe, ou l'image du monde à la Renaissance* (Paris: Albin Michel, 1991); Bernhard Klein, *Maps and the Writing of Space in Early Modern England and Ireland* (Houndmills: Palgrave, 2001); Andrew Hiscock, *The Uses of This World: Thinking Space in Shakespeare, Marlowe, Cary, and Jonson* (Cardiff: University of Wales Press, 2004); John Rennie Short, *Making Space: Revisioning the World, 1475–1600* (Syracuse: Syracuse University Press, 2004); Adam Max

Cohen, *Shakespeare and Technology: Dramatizing Early Modern Technological Revolutions* (London: Palgrave, 2006); and Angus Fletcher, *Time, Space, and Motion in the Age of Shakespeare* (Cambridge, MA: Harvard University Press, 2007). For a discussion of how Shakespeare was read as a geographer, see Robert J. Mayhew, "Was William Shakespeare an Eighteenth-Century Geographer? Constructing Histories of Geographical Knowledge," *Transactions of the Institute of British Geographers*, n.s., 23, no. 1 (1998): 21–37.

11. For a sampling of the literature on this theme, see Steven Mullaney, *The Place of the Stage: License, Play, and Power in Renaissance England* (Chicago: University of Chicago Press, 1988); Russell West, *Spatial Representations and the Jacobean Stage: From Shakespeare to Webster* (London: Palgrave, 2002); Tiffany Stern, *Making Shakespeare: From Stage to Page* (London: Routledge, 2004); Henry S. Turner, *The English Renaissance Stage: Geometry, Poetics, and the Practical Spatial Arts, 1580–1630* (Oxford: Oxford University Press, 2006); D. J. Hopkins, *City/Stage/Globe: Performance and Space in Shakespeare's London* (New York: Routledge, 2008); and Adam Zucker, *Places of Wit in Early Modern English Comedy* (Cambridge: Cambridge University Press, 2011); Ina Habermann and Michelle Witen, eds., *Shakespeare and Space: Theatrical Explorations of the Spatial Paradigm* (London: Palgrave, 2016).

12. Stuart Elden, *The Birth of Territory* (Chicago: University of Chicago Press, 2013); Elden, *Terror and Territory: The Spatial Extent of Sovereignty* (Minneapolis: University of Minnesota Press, 2009). For a succinct summary of the conceptual argument, see Elden, "Land, Terrain, Territory," *Progress in Human Geography* 34, no. 6 (2010): 799–817.

13. Space, of course, is anything but simple. My thinking on this has been shaped, above all, by Henri Lefebvre. For a discussion, see Stuart Elden, *Understanding Henri Lefebvre: Theory and the Possible* (London: Continuum, 2004), esp. chaps. 5 and 6; and Neil Brenner and Stuart Elden, "Henri Lefebvre on State, Space, Territory," *International Political Sociology* 3, no. 3 (2009): 353–77.

14. Michel Foucault, *Surveiller et punir: Naissance de la prison* (Paris: Gallimard/Tel, 1975), 31, 34, 39; Foucault, *Discipline and Punish: The Birth of the Prison*, trans. Alan Sheridan (London: Penguin, 1977), 24, 26, 30.

15. See, among other references, Michel Foucault, "The Meshes of Power," trans. Gerald Moore, in Jeremy W. Crampton and Stuart Elden, eds., *Space, Knowledge, and Power: Foucault and Geography* (Aldershot: Ashgate, 2007), 153–62; Foucault, *Security, Territory, Population: Lectures at the Collège de France 1977–1978*, trans. Graham Burchell (Basingstoke: Palgrave Macmillan, 2007), 289–90, 312; Foucault, "The Political Technology of Individuals," in *Power: Essential Works*, vol. 3, ed. James D. Faubion (London: Penguin, 2000), 403–17.

16. For a discussion along these lines, see Stuart Elden, "Governmentality, Calculation, Territory," *Environment and Planning D: Society and Space* 25, no. 3 (2007): 562–80; and Elden, "How Should We Do the History of Territory?," *Territory, Politics, Governance* 1, no. 1 (2013): 5–20. I discuss Foucault's work in detail in *Foucault's Last Decade* (Cambridge: Polity, 2016); and *Foucault: The Birth of Power* (Cambridge: Polity, 2017).

17. Elden, *The Birth of Territory*, 17; and the coda in this book.

18. On his language generally, see David Crystal, *Think on My Words: Exploring Shakespeare's Language* (Cambridge: Cambridge University Press, 2008), esp. chap. 7; and Frank Kermode, *Shakespeare's Language* (London: Penguin, 2000).

19. Gottfried Leibniz, "Extrait d'une Lettre à l'Auteur du *Journal des Sçavans,*" in *Sämliche Schriften und Briefe,* ed. Leibniz-Archiv Hannover (Darmstadt: Otto Reichl, 1923–), ser. 4, 2:360.

20. Robert Cawdrey, *The First English Dictionary, 1604: Robert Cawdrey's A Table Alphabeticall* (Oxford: The Bodleian Library, 2007), 147.

21. Jean E. Howard, "Shakespeare, Geography, and the Work of Genre on the Early Modern Stage," *Modern Language Quarterly* 64, no. 3 (2003): 299–322, 299.

22. Kingsley-Smith, *Shakespeare's Drama of Exile,* 1; see Robert F. Gorman, "Revenge and Reconciliation: Shakespearean Models of Exile and Banishment," *International Journal of Refugee Law* 2, no. 2 (1990): 211–37.

23. Anthony Munday, Henry Chettle, et al., *Sir Thomas More,* ed. John Jowett (London: Arden Shakespeare, 2011), vi.85–87.

24. *Much Ado About Nothing,* ed. Claire McEachern, rev. ed. (2006; London: Arden Shakespeare, 2016), I.i.5–6.

25. *All's Well That Ends Well,* ed. G. K. Hunter (London: Arden Shakespeare, 1959), II.iii.269.

26. *Troilus and Cressida,* ed. David Bevington, rev. ed. (1998; London: Arden Shakespeare, 2015).

27. For an excellent analysis, see James Shapiro, *Contested Will: Who Wrote Shakespeare?* (London: Faber and Faber, 2010).

28. For a useful discussion of the idea of "author" in Shakespeare's time, see Jeffrey Knapp, *Shakespeare Only* (Chicago: University of Chicago Press, 2009); and Lene B. Petersen, *Shakespeare's Errant Texts: Textual Form and Linguistic Style in Shakespearean "Bad" Quartos and Co-authored Plays* (Cambridge: Cambridge University Press, 2010), esp. chap. 1. On the "text," see E. A. J. Honigmann, *The Stability of Shakespeare's Text* (Lincoln: University of Nebraska Press, 1965). There are multiple biographies, of which I found the following the most useful: E. A. J. Honigmann, *Shakespeare: The "Lost Years,"* 2nd ed. (1985; Manchester: Manchester University Press, 1998); Jonathan Bate, *Soul of the Age: The Life, Mind, and World of William Shakespeare* (London: Penguin, 2009); Stephen Greenblatt, *Will in the World: How Shakespeare Became Shakespeare* (London: Pimlico, 2005).

29. See Kenneth Muir, *Shakespeare as Collaborator* (London: Methuen, 1960); Jeffrey Masten, *Textual Intercourse: Collaboration, Authorship, and Sexualities in Renaissance Drama* (Cambridge: Cambridge University Press, 1997); Brian Vickers, *Shakespeare, Co-Author: A Historical Study of the Five Collaborative Plays* (Oxford: Oxford University Press, 2002); Stanley Wells, *Shakespeare and Co.: Christopher Marlowe, Thomas Dekker, Ben Jonson, Thomas Middleton, John Fletcher, and the Other Players in His Story* (London: Allen Lane, 2006); Hugh Craig and Arthur F. Kinney, *Shakespeare, Computers, and the Mystery of Authorship* (Cambridge: Cambridge University Press, 2009); the special issue of *Shakespeare Survey* 67 (2014); and Peter Kirwan, *Shakespeare and the Idea of Apocrypha: Negotiating the Boundaries of the Dramatic Canon* (Cambridge: Cambridge University Press, 2015). The most recent edition of these plays is William Shakespeare and Others, *Collaborative Plays,* ed. Jonathan Bate and Eric Rasmussen (London: Palgrave Macmillan, 2013). Will Sharpe's essay in that collection, "Authorship and Attribution," 641–745, is a very useful survey of debates and detailed discussion of the arguments for and against each included play.

30. Richard Proudfoot, "*The Reign of King Edward the Third* (1596) and Shakespeare," *Proceedings of the British Academy* 71 (1985): 169–85, 184; Proudfoot, *Double Falsehood, or, The Distressed Lovers*, ed. Brean Hammond (London: Arden Shakespeare, 2010).

31. Craig and Kinney, *Shakespeare, Computers, and the Mystery of Authorship*, chaps. 4 and 6; *Arden of Faversham*, ed. Martin White, rev. ed. (London: AC & Black, 2007); Thomas Kyd, *The Spanish Tragedy*, ed. Clara Calvo and Jésus Tronch (London: Arden Shakespeare, 2011).

32. Eric Sams, "Introduction," in Eric Sams, ed., *Shakespeare's Lost Play: Edmund Ironside* (London: Fourth Estate, 1985), 26.

33. Sams, "Introduction," 52. This claim is rejected in several accounts, notably Craig and Kinney, *Shakespeare, Computers, and the Mystery of Authorship*, chap. 5.

34. For useful discussions of debates about textual transmission, see Gary Taylor, "General Introduction," in Stanley Wells and Gary Taylor, with John Jowett and William Montgomery, *William Shakespeare: A Textual Companion* (Oxford: Clarendon Press, 1987), 1–68; Paul Werstine, "Narratives about Printed Shakespeare Texts: 'Foul Papers' and 'Bad' Quartos," *Shakespeare Quarterly* 41, no. 1 (Spring 1990): 65–86; Werstine, "Touring and the Construction of Shakespeare Textual Criticism," in Lawrie E. Maguire and Thomas L. Berger, eds., *Textual Formations and Reformations* (Newark: University of Delaware Press, 1998), 45–66; and Barbara A. Mowat, "The Problem of Shakespeare's Text(s)," in Maguire and Berger, *Textual Formations and Reformations*, 131–48.

35. For full discussions, see Stephen Greenblatt, *Renaissance Self-Fashioning: From More to Shakespeare* (Chicago: University of Chicago Press, 1980), chap. 5; Garrett A. Sullivan Jr., "Geography and Identity in Marlowe," in Patrick Cheney, ed., *The Cambridge Companion to Christopher Marlowe* (Cambridge: Cambridge University Press, 2004), 231–44; Jacques Lezra, "Geography and Marlowe," in Emily C. Bartels and Emma Smith, eds., *Christopher Marlowe in Context* (Cambridge: Cambridge University Press, 2013), 125–37.

36. Jonathan Bate, "Shakespeare's Islands," in Tom Clayton, Susan Brock, and Vicente Forés, eds., *Shakespeare and the Mediterranean: The Selected Proceedings of the International Shakespeare Association World Congress Valencia, 2001* (Newark: University of Delaware Press, 2004), 298–301, 290.

37. For instances of the word "territories," see *Tamburlaine the Great, Part I*, IV.iii.14; *Tamburlaine the Great, Part II*, I.i.166, in *Doctor Faustus and Other Plays*, ed. David Bevington and Eric Rasmussen (Oxford: Oxford University Press, 1995).

38. *Tamburlaine the Great, Part I*, IV.iv.78–81. For a discussion of the calculative and cartographic geographies of this play, see Garrett A. Sullivan Jr., "Space, Measurement, and Stalking *Tamburlaine*," *Renaissance Drama* 28 (1997): 3–27; and Crystal Bartolovich, "Putting *Tamburlaine* on a (Cognitive) Map," *Renaissance Drama*, n.s., 28 (1997): 29–72.

39. See Bevington and Rasmussen, notes to *Doctor Faustus and Other Plays*, 414.

40. Michael Drayton, *Poly-Olbion: A Choreographicall Description of all the Tracts, Rivers, Mountains, Forests and Other Parts of this Renowned Isle of Great Britain* (London: Iohn Marriott, John Grismand and Thomas Dewe, 1622). On this text, see the Poly-Olbion Project at http://poly-olbion.exeter.ac.uk/.

CHAPTER ONE

1. *Henry VI, Part 1,* ed. Edward Burns (London: Arden Shakespeare, 2000), V.ii.167.

2. *Henry VI, Part 1,* V.ii.178–80.

3. *Henry VI, Part 1,* V.iii.102–12.

4. *Henry VI, Part 1,* V.iii.138–43.

5. Michael Taylor, note to *Henry VI, Part One,* ed. Michael Taylor (Oxford: Oxford University Press, 2003), 242.

6. *Henry VI, Part 2,* ed. Ronald Knowles (London: Arden Shakespeare, 1999), I.i.50–51.

7. *Henry VI, Part 2,* III.i.84–85.

8. *Henry VI, Part 2,* III.ii.245.

9. *Henry VI, Part 3,* ed. John D. Cox and Eric Rasmussen (London: Arden Shakespeare, 2001), I.i.107–9.

10. *Henry VI, Part 3,* I.i.110.

11. *Henry VI, Part 3,* I.i.111–12.

12. See Stephen Greenblatt, *Learning to Curse: Essays in Early Modern Culture* (New York: Routledge, 1990), 168. See also Thomas Cartelli, "Jack Cade in the Garden: Class Consciousness and Class Conflict in 2 *Henry VI,*" in Richard Burt and John Michael Archer, eds., *Enclosure Acts: Sexuality, Property, and Culture in Early Modern England* (Ithaca: Cornell University Press, 1994), 48–67; and Sullivan, *The Drama of Landscape,* chap. 6, esp. 226–28.

13. *Richard III,* ed. James R. Siemon (London: Arden Shakespeare, 2009), V.v.19–28, 35–41.

14. Note to *The Norton Shakespeare based on the Oxford Edition,* ed. Stephen Greenblatt, Walter Cohen, Jean E. Howard, and Katharine Eisaman Maus (New York: W.W. Norton, 1997), 595.

15. Jacques Lezra, *Wild Materialism: The Ethic of Terror and the Modern Republic* (New York: Fordham University Press, 2010), 72–79. Most of this text is reproduced from Lezra, "*Phares,* or Divisible Sovereignty," *Religion and Literature* 28, no. 3 (2006): 13–39.

16. Lezra, *Wild Materialism,* 72.

17. *The Tragedy of King Richard the Third: Parallel Texts of the First Quarto and the First Folio with Variants of the Early Quartos,* ed. Kristian Smidt (Oslo: Universitetforlaget, 1969), 220. Its spelling is modernized in the critical edition: *The First Quarto of King Richard III,* ed. Peter Davison (Cambridge: Cambridge University Press, 1996).

18. Lezra, *Wild Materialism,* 73.

19. Lezra, *Wild Materialism,* 73.

20. Lezra, *Wild Materialism,* 73–74.

21. Lezra, *Wild Materialism,* 75.

22. Lezra, *Wild Materialism,* 74.

23. *Henry VI, Part 2,* III.i.84–85.

24. See also *Richard III,* II.ii.47–48: "so much interest have I in thy sorrow / As I had title in thy noble husband."

25. *King Lear,* ed. R. A. Foakes (London: Arden Shakespeare, 1997), I.i.33.

26. *King Lear,* I.i.35–43.

27. The Arden Second Series (ed. Kenneth Muir [London: Routledge, 1972]) and Third Series (ed. R. A. Foakes) are conflated, as is the version edited by George Hunter, London: Penguin, 2005, though this tends to prefer the Folio text (see "An Account of the Text," 134). *King Lear,* ed. Stanley Wells (Oxford: Oxford University Press, 2000), is based on the Folio, while *The Tragedy of King Lear,* ed. Jay L. Halio (Cambridge: Cambridge University Press, 1992), is based on the Quarto. *The Norton Shakespeare* provides parallel texts, along with a "conflated" version.

28. See Steven Urkowitz, *Shakespeare's Revision of King Lear* (Princeton: Princeton University Press, 1980); Gary Taylor and Michael Warren, eds., *The Division of the Kingdoms: Shakespeare's Two Versions of "King Lear"* (Oxford: Clarendon Press, 1983); Mark Jay Mirsky, *The Absent Shakespeare* (Rutherford: Fairleigh Dickinson University Press, 1994), chaps. 2 and 3; and R. A. Foakes, *Hamlet versus Lear: Cultural Poetics and Shakespeare's Art* (Cambridge: Cambridge University Press, 1993), chaps. 4 and 7. For a critique, see Sidney Thomas, "The Integrity of *King Lear,*" *Modern Language Review* 90, no. 3 (1995): 572–84; and Brian Vickers, *The One King Lear* (Cambridge, MA: Harvard University Press, 2016).

29. I have used the parallel text in *The Complete King Lear, 1608–1623,* ed. Michael Warren (Berkeley: University of California Press, 1989), vol. 1, but have modernized the spelling.

30. On this generally, see Frederick T. Flahiff, "Lear's Map," *Cahiers Élisabéthain: Études sur la Pré-renaissance et la Renaissance Anglaises* 30 (1986): 17–33; Terence Hawkes, "Lear's Maps," in *Meaning by Shakespeare* (London: Routledge, 1992), chap. 6; Gillies, *Shakespeare and the Geography of Difference,* chap. 2; Gillies, "The Scene of Cartography in *King Lear,*" in Andrew Gordon and Bernhard Klein, eds., *Literature, Mapping, and the Politics of Space in Early Modern Britain* (Cambridge: Cambridge University Press, 2001), 109–37; Philip Armstrong, "Spheres of Influence: Cartography and the Gaze in Shakespearean Tragedy and History," *Shakespeare Studies* 23 (1997): 39–70; and Sullivan, *The Drama of Landscape,* chap. 3.

31. *King Lear,* I.i.56–57.

32. *King Lear,* I.i.69–70, 72–74. A variant noted by Muir substitutes "spacious" for "precious"; see Muir, note to *King Lear,* 7.

33. Henry S. Turner, "*King Lear* Without: The Heath," *Renaissance Drama* 28 (1997): 161–83, 172.

34. *King Lear,* I.i.63–67. The words "and with champains riched, / With plenteous rivers" are not in the Quarto.

35. *King Lear,* I.i.79–82.

36. *King Lear,* I.i.83–85. These lines are not in the Quarto.

37. *King Lear,* I.i.85–86. Emily W. Leider, "Plainness of Style in *King Lear,*" *Shakespeare Quarterly* 21, no. 1 (1970): 45–53, 46, suggests that "space, validity, and pleasure" is, like "rule, / Interest of territory, cares of state," a grouping of "three abstractions," which she interprets as related to the "three requests . . . responses . . . bequests, a triple division of land."

38. *King Lear,* I.i.87–90.

39. Turner, "*King Lear* Without," 178; Peter L. Rudnytsky, "'The Dark and Vicious Place': The Dread of the Vagina in *King Lear,*" *Modern Philology* 96, no. 3 (1999): 291–311;

and Edward Grant, *Much Ado About Nothing: Theories of Space and Vacuum from the Middle Ages to the Scientific Revolution* (Cambridge: Cambridge University Press, 1981).

40. *King Lear,* I.i.93.

41. *King Lear,* I.i.99–100.

42. *King Lear,* I.i.101–2. See also *Othello,* ed. E. A. J. Honigmann (London: Arden Shakespeare, 1997), I.iii.180–89, where Desdemona declares a "divided duty" between her father and her husband.

43. *King Lear,* I.i.109.

44. *King Lear,* I.i.1–2.

45. *King Lear,* I.i.3–6.

46. James Shapiro, *1606: William Shakespeare and the Year of Lear* (London: Faber & Faber, 2015).

47. Gillies, "The Scene of Cartography," 116. Prince Henry died before his father, in 1612, with his brother Charles becoming heir. On the wider context, see Alvin Kernan, *Shakespeare, the King's Playwright: Theatre in the Stuart Court, 1603–1613* (New Haven: Yale University Press, 1995).

48. *King Lear,* I.i.43–44.

49. Harry V. Jaffa, "The Limits of Politics: *King Lear* I.I," in Allan Bloom with Harry V. Jaffa, *Shakespeare's Politics* (New York: Basic Books, 1964), 113–45, at 122. There is a similar division into three proposed by the forces aligned against the King in *Henry IV, Part 1* (see chapter 7).

50. Jaffa, "The Limits of Politics," 123–24.

51. *King Lear,* I.i.128–29.

52. *King Lear,* I.i.150.

53. William Dodd, "Impossible Worlds: What Happens in King Lear 1.1?," *Shakespeare Quarterly* 50, no. 4 (1999): 477–507, 504. The Quarto has "reuerse thy doome" instead, which Dodd suggests is more appropriate in this context (504 n. 105). That line is repeated in the Quarto some fifteen lines later, whereas the Folio now has "revoke thy gift."

54. *King Lear,* I.i.244–46.

55. *King Lear,* I.i.249–50.

56. *King Lear,* I.i.243, 258–59.

57. *King Lear,* I.i.264–68.

58. Francis Barker, *The Culture of Violence: Tragedy and History* (Chicago: University of Chicago Press, 1993), 5; and more generally 3–7.

59. Martin Orkin, *Shakespeare against Apartheid* (Craighall: Ad. Donker, 1987), chap. 4; Nicholas Visser, "Shakespeare and Hanekom, *King Lear* and Land," *Textual Practice* 11, no. 1 (1997): 25–37.

60. *Ran,* directed by Akira Kurosawa, 1985. On *Ran,* see James Goodwin, *Akira Kurosawa and Intertextual Cinema* (Baltimore: Johns Hopkins University Press, 1994), 191–216.

61. Visser, "Shakespeare and Hanekom, *King Lear* and Land," 29.

62. Margreta de Grazia, *Hamlet without Hamlet* (Cambridge: Cambridge University Press, 2007), 44.

63. Flahiff, "Lear's Map," 21, suggests that the very name Gloucester, deriving from the Saxon, meant "to cut or separate—division." Flahiff's source is Samuel D. Rudder, *A*

New History of Gloucestershire (Cirencester: Samuel Rudder, 1779), 19, but this makes it clear that "Gloucester" means "a handsome city," and it is "shire," which is a division. On "Gloucester," see Rudder, 82–83. It has equally been suggested that the names Ed-gar and Ed-mund point to God and the world, respectively. See Zdravko Plannic, "'. . . this scattered kingdom': A Study of *King Lear*," *Interpretation* 29, no. 2 (2001–2): 171–85, at 173.

64. *King Lear*, I.ii.16, 181. This reading is indebted to Visser, "Shakespeare and Hanekom, *King Lear* and Land," 29–30.

65. *King Lear*, II.i.57.

66. *King Lear*, II.i.67.

67. *King Lear*, II.i.79–85.

68. It is worth noting that "space" appears only four times in the play: it is used once each by Goneril and Lear in the opening scene in passages quoted above; and once each from the mouths of Edgar (IV.vi.266) and Edmund (V.iii.54). On Lear and Gloucester's failures to read their relations, see Greenblatt, *Learning to Curse*, chap. 5.

69. Foakes, note to *King Lear*, 221.

70. *King Lear*, I.iv.130.

71. *King Lear*, I.iv.131–32.

72. Paul W. Kahn, *Law and Love: The Trials of King Lear* (New Haven: Yale University Press, 2000), 35.

73. *King Lear*, I.iv.137, 139.

74. *King Lear*, I.iv.141–43.

75. *King Lear*, I.iv.148–54.

76. *King Lear*, IV.ii.57. While "noiseless" probably means unprepared, Muir, note to *King Lear*, 147, suggests that it means the drum has not yet sounded, as the previous line "Where's thy drum?" suggests, war has not been engaged.

77. *King Lear*, V.i.25.

78. *King Lear*, III.vii.2–3; IV.ii.4. In III.iii.12–13 the Folio has "There is part of a power already footed" in place of the Quarto's "already landed." See also III.vii.45: "footed in the kingdom." See Foakes, *Hamlet versus Lear*, 106–7; Mirsky, *The Absent Shakespeare*, 19–20; and above all Gary Taylor, "The War in 'King Lear,'" *Shakespeare Survey* 33 (1980): 27–34.

79. These questions are posed by Kenneth Muir, "Introduction," in *King Lear*, ed. Muir, xiii-lviii, xlvi.

80. *King Lear*, IV.iv.23–29.

81. Muir, "Introduction," *King Lear*, xlvi.

82. Muir, "Introduction," *King Lear*, xxx.

83. W. W. Greg, "Time, Place, and Politics in 'King Lear,'" *Modern Language Review* 35, no. 4 (1940): 431–46, 439–40.

84. Greg, "Time, Place, and Politics in 'King Lear,'" 441.

85. *King Lear*, III.iii.9–11 (Gloucester); III.v.10–12 (Edmund/Cornwall); III.vii.2–3 (Cornwall).

86. Greg, "Time, Place, and Politics in 'King Lear,'" 443 n. 1.

87. Greg, "Time, Place, and Politics in 'King Lear,'" 444.

88. Greg, "Time, Place, and Politics in 'King Lear,'" 445.

89. *King Lear,* IV.vii.76.

90. The Quarto has no scene divisions, but it would come between scenes ii and iii of the Folio, and appears as scene iii in most composite editions.

91. *King Lear,* III.i.19–25. The Quarto text is in a note to *King Lear,* ed. Foakes, 260.

92. Taylor, "The War in 'King Lear,'" 32.

93. Muir, "Introduction," *King Lear,* xiv.

94. Taylor, "The War in 'King Lear,'" 31.

95. Geoffrey of Monmouth, *The History of the Kings of Britain,* trans. Lewis Thorpe (Harmondsworth: Penguin, 1966), II.11.

96. On the links and differences between the two texts, see Mark Allen McDonald, *Shakespeare's "King Lear" with "The Tempest": The Discovery of Nature and the Recovery of Classical Natural Right* (Lanham, MD: University Press of America, 2004), 221.

97. There are several appendices to the Muir edition, for instance, providing relevant passages.

98. Thomas Norton and Thomas Sackville, *Gorboduc, or, Ferrex and Porrex,* ed. Irby B. Cauthen Jr. (1562; Lincoln: University of Nebraska Press, 1970). For a comprehensive sampling, see Russell A. Fraser and Norman Rabkin, eds., *Drama of the English Renaissance I: The Tudor Period* and *II: The Stuart Period* (New York: Macmillan, 1976).

99. Franco Moretti, *Signs Taken for Wonders: Essays in the Sociology of Literary Forms,* trans. Susan Fischer, David Forgacs, and David Miller (London: Verso, 1983), 43; see Irby B. Cauthen Jr., "Introduction," in *Gorboduc,* xi-xxx, xiii.

100. Moretti, *Signs Taken for Wonders,* 50.

101. *Gorboduc,* V.ii.

102. *Gorboduc,* "The Argument of the Tragedy."

103. Geoffrey of Monmouth, *The History of the Kings of Britain,* II.xvi; see Cauthen, "Introduction," *Gorboduc,* xiv.

104. Foakes, *Hamlet versus Lear,* 224.

105. Muir, "Introduction," *King Lear,* xviii.

106. Dodd, "Impossible Worlds," 481; see Christopher Haigh, *Elizabeth I,* 2nd ed. (London: Longman, 2000), 24–26, 111, 160–61.

107. Dodd, "Impossible Worlds," 482.

108. Dodd, "Impossible Worlds," 484–86. This provides a detailed and extensively referenced discussion of the relations, similarities, and differences.

109. Orkin, *Shakespeare against Apartheid,* 147. The passage, discussed above, is in *King Lear,* I.iv.137–43. On this generally, see Gary Taylor, "Monopolies, Show Trials, Disaster, and Invasion: *King Lear* and Censorship," in Taylor and Warren, *The Division of the Kingdoms,* 75–119.

110. Orkin, *Shakespeare against Apartheid,* 165–66.

111. http://www.parliament.uk/about/living-heritage/transformingsociety/town country/landscape/overview/enclosingland/.

112. *King Lear,* I.ii.97.

113. *King Lear,* II.ii.467–71.

114. *King Lear,* III.i.4–7.

115. *King Lear,* III.iv.112–16.

116. *King Lear,* IV.iv.15–17.

117. *King Lear,* V.iii.259.

118. *Richard III,* I.ii.63–65.

119. *Richard III,* IV.iv.75–76.

120. Some other aspects, particularly concerning the spaces of Britain, are discussed in Plannic, "'. . . this scattered kingdom.'"

121. Turner, "*King Lear* Without," 165.

122. Shapiro, *1606,* 58.

123. Shapiro, *1606,* 61.

124. Flahiff, "Lear's Map," 24.

125. See Elden, *The Birth of Territory.*

126. Leider, "Plainness of Style in *King Lear,*" 46.

127. For a fuller discussion, see Stuart Elden, "Place Symbolism and Land Politics in *Beowulf,*" *Cultural Geographies* 16, no. 4 (2009): 447–63.

128. Paul Delany, "King Lear and the Decline of Feudalism," *PMLA* 92, no. 3 (1977): 429–40; Visser, "Shakespeare and Hanekom, *King Lear* and Land."

129. This is suggested by Richard Halpern, *The Poetics of Primitive Accumulation: English Renaissance Culture and the Genealogy of Capital* (Ithaca: Cornell University Press, 1991), chap. 6; and see Dodd, "Impossible Worlds."

CHAPTER TWO

1. G. R. Hibbard, "Introduction," in *Hamlet,* ed. G. R. Hibbard (Oxford: Oxford University Press, 1987), 35.

2. On this, see Hibbard, "Introduction," *Hamlet,* 33–36.

3. English texts can be found in Geoffrey Bullough, *Narrative and Dramatic Sources of Shakespeare* (London: Routledge & Kegan Paul, 1966), 7:60–79, 81–124.

4. A detailed discussion of these sources can be found in Hibbard, "Introduction," *Hamlet,* 6–12; Ann Thompson and Neil Taylor, "Introduction," in *Hamlet,* ed. Ann Thompson and Neil Taylor (London: Arden Shakespeare, 2006), 64–70; and Philip Edwards, "Introduction," in *Hamlet, Prince of Denmark* (1985; Cambridge: Cambridge University Press, 2003), 1–2; and above all Margrethe Jolly, *The First Two Quartos of "Hamlet": A New View of the Origins and Relationship of the Texts* (Jefferson, NC: Marfarlane, 2014), chaps. 2–5. Thompson and Taylor's text was reissued in 2016, with some additions and differences in pagination. References are to the 2006 edition unless otherwise stated.

5. See, among other discussions, Edwards, "Introduction," *Hamlet,* 2–3; Thompson and Taylor, "Introduction," *Hamlet,* 44–48; and, especially, Jolly, *The First Two Quartos of "Hamlet,"* 11–14, 143–76.

6. Eric Sams, *The Real Shakespeare: Retrieving the Early Years, 1564–1594* (New Haven: Yale University Press, 1995), 121–35; Terri Bourus, *Young Shakespeare's Young Hamlet: Print, Piracy, and Performance* (London: Palgrave, 2014). Jolly, *The First Two Quartos of "Hamlet,"* tentatively supports this view in her conclusion.

7. The title page is reproduced in Thompson and Taylor, "Introduction," *Hamlet,* 77. I have modernized the spelling.

8. See Edwards, "Introduction," *Hamlet,* 9.

9. Thompson and Taylor, "Introduction," *Hamlet,* 80 and 80 n. 1.

10. John Dover Wilson, *The Manuscript of Shakespeare's "Hamlet" and the Problems of Its Transmission: An Essay in Critical Bibliography,* vol. 1, *The Texts of 1605 and 1623* (Cambridge: Cambridge University Press, 1934), xv. More generally on editing and variants of this text, see George Ian Duthie, *The "Bad" Quarto of "Hamlet": A Critical Study* (Cambridge: Cambridge University Press, 1941); Ann Thompson and Neil Taylor, "Appendix II," *Hamlet;* the special issue of *Studies in Bibliography* 7 (1955); and J. M. Nosworthy, *Shakespeare's Occasional Plays: Their Origin and Transmission* (London: Edward Arnold, 1965), chaps. X–XII; and Jolly, *The First Two Quartos of "Hamlet."*

11. See, for example, Wilson, *The Manuscript of Shakespeare's "Hamlet,"* 1:41; and vol. 2, *Editorial Problems and Solutions,* 177.

12. Edwards, "Introduction," *Hamlet,* esp. 8–19. His diagram of textual transmission on p. 31 is also helpful.

13. *Hamlet,* ed. Harold Jenkins (London: Arden Shakespeare, 1982).

14. Accordingly, references are to the Second Quarto, found in the Arden Third Series edition, ed. Thompson and Taylor, unless otherwise noted. The First Quarto and First Folio texts in modernized spelling can be found in *Hamlet: The Texts of 1603 and 1623,* ed. Ann Thompson and Neil Taylor (London: Arden Shakespeare, 2006). I have also used the very helpful *The Three-Text Hamlet: Parallel Texts of the First and Second Quartos and First Folio,* ed. Paul Bertram and Bernice W. Kliman (New York: AMS Press, 1991); and *The First Quarto of Hamlet,* ed. Kathleen O. Irace (Cambridge: Cambridge University Press, 1998).

15. *Hamlet,* I.i.12–13. For a pun on "ground" as reason and place, see V.i.151–52: "Upon what ground? / Why, here in Denmark."

16. *Hamlet,* IV.iv.17; V.i.72.

17. Edwards, note to *Hamlet,* 89.

18. *Hamlet,* I.i.45–48.

19. *Hamlet,* First Quarto, 1.36–38.

20. *Hamlet,* I.i.40.

21. *Hamlet,* I.i.42, 43.

22. See Hibbard, "Introduction," *Hamlet,* 40. On this question more generally, see Stephen Greenblatt, *Hamlet in Purgatory* (Princeton: Princeton University Press, 2001); Poole, *Supernatural Environments in Shakespeare's England,* chap. 3.

23. *Hamlet,* I.i.59–62.

24. *Love's Labour's Lost,* ed. H. R. Woudhuysen (London: Arden Shakespeare, 1998), V.ii.571–72. See *Love's Labour's Lost 1598: Shakespeare's Quarto Facsimile's No 10 (*Oxford: Clarendon Press, 1957); and *The Norton Facsimile: The First Folio of Shakespeare, Based on Folios in the Folger Shakespeare Library Collection,* prepared by Charlton Hinman, 2nd ed. (London: W.W. Norton, 1996). The line is delivered by Costard, and it is likely that "close-stool" is a seat in a privy, and the pole-axe is phallic; it thus parodies a lion on a throne with a battle-axe—Alexander the Great's coat of arms. See notes to *Love's Labour's Lost,* ed. William C. Carroll (Cambridge: Cambridge University Press, 2009), 163; Woudhuysen, note to *Love's Labour's Lost,* 276.

25. Thompson and Taylor, note to *Hamlet,* 154. One edition emends to "steelèd," but no explanation is given, and this would require all three texts to be seriously corrupt. See *Hamlet,* ed. Jonathan Bate and Eric Rasmussen (Houndmills: Macmillan, 2008), 30.

26. Irace, "Textual Notes," *The First Quarto of Hamlet,* 94.

27. Thompson and Taylor, note to *Hamlet,* 154.

28. *Hamlet,* II.ii.63; II.ii.75; IV.iv.22; V.ii.360.

29. Thompson and Taylor, note to *Hamlet,* 154.

30. T. J. B. Spencer, "Commentary," in *Hamlet,* ed. T. J. B. Spencer (London: Penguin, 2005), 187, thinks this "surprising choice" might be acceptable in "real life," "but in fiction we expect things to be more carefully arranged." See also Edwards, "Introduction," *Hamlet,* 70, where he says the name is "even more perplexing" than the French name for a Norwegian king.

31. Edwards, "Introduction," *Hamlet,* 71.

32. Spencer, "Commentary," *Hamlet,* 176–77.

33. *The First Quarto of Hamlet,* 36 and n.

34. Hibbard, note to *Hamlet,* 148. Hibbard notes that *The Rape of Lucrece* 176 (in *Shakespeare's Poems,* ed. Katherine Duncan-Jones and H. R.Woudhuysen [London: Arden Shakespeare, 2007]) is the nearest parallel in Shakespeare: "His falchion on a flint he softly smiteth." This interpretation is also the one given by Edwards, note to *Hamlet,* 90.

35. *Hamlet,* I.i.67–68. On the eruption and disease, see de Grazia, *Hamlet without Hamlet,* 30, 79–80.

36. *Hamlet,* I.i.69–78.

37. *Hamlet,* I.i.78–106. The First Quarto somewhat compresses this speech; it is much the same in the Folio.

38. See Simon Critchley and Jamieson Webster, *The Hamlet Doctrine* (London: Verso, 2013), 49.

39. *Hamlet,* I.i.185. In the First Quarto he is "gallant" (2.100); in the Folio "goodly."

40. "Heraldry" appears as "heraldy" (as spelled in the Second Quarto) in some editions, notably Spencer's Penguin text; it is spelled "heraldrie" in both First Quarto and Folio.

41. Hibbard, note to *Hamlet,* 149, suggests that these were Norwegian "crown lands, in which the king is his own tenant in chief, and from which he receives the revenues," rather than the whole kingdom: "Shakespeare is thinking in terms of his own England."

42. *Hamlet,* V.i.139.

43. *King John,* I.i.177. See chapter 3.

44. Edwards, note to *Hamlet,* 91, see his "Introduction," 41; and *Hamlet: The Texts of 1603 and 1623,* 182. Wilson, *The Manuscript of Shakespeare's "Hamlet,"* 150, sees this as one of the Folio's "misreadings or miscorrections."

45. *Hamlet,* V.i.71–72. This passage is in the Folio too; in the First Quarto it reads: "See how the slave jowls their heads against the earth."

46. *Hamlet,* First Quarto, 16.45–52.

47. Thompson and Taylor, note to *Hamlet: The Texts of 1603 and 1623,* 157; Jolly, *The First Two Quartos of "Hamlet,"* 63.

48. *Hamlet,* V.i.93–105.

49. For a discussion, see de Grazia, *Hamlet without Hamlet,* 141–42.

50. Thompson and Taylor, notes to *Hamlet,* 416–17; B. J. Sokol and Mary Sokol, *Shakespeare's Legal Language: A Dictionary* (London: Athlone, 2000), 316. See *Henry IV, Part 1,* ed. David Scott Kastan (London: Arden Shakespeare, 2002), I.ii.42–43: "thy quips

and thy quiddities"; and various instances of "quillets," notably *Henry VI, Part 1*, II.iv.17: "these nice sharp quillets of the law"; and *Timon of Athens*, ed. Anthony B. Dawson and Gretchen E. Minton (London: Arden Shakespeare, 2008), IV.iii.153–55: "Crack the lawyer's voice / That he may never more false title plead / Nor sound his quillets shrilly."

51. Thompson and Taylor, note to *Hamlet*, 417; Hibbard, note to *Hamlet*, 325. The date may suggest the text was revised shortly before being printed as the Second Quarto, though the phrase was common.

52. Thompson and Taylor, note to *Hamlet*, 417. The best discussion of these remains Paul S. Clarkson and Clyde T. Warren, *The Law of Property in Shakespeare and the Elizabethan Drama* (1942; New York: Gordian Press, 1968), pt. I, especially chap. IX; also invaluable is Sokol and Sokol, *Shakespeare's Legal Language*; and see also Paul Raffield, *Shakespeare's Imaginary Constitution: Late Elizabethan Politics and the Theatre of Law* (Oxford: Hart, 2010). For a suggestion that some of the legal aspects owe much to debates about rhetoric, see Quentin Skinner, *Forensic Shakespeare* (Oxford: Oxford University Press, 2014).

53. Edwards, note to *Hamlet*, 229; Hibbard, note to *Hamlet*, 326. See chapter 7 on *Henry IV, Part 1*.

54. Edwards, note to *Hamlet*, 229.

55. *Hamlet*, First Folio, V.i.96–110.

56. Spencer, "Commentary," *Hamlet*, 295, says Hamlet achieves "four puns on the word *fine*," but it could be more.

57. Thompson and Taylor, note to *Hamlet*, 417.

58. *Hamlet*, III.ii.313–14; de Grazia, *Hamlet without Hamlet*, 157. The Second Quarto has 'diſeaſd'; the Folio 'diſeas'd.'

59. *Hamlet*, I.i.85–86. This is the same in all texts.

60. Linda Charnes, *Hamlet's Heirs: Shakespeare and the Politics of a New Millennium* (London: Routledge, 2006), 61.

61. See Sokol and Sokol, *Shakespeare's Legal Language*, 110–12.

62. *Hamlet*, V.i.112–13, 158–65.

63. Virgil, *Aeneid* 1.1.365–68; Livy, *Ab urbe condita*, 34.62; Marlowe, *Dido, Queen of Carthage*, in Fredson Bowers, *The Complete Works of Christopher Marlowe*, vol. 1, 2nd ed. (Cambridge: Cambridge University Press, 1981), IV.ii.13. On hides generally, see Elden, *The Birth of Territory*, 70–71, 122, 155; de Grazia, *Hamlet without Hamlet*, 35, 146, 234 n. 59.

64. *Hamlet*, I.ii.17–25. This is not in the First Quarto.

65. *Hamlet*, I.ii.25–39. The First Quarto compresses this speech.

66. *Hamlet*, I.ii.8; not in First Quarto; same in Folio.

67. Spencer, "Commentary," *Hamlet*, 179.

68. Spencer, "Commentary," *Hamlet*, 178.

69. *Hamlet*, II.ii.60–80. There are minor variations in the First Quarto and Folio.

70. *Hamlet*, II.ii.80–82. The First Quarto has "we'll read and answer these his articles," implying a more legal sense than "business."

71. *Hamlet*, IV.iv.1–7. A shorter variant of this speech is in the First Quarto.

72. *Hamlet*, IV.iv.8. This scene is greatly compressed in the First Quarto, and Hamlet does not meet the army.

73. *Hamlet,* IV.iv.9–13.

74. *Hamlet,* IV.iv.14–15.

75. Spencer, "Commentary," *Hamlet,* 270.

76. Thompson and Taylor, note to *Hamlet,* 368.

77. *Hamlet,* IV.iv.16–21. On the possible topical allusion of this campaign, see Jolly, *The First Two Quartos of "Hamlet,"* 20.

78. *Hamlet,* IV.iv.22–23.

79. Critchley and Webster, *The Hamlet Doctrine,* 31.

80. *Hamlet,* IV.iv.61–64.

81. *Hamlet,* IV.iv.24, 59.

82. Spencer, "Commentary," *Hamlet,* 270; Hibbard, note to *Hamlet,* 363.

83. *Hamlet,* V.i.187–201.

84. *Hamlet,* V.i.202. Both passages appear in the First Quarto and the Folio, though the Caesar part is reduced in the First Quarto.

85. *Hamlet,* IV.iv.46.

86. *Hamlet,* IV.iv.64–65.

87. *Hamlet,* V.ii.197–202. This passage appears in the Folio, and in a slightly different form in the First Quarto.

88. Wilson, *The Manuscript of Shakespeare's "Hamlet,"* 31.

89. *Hamlet,* V.ii.333.

90. *Hamlet,* V.ii.334–36. Not in First Quarto, very similar in Folio. Osric had gained land himself, which Hamlet notes and mocks: "He hath much land, and fertile . . . spacious in the possession of dirt" (V.ii.72, 74–75). See de Grazia, *Hamlet without Hamlet,* 146–47.

91. *Hamlet,* V.ii.360.

92. *Hamlet,* V.ii.336–42. The passage is different in the First Quarto, and Fortinbras is not named; the passage is the same in the Folio.

93. See Julia Reinhard Lupton, *Thinking with Shakespeare: Essays on Politics and Life* (Chicago: University of Chicago Press, 2011), chap. 2, which discusses this question in dialogue with Carl Schmitt, *Hamlet or Hecuba: The Intrusion of Time into the Play,* trans. David Pan and Jennifer R. Rust (New York: Telos, 2009). On this text, see Carlo Galli, "*Hamlet:* Representation and the Concrete," trans. Adam Sitze and Amanda Minervini, in Graham Hammill and Julia Reinhard Lupton, eds., *Political Theology and Early Modernity* (Chicago: University of Chicago Press, 2012), 60–83.

94. *Hamlet,* V.ii.385–86.

95. *Hamlet,* V.ii.354–85.

96. *Hamlet,* V.ii.372–74. The first line does not appear in the First Quarto.

97. Edwards, "Introduction," *Hamlet,* 41.

98. De Grazia, *Hamlet without Hamlet,* 2, 43; see 3, 156.

99. *Hamlet,* I.ii.179–80. Variants in other texts. As Poole, *Supernatural Environments in Shakespeare's England,* 130, points out, *Hamlet* begins with "inward movement" to Elsinore, and then "turns to an outward movement."

100. *Hamlet,* IV.iii.19–21. The second line is not in the First Quarto. See de Grazia, *Hamlet without Hamlet,* 71–72.

101. *Hamlet,* I.ii.50–56. This is rather different in the First Quarto; minor variants in Folio.

102. *Hamlet*, I.ii.59–60. These specific words are not in the First Quarto or the Folio.

103. *Hamlet*, I.iii.20–24. Not in First Quarto; variants in Folio.

104. See Spencer, "Commentary," *Hamlet*, 195.

105. *Hamlet*, IV.v.88.

106. *Hamlet*, IV.v.101.

107. *Hamlet*, IV.v.106. Much of this scene is missing from the First Quarto; it appears as Act IV.I in the Folio.

108. See *Hamlet*, Folio, V.ii.77–78.

109. *Hamlet*, IV.vii.80–104; see de Grazia, *Hamlet without Hamlet*, 63–64.

110. *Hamlet*, III.i.163–64.

111. *Hamlet*, V.ii.39; see de Grazia, *Hamlet without Hamlet*, 62–63, 141.

112. *Hamlet*, III.iii.1–7. Not in First Quarto.

113. *Hamlet*, Folio, III.iii.5–7.

114. *Hamlet*, IV.iii.56–66. The First Quarto is much briefer; some variants in Folio.

115. *Hamlet*, V.ii.19–25. Not in First Quarto.

116. *Hamlet*, III.iv.202; not in Folio; see Hibbard, note to *Hamlet*, 361.

117. These events are narrated in a letter to Horatio, in *Hamlet*, IV.vi.13–28.

118. Thompson and Taylor, "Introduction," *Hamlet*, 29.

119. De Grazia, *Hamlet without Hamlet*, 61–62.

120. Edwards, "Introduction," *Hamlet*, 75.

121. "In Performance: The RSC and Beyond," in *Hamlet*, ed. Bate and Rasmussen, 217.

122. Charnes, *Hamlet's Heirs*, 27. This argument is also made by Stuart Kurland, "Hamlet and the Scottish Succession," *Studies in English Literature, 1500–1900* 34, no. 2 (1994): 279–300. See Lilian Winstanley, *Hamlet and the Scottish Succession* (Cambridge: Cambridge University Press, 1921); and Schmitt, *Hamlet or Hecuba*, 53–58. Winstanley is good on the broad historical background at the time, but it is questionable that the explanation is quite that straightforward.

123. *Hamlet*, Folio, II.ii.242, 244–45. Edwards, note to *Hamlet*, 140, suggests this cut may be due to King James's consort, Anne of Denmark. Hibbard, "Introduction," *Hamlet*, 112, suggests it is to bring out more clearly the duplicity of Rosencrantz and Guildenstern. See Thompson and Taylor, "Appendix 1," *Hamlet*, 466–67 and note on p. 466.

124. Thompson and Taylor, "Appendix 5," *Hamlet*, 565; "Additions and Reconsiderations," in *Hamlet*, ed. Ann Thompson and Neil Taylor, rev. ed. (London: Arden Shakespeare, 2016), 157.

125. *Hamlet*, I.iv.17–18.

126. *Hamlet*, I.iv.28.

127. *Hamlet*, V.i.213.

128. Franco Moretti, "Network Analysis, Plot Analysis," *New Left Review* 68 (2011): 80–102; see also the longer version published as Literary Lab Pamphlet 2, May 1, 2011, http://litlab.stanford.edu/LiteraryLabPamphlet2.pdf. On the positivist relapse of Moretti's approach, see Federico Italiano, *Translation and Geography* (London: Routledge, 2016), 9.

129. Thompson and Taylor, "Additions and Reconsiderations," *Hamlet*, 157–58, at 158. For a longer discussion of his role in the play, see Christopher Warley, "Specters of Horatio," *English Literary History* 75, no. 4 (2008): 1023–50.

130. Jonathan Bate, "Introduction," in *Macbeth*, ed. Jonathan Bate and Eric Rasmussen (London: RSC/Palgrave, 2009), 6; see Clark and Mason, note to *Macbeth*, 130. For a reading of the play in the context of the time, including the failed "gunpowder plot," the accession of King James, and attitudes to witchcraft, see Stephen Greenblatt, "Shakespeare Bewitched," in Tetsuo Kishi, Roger Pringle, and Stanley Wells, eds., *Shakespeare and Cultural Traditions: The Selected Proceedings of the International Shakespeare Association World Congress, Tokyo, 1991* (Newark: University of Delaware Press, 1994), 17–42; Garry Wills, *Witches and Jesuits: Shakespeare's Macbeth* (New York: Oxford University Press, 1995); Mary Floyd-Wilson, *Occult Knowledge, Science, and Gender on the Shakespearean Stage* (Cambridge: Cambridge University Press, 2013); Rebecca Lemon, *Treason by Words: Literature, Law, and Rebellion in Shakespeare's England* (Ithaca: Cornell University Press, 2006), esp. chap. 4; and John Kerrigan, *Archipelagic English: Literature, History, and Politics, 1603–1707* (Oxford: Oxford University Press, 2008), chap. 2. On the historical figure, see Nick Aitchison, *Macbeth: Man and Myth* (Stroud: Sutton Publishing Limited, 1999).

131. Poole, *Supernatural Environments in Shakespeare's England*, 139–40. On the supernatural in this period, the classic study is Keith Thomas, *Religion and the Decline of Magic: Studies in Popular Beliefs in Sixteenth- and Seventeenth-Century England* (London: Penguin, 1971).

132. *Macbeth*, ed. Sandra Clark and Pamela Mason (London: Arden Shakespeare, 2015), I.ii.7–15.

133. Notes to *Macbeth*, ed. Kenneth Muir (1951; London: Methuen, 1984), 6; *Macbeth*, ed. Clark and Mason, 131.

134. *Macbeth* ,V.vii.18–19.

135. Clark and Mason, note to *Macbeth*, 131; A. R. Braunmuller, note to *Macbeth*, ed. A. R. Braunmuller (1997; Cambridge: Cambridge University Press, 2008), 120–21.

136. *Macbeth*, I.ii.15–23.

137. *Macbeth*, I.ii.31–33.

138. On this scene, see J. M. Nosworthy, "The Bleeding Captain Scene in *Macbeth*," *Review of English Studies* 22, no. 86 (1946): 126–30, which suggests that it may be cut; and Harry Berger Jr., "The Early Scenes of *Macbeth*: Preface to a New Interpretation," *English Literary History* 47, no. 1 (1980): 1–31.

139. *Macbeth*, I.ii.49–63.

140. *Macbeth*, I.ii.65–68.

141. *Macbeth*, I.iii.72–73, 109–10. See Clark and Mason, notes to *Macbeth*, 135, 143; Kerrigan, *Archipelagic English*, 101.

142. *Macbeth*, I.iii.39.

143. Braunmuller, "Introduction," *Macbeth*, 37, fig. 5.

144. Muir, note to *Macbeth*, 5.

145. Brian Morris, "The Kingdom, the Power, and the Glory in *Macbeth*," in John Russell Brown, ed., *Focus on Macbeth* (London: Routledge & Kegan Paul, 1982), 30–53, 34.

146. Poole, *Supernatural Environments in Shakespeare's England*, 140.

147. While modern editions, which want Macbeth to be Bellona's bridegroom, conflate the battles, support for this reading comes from *A New Variorum Edition of Shakespeare*, vol. 2, *Macbeth*, ed. Horace Howard Furness Jr. (Philadelphia: J.B. Lippincott, 1901), 27n; Edwin Litchfield, "Macbeth, I. ii," *Notes and Queries*, September 10, 1892, 204. It is also

what is in Shakespeare's source, *Holinshed's Chronicles: Volume II, Historie of Scotland,* 2nd ed. (1587), 168–172. In at least one stage production the text was changed to identify Macbeth: "lapped in proof" is replaced with "brave Macbeth." *Macbeth,* directed by Trevor Nunn, 1979. The *Chronicles* were named after their editor, Raphael Holinshed, though the authors were a large group. On their use by Shakespeare, see W. G. Boswell-Stone, *Shakespeare's Holinshed: The Chronicle and the Historical Plays Compared,* 2nd ed. (London: Chatto and Windus, 1907); *Holinshed's Chronicles as Used in Shakespeare's Plays,* ed. Allardyce Nicoll and Josephine Nicoll (London: Dent, 1927); and more generally Annabel Patterson, *Reading Holinshed's Chronicles* (Chicago: University of Chicago Press, 1994).

148. *Macbeth,* I.iii.71.

149. *Macbeth,* I.iii.72–73.

150. *Macbeth,* I.iii.109–10.

151. *Macbeth,* I.iii.147–49. A similar line is said by the recently crowned Henry V: "This new and gorgeous garment, majesty / Sits not so easy on me as you think" (*Henry IV, Part 2,* ed. James C. Bulman [London: Arden Shakespeare, 2016], V.ii.44–45).

152. *Macbeth,* V.ii.15–16.

153. *Macbeth,* V.ii.20–22.

154. Clark and Mason, note to *Macbeth,* 135, 143.

155. *Macbeth,* I.iii.110–17.

156. *Macbeth,* I.ii.31.

157. *Macbeth,* I.iii.105; see Braunmuller, note to *Macbeth,* 131.

158. *Macbeth,* I.iii.118–19.

159. *Macbeth,* I.iii.73–75.

160. *Macbeth,* I.iii.136–39.

161. *Macbeth,* I.iii.144–45.

162. *Macbeth,* I.iv.35–43.

163. *Macbeth,* I.iv.48–50. Kenneth Muir, "Introduction," *Macbeth,* xxxxvii-iii, notes that in Holinshed's *Chronicles,* Macbeth may have had a legitimate grievance against Duncan for being passed over in this way, because of Lady Macbeth's previous marriage. But Shakespeare cuts this from his narrative, as other matters, to compress the story and take away any basis for the crime. For a fuller discussion of how Shakespeare used and departed from his sources, see David Norbrook, "Macbeth and the Politics of Historiography," in Kevin Sharpe and Steven N. Zwicker, eds., *Politics of Discourse: The Literature and History of Seventeenth-Century England* (Berkeley: University of California Press, 1987), 78–116; and on the context more generally, Henry R. Paul, *The Royal Play of Macbeth* (1948; New York: Octagon, 1978); Michael Hawkins, "History, Politics, and Macbeth," in Brown, *Focus on Macbeth,* 155–88.

164. *Macbeth,* II.iii.138–39.

165. *Macbeth,* II.iv.25–27.

166. *Macbeth,* II.iv.30–32. Sovereignty is an important theme in the play. For a discussion, see John Drakakis, "*Macbeth* and 'Sovereign Process,'" in John Drakakis and Dale Townshend, eds., *Macbeth: A Critical Reader* (London: Arden Shakespeare, 2013), 123–52.

167. *Macbeth,* II.iv.36.

168. *Macbeth,* IV.i.70–71.

169. *Macbeth,* III.ii.46–51.

170. See Clark and Mason, notes to *Macbeth,* 214; Nicholas Brooke, notes to *Macbeth,* ed. Nicholas Brooke (Oxford: Oxford University Press, 1990), 150–51; Hunter, "Commentary," *Macbeth,* 129. On the pale, see the coda below.

171. See *Macbeth,* III.vi.24–29.

172. *Macbeth,* III.i.29–34.

173. *Macbeth,* IV.i.141.

174. *Macbeth,* III.vi.27–39.

175. Carol Chillington Rutter, "Introduction," *Macbeth,* lxvii; *Macbeth,* V.ii.2; see Furness, note to *Macbeth,* 288.

176. Clark and Mason, "Introduction," *Macbeth,* 69; and their "Appendix I," *Macbeth,* 323. The relevant passages are III.vi and IV.i.135–55. See also Nicholas Brooke, "Introduction," in *Macbeth,* ed. Nicholas Brooke (Oxford: Oxford University Press, 1990), 51–53. The fullest discussion and an attempt at a resolution can be found in A. R. Braunmuller, "Textual Analysis," in *Macbeth,* 277–78.

177. For example, Muir; see his note to *Macbeth,* 103, which puts "their" down to a printer thinking "the king referred to was Edward the Confessor"; and "Commentary," *Macbeth,* ed. George Hunter (London: Penguin, 2005), 136.

178. For discussions, see Clark and Mason, note to *Macbeth,* 233; Brooke, "Introduction," *Macbeth,* 51–53.

179. For a discussion, see Braunmuller, "Textual Analysis," *Macbeth,* 271–75; Clark and Mason, "Appendix 1," *Macbeth,* esp. 325–36. *Macbeth* is included in Thomas Middleton, *The Collected Works,* ed. Gary Taylor and John Lavagnino (Oxford: Clarendon Press, 2007) as a text by Shakespeare, "adapted by Thomas Middleton," which is taken to mean both cuts and additions. For a fuller discussion, see *Thomas Middleton and Early Modern Textual Culture: A Companion to the Collected Works,* ed. Gary Taylor and John Lavagnino (Oxford: Clarendon Press, 2007), 690–703. A contrary view is advanced by J. M. Nosworthy, *Shakespeare's Occasional Plays: Their Origin and Transmission* (London: Edward Arnold, 1965), 24–44, which suggests just the songs are by Middleton, with later additions made by Shakespeare after 1608–9. This fits generally with his claim of Shakespeare's revision of other plays, including *Hamlet.*

180. *Macbeth,* IV.iii.35–37.

181. *Macbeth,* IV.iii.39–41; see Braunmuller, note to *Macbeth,* 222.

182. *Macbeth,* IV.iii.43–44.

183. *Macbeth,* IV.iii.190–91.

184. The references to "gracious Duncan" and "gracious father" are III.i.65 (by Macbeth); III.vi.3, 10 (both by Lennox).

185. *Macbeth,* V.iii.30.

186. *Macbeth,* V.v.29.

187. *Macbeth,* IV.iii.134; see 191.

188. Rutter, "Introduction," *Macbeth,* ed. Hunter, lxvii.

189. *Macbeth,* IV.i.92–94.

190. *Macbeth,* IV.i.94–96.

191. *Macbeth,* III.iv.121.

192. Robert Pogue Harrison, *Forests: The Shadow of Civilization* (Chicago: University of Chicago Press, 1992), 103; Clark and Mason, note to *Macbeth,* 242.

193. *Macbeth,* V.iv.4–7.

194. *Macbeth,* V.v.34, 37.

195. *Macbeth,* V.v.44–45.

196. *Macbeth,* V.ix.32–33.

197. *Macbeth,* V.ii.7.

198. Rutter, "Introduction," *Macbeth,* lxviii; Hawkins, "History, Politics, and Macbeth," 182.

199. *Macbeth,* II.iii.141–42.

200. *Macbeth,* III.i.3.

201. *Macbeth,* III.i.4–6.

202. *Macbeth,* IV.i.111–20.

203. On this political context, see Leah S. Marcus, *Puzzling Shakespeare: Local Reading and Its Discontents* (Berkeley: University of California Press, 1988), chap. 4, esp. 116–48; and Ros King, *Cymbeline: Constructions of Britain* (Aldershot: Ashgate, 2005), chap. 2.

204. Braunmuller, note to *Macbeth,* 259.

205. E. B. Lyle, "'The Twofold Balls and Treble Sceptres' in *Macbeth,*" *Shakespeare Quarterly* 28, no. 4 (1977): 516–19. See Clark and Mason, note to *Macbeth,* 244. Hunter, "Commentary," *Macbeth,* 141, suggests the treble are England, Scotland, and Ireland. For a discussion, see Arthur F. Kinney, "Scottish History, the Union of the Crowns, and Right Rule: The Case of Shakespeare's *Macbeth,*" in Jean R. Brink and William F. Gentrup, eds., *Renaissance Culture in Context: Theory and Practice* (Aldershot: Scolar Press, 1993), 18–53; and his "Shakespeare's *Macbeth* and the Question of Nationalism," in Vincent Newey and Ann Thompson, eds., *Literature and Nationalism* (Liverpool: Liverpool University Press, 1991), 56–75. More generally, see Christopher Highley, *Shakespeare, Spenser, and the Crisis in Ireland* (Cambridge: Cambridge University Press, 2009). There are also some ambiguities in the use of "dominions," "kingdom," and "realms" in *Henry VIII,* II.iv.14; II.iv.191; II.iv.194. See Willy Maley, "Postcolonial Shakespeare: British Identity Formation and *Cymbeline,*" in Jennifer Richards and James Knowles, eds., *Shakespeare's Late Plays: New Readings* (Edinburgh: Edinburgh University Press, 1999), 155 (this essay is also reprinted in his *Nation, State, and Empire in English Renaissance Literature* [London: Palgrave, 2003]).

206. George Walton Williams, "Macbeth: King James' Play," *South Atlantic Review* 47, no. 2 (1982): 12–21, 14.

207. Clark and Mason, "Introduction," *Macbeth,* 82. The last speech is from Malcolm: "We shall not spend a large expanse of time / . . . / Which would be planted newly with the time, / . . . / We will perform in measure, time and place" (V.ix.26–39).

208. *Macbeth,* V.v.16–27.

209. *Hamlet,* I.iv.28.

210. *Macbeth,* III.ii.50–51.

211. James R. Siemon, "Landlord not King: Agrarian Change and Interarticulation," in Burt and Archer, eds., *Enclosure Acts,* 22.

212. *The First Part of the Contention 1594* (London: Malone Society, 1985), 33 (folio E1).

213. Christopher Marlowe, *Edward II,* in *Doctor Faustus and Other Plays,* II.ii.163–64.

CHAPTER THREE

1. For a discussion, see Ivor John, "Introduction," in *The Life and Death of King John,* ed. Ivor B. John (London: Methuen, 1907), xxviii-xxxii (1593–95); Eugene Giddens, "Introduction," in *King John,* ed. Robert Smallwood (1974; London: Penguin, 2015), xxii (1596); A. R. Braunmuller, "Introduction," in *The Life and Death of King John,* ed. A. R. Braunmuller (Oxford: Oxford University Press, 1989), 2–15 (1595–96). MacDonald Jackson, "Pause Patterns in Shakespeare's Verse," *Literary and Linguistic Computing* 17, no. 1 (2002): 37–46, 41, 45, suggests that it correlates with plays dated to around 1594–95, including *Romeo and Juliet* and *Richard II,* and that if it is earlier, then the dates of those plays must be incorrect.

2. Emrys Jones, "Reclaiming Early Shakespeare," *Essays in Criticism* 51 (2001): 35–50, 41, dates it to around 1590; E. A. J. Honigmann, "Introduction," in *King John,* ed. E. A. J. Honigmann (1954; London: Arden Shakespeare, 2007), xix, lviii, to 1590–91. As well as comparing multiple modern editions I have also consulted *The Norton Facsimile.*

3. For samplings of the critical literature on the play, see Joseph Candido, ed., *King John: The Critical Tradition* (London: Athlone, 1996); and Deborah T. Curren-Aquino, ed., *King John: New Perspectives* (Newark: University of Delaware Press, 1989). For good overall discussions of the play, see Adrien Bonjour, "The Road to Swinstead Abbey: A Study of the Sense and Structure of *King John,*" *English Literary History* 18, no. 4 (1951): 253–74; and David Womersley, "The Politics of Shakespeare's *King John,*" *Review of English Studies,* n.s., 40, no. 160 (1989): 497–515; A. J. Piesse, "*King John:* Changing Perspectives," in Michael Hattaway, ed., *The Cambridge Companion to Shakespeare's History Plays* (Cambridge: Cambridge University Press, 2002), 126–40.

4. Brian Boyd, "*King John* and *The Troublesome Raigne:* Sources, Structure, Sequence," *Philological Quarterly* 74, no. 1 (1995): 37–56, 38.

5. See Saccio, *Shakespeare's English Kings,* 201.

6. Giddens, "Introduction," *King John,* xxiv, notes that Richard had named John as his successor in his will, but that Shakespeare barely hints at this. See the exchange between Eleanor and Constance at II.i.191–94.

7. Robert Smallwood, "Commentary," in *King John,* ed. Smallwood, 134 nn. 9–11.

8. It is not mentioned in George Peele, *The Troublesome Reign of John, King of England,* ed. Charles R. Forker (Manchester: Manchester University Press, 2011) either, but has an important role in Holinshed's *Chronicles,* which is the key source for both plays.

9. These questions are shaped by Jonathan Bate, "Introduction," in *King John and Henry VIII,* ed. Jonathan Bate and Eric Rasmussen (London: RSC/Macmillan, 2012), 3.

10. L. A. Beaurline, "Appendix: Date, Sources, and *The Troublesome Reign,*" in *King John,* ed. L. A. Beaurline (Cambridge: Cambridge University Press, 1990), 200. See Robert Lane, " 'The Sequence of Posterity': Shakespeare's *King John* and the Succession Controversy," *Studies in Philology* 92, no. 4 (1995): 460–81.

11. Braunmuller, "Introduction," *King John,* 76. See also his "*King John* and Historiography," *English Literary History* 55, no. 2 (198): 309–32; and on the Bastard, Edward Gieskes, " 'He Is but a Bastard to the Time: Status and Service in *The Troublesome Raigne of John* and Shakespeare's *King John,*" *English Literary History* 65, no. 4 (1998): 779–98.

12. *King John*, I.i.3–4 (see *Henry V*, ed. T. W. Craik [London: Arden Shakespeare, 1995], II.iv.79: "borrowed glories"). See Braunmuller, note to *The Life and Death of King John*, ed. Braunmuller, 120.

13. *King John*, I.i.5.

14. *King John*, I.i.39–41.

15. *King John*, I.i.110–11, 113, 116–17.

16. *King John*, II.i.118–21.

17. *King John*, II.i.350–51; 480.

18. *King John*, II.ii.55–56, 58–61.

19. *King John*, III.i.25–27.

20. *King John*, III.i.247–48.

21. *King John*, IV.ii.208–14.

22. *Richard II*, ed. Charles R. Forker (London: Arden Shakespeare, 2002), I.iii.213–15.

23. *King John*, IV.iii.148.

24. *King John*, IV.ii.1–16.

25. *The Life and Death of King John*, ed. Braunmuller, 212 n. 10; see Braunmuller, "Introduction," *The Life of King John*, 48–49.

26. *King John*, IV.II.35.

27. *King John*, IV.ii.40–41; see Braunmuller, "Introduction," *The Life of King John*, 50.

28. *King John*, V.i.1–4.

29. Smallwood, "Commentary," *King John*, 272 n. 3.

30. For a much fuller discussion, see Elden, *The Birth of Territory*, chap. 5.

31. Outside the play there may have been others. What I am calling his second may actually have been his fourth coronation. See L. A. Beaurline, "Supplementary Notes," in *King John*, ed. Beaurline, 177. For a more general discussion, see Stuart Elden, "Foucault and Shakespeare: Ceremony, Theatre, Politics," *Southern Journal of Philosophy* 55, Spindel Supplement S1 (2017): 153–72.

32. *King John*, V.i.62–65.

33. *King John*, V.i.66–69.

34. Alexander Schmidt, *Shakespeare-Lexicon: A Complete Dictionary of All the English Words, Phrases, and Constructions in the Works of the Poet*, 3rd ed., rev. Gregor Sarrazin (Berlin: Georg Reimer, 1902), 2:1193; Louis Ure, *A Concordance to the Shakespeare Apocrypha* (Hildesheim: Georg Olms, 1987), 3:1269. For a general discussion, see Alfred Hart, "Vocabularies of Shakespeare's Plays," *Review of English Studies* 19, no. 74 (1943): 128–40.

35. *As You Like It*, III.i.7–8.

36. *King John*, I.i.7–15; reading the place-names as the Braunmuller edition.

37. See Braunmuller, "Introduction," *The Life and Death of King John*, 19–37; Beaurline, "Textual Analysis," in *King John*, ed. Beaurline, 184–93.

38. *The Life and Death of King John*, ed. John, 4 n. 10.

39. *A New Variorum Edition of Shakespeare, The Life and Death of King John*, ed. H. H. Furness Jr. (Philadelphia and London: J.P. Lippincott Company, 1999), 20.

40. John Dover Wilson, "Introduction," *in King John*, ed. John Dover Wilson (Cambridge: Cambridge University Press, 1936), xxv.

41. C. T. Onions, *A Shakespeare Glossary,* enl. and rev. Robert D. Eagles (Oxford: Clarendon Press, 1986), 282; see also David Crystal and Ben Crystal, *Shakespeare's Words: A Glossary and Language Companion* (London: Penguin, 2002), 447.

42. "Peele," *The Troublesome Reign of John,* Part I.ii.1–2. I have used the Manchester edition for quotations from the play, but have also consulted *The Troublesome Raigne of Iohn King of England,* ed. Charles Praetorius, 2 vols. (London: C. Pretorius, 1888); and *"The Troublesome Reign of King John," Being the Original of Shakespeare's "Life and Death of King John,"* ed. F. J. Furnivall and John Munro (London: Chatto & Windus, 1913).

43. Wilson, "Introduction," *King John,* xxv.

44. *The Life and Death of King John,* ed. Braunmuller, 120 n. 10, citing William Aldis Wright, *King John* (Oxford: Clarendon Press, 1886), 81: "*Territories* appears to be used here in the sense of feudal dependencies."

45. Honigmann, note to *King John,* 4 n. 10.

46. Smallwood, "Commentary," *King John,* 135.

47. *King John and Henry VIII,* 24 n. 10.

48. *King John,* ed. Beaurline, 63 n. 10.

49. Ramon Jiménez, "*The Troublesome Raigne of John, King of England:* Shakespeare's First Version of *King John,*" *The Oxfordian* 12 (2010): 21–55. See also E. A. J. Honigmann, "Shakespeare's Self-Repetitions and *King John,*" *Shakespeare Survey* 53 (2000): 175–83.

50. John Munro, "Introduction," in *"The Troublesome Reign of King John," Being the Original of Shakespeare's "Life and Death of King John,"* ed. F. J. Furnivall and John Munro (London: Chatto & Windus, 1913), xi.

51. Ivor B. John, "Introduction," *The Life and Death of King John,* x.

52. See also E. A. J. Honigmann, *Shakespeare's Impact on His Contemporaries* (London: Macmillan, 1982), 56–66, 78–88.

53. Guy Hamel, "*King John* and *The Troublesome Raigne:* A Reexamination," in Curren-Aquino, *King John,* 41–62, 41; see Philip Edwards, *Threshold of a Nation: A Study of English and Irish Drama* (Cambridge: Cambridge University Press, 1979), 115.

54. Beaurline, "Appendix," 194–210. For a good discussion, see Boyd, "*King John* and *The Troublesome Raigne*"; Rosalind King, "The Case for the Early Canon," in John Batchelor, Tom Cairn, and Claire Lamont, eds., *Shakespearean Continuities: Essays in Honour of E. A. J. Honigmann* (Houndmills: Basingstoke, 1997), 108–22; and John Klause, "New Sources for Shakespeare's *King John:* The Writings of Robert Southwell," *Studies in Philology* 98, no. 4 (2001): 401–27.

55. Braunmuller, "Introduction," *King John,* 4.

56. *The Troublesome Reign of John,* Part II.iii.224–25; see Honigmann, note to *King John,* 4 n. 10.

57. Hamel, "*King John* and *The Troublesome Raigne,*" 60.

58. *The Troublesome Reign of John,* Part I.ii.2; xiii.30; xiii.64; Part II.iii.225.

59. *The Troublesome Reign of John,* Part I.i.33–35.

60. Smallwood, "Commentary," *King John,* 135.

61. *King John,* V.ii.128–36.

62. Wilson suggests "*territories.* So F." (*King John,* ed. Wilson, 181 n. 136); i.e., dependent on the Folio, but there is no Quarto edition, and all other editions have the same. The meaning of his note is thus unclear, except if it is to mark that "his Territories" is capitalized.

63. *Coriolanus,* ed. Peter Holland (London: Arden Shakespeare, 2013), IV.v.35; IV.vi.40; IV.vi.78. For a fuller discussion, see chapter 8.

64. *Henry VI, Part 1,* V.ii.167; V.iii.139. See chapter 1.

65. *Richard II,* I.iii.139. On this play, see chapter 4.

66. *The Two Gentlemen of Verona,* ed. William C. Carroll (London: Arden Shakespeare, 2004), III.i.163. See chapter 9.

67. *As You Like It,* III.i.8. See chapter 9.

68. *Edward III,* V.i.232. See chapter 5.

69. *Henry VI, Part 2,* III.ii.245; III.i.84–85. See chapter 1.

70. *King Lear,* I.i.50. As shown in chapter 1, this only appears in the Folio version of the text.

71. Hamel, "*King John* and *The Troublesome Raigne,*" 47.

72. *King John,* II.i.19–31.

73. Honigmann, note to *King John,* ed. Honigmann, 22.

74. Smallwood, "Commentary," *King John,* 159 n. 23.

75. *King John,* II.i.147–50.

76. See "Appendix A: Speech-Prefixes in Act II," *The Life and Death of King John,* ed. Braunmuller, 271–72. For detailed comparison, see Wells and Taylor, *A Textual Companion,* 319.

77. See Smallwood, "Commentary," *King John,* 166.

78. *King John,* II.i.151–54.

79. *King John,* II.i.484–89; reading "Poitiers" with the Braunmuller edition.

80. *King John,* II.i.551–53. The city Angiers is sometimes confused with the province Anjou in the Folio text.

81. *King John,* II.i.527–30; reading "Poitiers" with the Braunmuller edition.

82. *King John,* Ii.i.561–64.

83. *The Life and Death of King John,* ed. Braunmuller, 167 n. 562.

84. *King John,* II.ii.3, 34

85. *King John,* III.i.73–86.

86. *The Life and Death of King John,* ed. Braunmuller, 177 n. 147; *King John,* ed. Honigmann, 63 nn. 73–74; Smallwood, "Commentary," 205.

87. Pope Innocent III, "Per Venerabilem," in *Quellen zur Geschichte des Papsttums und des Römanischen Katholizismus,* ed. D. Carl Mirbt, 4th ed. (Tübingen: J. C. B. Mohr, 1924), 176; trans. in Brian Tierney, *The Crisis of Church and State, 1050–1300* (Toronto: University of Toronto Press, 1988), 136.

88. Henry VIII, *State Papers Published under the Authority of Her Majesty's Commission* (London: George E. Eyre and William Spottiswoode, 1849), 7:261–62.

89. *King John,* III.i.88–97.

90. Smallwood, "Commentary," *King John,* 205–6 nn. 164–67.

91. *King John,* III.i.102–5.

92. *King John,* IV.II.106–12.

93. *King John,* IV.II.128.

94. Generally, see Gwilym Jones, *Shakespeare's Storms* (Manchester: Manchester University Press, 2015).

95. *King John,* IV.II.137–39.

96. *King John,* V.i.17–21.

97. *King John,* III.iii.1–3.

98. *The Life and Death of King John,* ed. Braunmuller, 195 n. 2.

99. The claim about the *Henry VI* plays can be found in Anthony B. Dawson and Paul Yachnin, "Introduction," in *Richard II,* ed. Anthony B. Dawson and Paul Yachnin (Oxford: Oxford University Press, 2011), 57 n. 1; and Jones, "Reclaiming Early Shakespeare," 40.

100. On this subplot, and the relation between *King John* and *King Lear,* see Fionnuala O'Neill, "Toward Tyranny: Geopolitics and Genre, A Response to Stuart Elden," *Law and Literature* 25, no. 2 (2013): 166–74.

101. *King John,* I.i.69–70, 73.

102. *King John,* I.i.91, 93, 97, 115, 129, etc. "Land" appears twenty-nine times in the play; "lands" as a noun once; "landed" and "landless" once each; and "landed" and "lands" as verbs once each.

103. *King John,* I.i.128–29.

104. *King John,* I.i.163–64.

105. *King John,* I.i.176–77.

106. *King John,* I.i.182–83.

107. Honigmann, "Introduction," *King John,* lxviii.

108. For a very good discussion, see Braunmuller, "Introduction," *King John,* 50–58. See also the exchange between Baptista and Petruccio in *The Taming of the Shrew,* ed. Barbara Hodgdon (London: Arden Shakespeare, 2010), II.i.115–26. There they discuss the dowry that Katherina will bring with her into marriage, and the planned inheritance if she predeceases her husband. Both concern lands and other property, and are structured around legal terms: "specialties" are "explicit contracts"; "covenants" are "formal financial agreements or compacts." See Hodgdon's notes, 199.

109. *King John,* II.i.265–66, 366.

110. *King John,* IV.II.144–45.

111. *King John,* III.II.50–51.

112. *King John,* IV.i.32.

113. *King John,* IV.ii.203.

114. *King John,* V.iii.17.

115. See James L. Calderwood, "Commodity and Honour in *King John,*" *University of Toronto Quarterly* 29, no. 3 (1960): 341–56.

116. *King John,* II.i.573–75, 582.

117. Smallwood, "Commentary," *King John,* 193 n. 574.

118. Bate, "Introduction," *King John and Henry VIII,* 2.

119. *King John,* V.ii.73–74.

120. *King John,* V.II.78–82, 93–96.

121. *King John,* V.ii.135–36.

122. *King John,* V.ii.160, 162.

123. *King John,* V.iii.9–11.

124. *King John,* V.v.3–4, 7–8.

125. *King John,* V.v.10–13.

126. *King John,* V.vii.82–86.

127. *King John,* V.v.39–42.

128. *King John,* V.vii.61–63

129. Braunmuller, "Introduction," *The Life of King John,* 18.

130. *King John,* I.i.48–49. Braunmuller, *The Life of King John,* 122 n. 49, indicates that there may be a "turned" *n,* meaning "expeditions" not "expeditious," but he thinks the speed is implied and the sense is sound.

131. *King John,* III.ii.16–21. Scene division is different in some editions.

132. Smallwood, "Commentary," *King John,* 194, 218.

133. *King John,* III.ii.22–23.

134. *King John,* III.ii.141–42.

135. *King John,* III.iii.171–73.

136. See Giddens, "Introduction," xxxix.

137. *King John,* IV.iii.140–54.

138. Smallwood, "Commentary," *King John,* 269 n. 149. See also *Coriolanus,* IV.v.213; *Henry IV, Part 1,* I.i.97–98.

139. See notes in *King John,* ed. Honigmann, 117; notes in *The Life and Death of King John,* ed. Braunmuller, 237–38.

140. *King John,* V.ii.89.

141. *King John,* IV.ii.243–48.

142. Honigmann, "Introduction," *King John,* lxii, notes that "body-images are especially important in *John,* because of their variety and profusion," with several words appearing more frequently than in other plays, and some not far behind their occurrence in other plays. He notes too that *King John* is only 2,600 lines long, so the frequency is remarkable.

143. *King John,* V.vii.34–43.

144. *King John,* II.i.84–85.

145. *King John,* II.i.89–92. These speeches are part of a highly ritualized rhetoric of the opening exchanges; see Giddens, "Introduction," xxvii-xxviii.

146. See also *King John,* II.i.202: a line in two halves, spoken by Philip and John: " 'Tis France, for England"; "England, for itself."

147. *King John,* V.vii.102.

148. *King John,* V.vii.111–18.

149. Jean Bodin, *Les six livres de la république,* ed. Christiane Frémont, Marie-Dominique Couzinet, and Henri Rochais, 6 vols. (Paris: Fayard, 1986).

150. Jean Bodin, *De republica libri sex* (Paris: J. de Puys, 1586).

151. Bodin, *Les six livres de la république,* Book I, Chapter viii.

152. Bodin, *Les six livres de la République,* Book I, Chapter x.

153. R. W. Desai, " 'What means Sicilia? He something seems unsettled': Sicily, Russia, and Bohemia in 'The Winter's Tale,' " *Comparative Drama* 30, no. 3 (1996): 311–24, 320, suggests it is "equally [as] likely" as Shakespeare's knowledge of Montaigne and Machiavelli.

154. Jean Bodin, *The Six Bookes of a Commonweale: A Facsimile Reprint of the English Translation of 1606,* trans. Richard Knolles, ed. Kenneth Douglas McRae (Cambridge, MA: Harvard University Press, 1962). On this edition, see Elden, *The Birth of Territory,* 261–65.

155. Leibniz, "Extrait d'une Lettre à l'Auteur du *Journal des Sçavans,*" ser. 4, 2:360.

CHAPTER FOUR

1. On readings of the play, see Charles R. Forker, *Shakespeare: The Critical Tradition: Richard II* (London: Athlone, 1998).

2. *Richard II,* I.iii.125–26.

3. *Richard II,* I.iii.139–43.

4. Stanley Wells, "Commentary," *Richard II,* ed. Stanley Wells (London: Penguin, 2008), 141.

5. *Richard II,* I.iii.148–53.

6. Forker, "Introduction," *Richard II,* 65.

7. *Richard II,* I.i.196.

8. *Richard II,* I.iii.152–53.

9. The reason for the dispute is not immediately clear. It concerned the death of Thomas of Woodstock, Duke of Gloucester and Richard's uncle, who had served as Lord Protector. This becomes clearer in Act I, scene ii, so some modern productions reverse the opening scenes. It has been suggested that Shakespeare's audience would have been familiar with the anonymous play *Thomas of Woodstock: or, Richard II, Part One,* ed. Peter Corbin and Douglas Sedge (Manchester: Manchester University Press, 2002) (see Corbin and Sedge, "Introduction," 38–39).

10. *Richard II,* I.iii.180–84.

11. *Richard II,* I.iii.188–90.

12. Andrew Gurr, "Introduction," in *King Richard II,* updated ed. (Cambridge: Cambridge University Press, 2003), 38–39. The references are to II.i.114; IV.i.180; IV.i.209; and V.vi.6. See also the "state and honour" given to Henry by the Duke of York in V.ii.40.

13. See *Richard II,* III.ii.72, 117; III.iv.102 (used of the Queen); IV.i.192; V.v.47.

14. *Richard II,* III.ii.163; IV.i.252; V.i.18.

15. *Richard II,* III.iv.27–28.

16. Forker, note to *Richard II,* 364.

17. *Richard II,* I.iii.209–12.

18. *Richard II,* I.iii.213–15. This reading is indebted to Forker, "Introduction," *Richard II,* 65–66.

19. *Richard II,* I.iii.218–24.

20. *Richard II,* I.iii.225–26.

21. Michel Foucault, *Histoire de la sexualité I: La volonté de savoir* (Paris: Gallimard, 1976), 177; trans. Robert Hurley as *The History of Sexuality I: The Will to Knowledge* (London: Penguin, 1978), 135.

22. Foucault, *Histoire de la sexualité I,* 181; *The History of Sexuality I,* 138.

23. *Richard II,* I.iii.46–51.

24. *Richard II,* I.iii.264.

25. *Richard II,* I.i.63–66.

26. Wells, "Commentary," *Richard II,* 128.

27. *Richard II,* I.iii.306–9.

28. *Richard II,* I.iv.42–52.

29. Forker, "Introduction," *Richard II,* 69.

30. Richard D. Altick, "Symphonic Imagery in *Richard II,*" *PMLA* 62 (1947): 339–65, 341. More generally, see Caroline M. Barron, "Richard II: Image and Reality," in Dillian Gordon, *Making & Meaning: The Wilton Diptych* (London: National Gallery, 1993), 13–19.

31. Forker, "Introduction," *Richard II,* 70.

32. In *Henry V,* V.ii.126–27, perhaps referring back to this claim, the King jokes that his bride-to-be might "think I had sold my farm to buy my crown."

33. *Richard II,* II.i.110.

34. See Wells, "Commentary," *Richard II,* 148; citing *Woodstock,* IV.i.181–94.

35. Forker, note to *Richard II,* 238.

36. See also Wells, "Commentary," *Richard II,* 148, citing a passage from Holinshed; and saying that again *Woodstock* discusses this.

37. Klinck, "Shakespeare's Richard II as Landlord and Wasting Tenant," 29.

38. *Richard II,* I.iv.59–65.

39. *Richard II,* II.i.31–68.

40. Phyllis Rackham, *Stages of History: Shakespeare's English Chronicles* (Ithaca: Cornell University Press, 1990), 101. See also Willy Maley, "'This Sceptred Isle': Shakespeare and the British Problem," in John J. Joughlin, ed., *Shakespeare and National Culture* (Manchester: Manchester University Press, 1997), 83–108.

41. Compare to *King Lear,* II.ii.188–89: "from low farms, / Poor pelting villages, sheep-cotes, and mills"; and a handful of other examples of "pelting" meaning small, such as *Measure for Measure,* II.ii.112: "For every pelting petty officer."

42. *Woodstock,* IV.i.147–48. The dating is based on the assumption that it was influenced by *Henry VI, Part 2,* but influenced Shakespeare's *Richard II* (Corbin and Sedge, "Introduction," 4; A. P. Rossiter, "Preface," *in Woodstock, a Moral History,* ed. A. P. Rossiter [London: Chatto & Windus, 1946], 16). It is, however, possible that *Richard II* predates *Woodstock.* In terms of authorship, Corbin and Sedge suggest that "Shakespeare is perhaps the one known dramatist in the 1590s whose dramatic style most closely resembles that of *Thomas of Woodstock.* . . . Any ascription of the play to Shakespeare or any other dramatist must, however, remain highly speculative; what is certain is that the level of dramatic skill deployed suggests an author (or authors) of considerable range and competence." Rossiter, "Preface," 73, is clear: "There is not the smallest chance that he was Shakespeare." On this play, see Sullivan, *The Drama of Landscape,* chap. 2.

43. *Woodstock,* IV.i.221–22.

44. Sullivan, *The Drama of Landscape,* 83.

45. *Richard II,* II.i.95–96.

46. *Richard II,* II.i.101–4.

47. Forker, note to *Richard II,* 252; see Wells, "Commentary," 153.

48. For a general discussion, see Sokol and Sokol, *Shakespeare's Legal Language*, 408–10.

49. Forker, note to *Richard II*, 252. For a full discussion, see Dennis R. Klinck, "Shakespeare's Richard II as Landlord and Wasting Tenant," *College Literature* 25, no. 1 (1998): 21–34; Siemon, "Landlord not King"; and William O. Scott, "'Like to a Tenement': Landholding, Leasing, and Inheritance in *Richard II*," in Constance Jordan and Karen Cunningham, eds., *The Law and Shakespeare* (Houndmills: Palgrave, 2010), 58–72.

50. *Richard II*, II.i.109–14.

51. *Woodstock*, V.iii.106–7. The other passages are IV.i.147 (cited above), 211, 244; V.i.89.

52. Peter Ure, "Introduction," in *Richard II*, ed. Peter Ure (London: Arden Shakespeare, 1956), xxxviii; see Corbin and Sedge, "Introduction," 7–8. Christopher Marlowe's *Edward II* is another potential inspiration. Forker, "Introduction," 116–18, 144–52, 159–64. Rossiter, "Preface," 53–65, disputes that Marlowe's play was an inspiration for *Woodstock*, suggesting that *Henry VI, Part 2* is much more important, 66–71.

53. See Wells, "Commentary," *Richard II*, 154.

54. *Richard II*, II.i.153–62.

55. See also "To forfeit all your goods, lands, tenements, / Chattels and whatsoever" (*Henry VIII*, ed. Gordon McMullan [London: Arden Shakespeare, 2000], III.ii.342–43).

56. *Richard II*, II.iii.190, 196, 198–99.

57. *Richard II*, II.i.261.

58. *Richard II*, II.i.201–5.

59. This reading is indebted to Wells, "Commentary," *Richard II*, 157; and Forker, note to *Richard II*, 262. The phrase "sue his livery" recurs in *Henry IV, Part 1*, IV.iii.62, where Hotspur mentions Bolingbroke's claim. For a broader discussion of legal issues in the play, see Donna Hamilton, "The State of Law in *Richard II*," *Shakespeare Quarterly* 34, no. 1 (1983): 5–17; and W. F. Bolton, "Ricardian Law Reports and *Richard II*," *Shakespeare Studies* 20 (1988): 53–65.

60. *Richard II*, II.i.209–10.

61. Forker, "Introduction," *Richard II*, 22.

62. *Richard II*, II.i.224–27.

63. *Richard II*, II.i.246–50.

64. Forker, "Longer Notes," *Richard II*, 489.

65. *Richard II*, II.ii.88–89.

66. *Richard II*, II.i.256–61.

67. *Richard II*, II.ii.80–83.

68. Forker, note to *Richard II*, 282.

69. *Richard II*, iii.i.4–26.

70. Ernst H. Kantorowicz, *The King's Two Bodies: A Study in Mediaeval Political Theology* (Princeton: Princeton University Press, 1957).

71. *Richard II*, V.v.31.

72. Kantorowicz, *The King's Two Bodies*, 26; for the reading generally, see 24–41; and for discussions, see David Norbrook, "The Emperor's New Body? Richard II, Ernst Kantorowicz, and the Politics of Shakespeare's Criticism," *Textual Practice* 10, no. 2 (1996):

329–57; Anselm Haverkamp, *Shakespearean Genealogies of Power* (London: Routledge, 2011), chap. 4.

73. *Richard II*, IV.i.289.

74. Ruth Nevo, *Tragic Form in Shakespeare* (Princeton: Princeton University Press, 1972), 74–76.

75. Forker, "Introduction," 67.

76. See Klinck, "Shakespeare's Richard II as Landlord and Wasting Tenant," 23.

77. Forker, "Introduction," 68.

78. *Richard II*, II.iii.108–12.

79. *Richard II*, II.iii.113–22. This is anticipated in an exchange at II.iii.69–70, where Bolingbroke refuses to answer to "Hereford" and requests to be addressed as "Lancaster."

80. *Richard II*, III.iii.39–41.

81. *Richard II*, III.iii.42–50.

82. *Richard II*, III.iii.91–100.

83. Forker, "Introduction," *Richard II*, 23.

84. *Richard II*, III.ii.95–96. See also the speech by Richard on "care," with multiple meanings, in the deposition scene IV.i.195–99.

85. *Richard II*, III.ii.109–11.

86. *Richard II*, III.ii.54–62.

87. *Richard II*, III.ii.123–25.

88. Forker, note to *Richard II*, 326.

89. *Richard II*, III.ii.127, 132, 134. See Wells, "Commentary," 177–78, on what may be imperfect revision by Shakespeare concerning the status of Bagot.

90. *Richard II*, III.ii.137–38.

91. *Richard II*, III.ii.145–63.

92. Wells, "Commentary," *Richard II*, 179.

93. *The Mirror for Magistrates: Edited from Original Texts in the Huntington Library*, ed. L. B. Campbell (New York: Barnes & Noble, 1938).

94. *Richard II*, III.iii.133–36.

95. *Richard II*, III.iii.147–57.

96. *Richard II*, III.iii.161–63.

97. *Richard II*, III.iii.190–91.

98. *Richard II*, III.iii.196–97.

99. *Richard II*, II.iii.165–67.

100. M. M. Reese, *The Cease of Majesty: A Study of Shakespeare's History Plays* (London: Edward Arnold, 1961), 252.

101. *Richard II*, III.i.22–23.

102. Forker, note to *Richard II*, 311.

103. Klinck, "Shakespeare's Richard II as Landlord and Wasting Tenant," 25.

104. On garden scene in *Richard II*, see James Siemon, *Word against Word: Shakespearean Utterance* (Amherst: University of Massachusetts Press, 2002), chap. 5. For parallels, see *Henry V*, V.ii.36–62. On the wider context, see Andrew McRae, *God Speed the Plough: The Representation of Agrarian England, 1500–1660* (Cambridge: Cambridge University Press, 1996).

105. *Richard II*, III.iv.33–47.

106. Anthony B. Dawson and Paul Yachnin, "Introduction," in *Richard II,* ed. Anthony B. Dawson and Paul Yachnin (Oxford: Oxford University Press, 2011), 30.

107. *Richard II,* III.iv.52–53.

108. *Richard II,* III.iv.55–66.

109. Wells, "Commentary," *Richard II,* 199.

110. *Richard II,* III.iv.138–48. See also the opening lines of *Henry IV, Part 1,* I.i.1–9.

111. *Richard II,* IV.i.178–81.

112. Forker, note to *Richard II,* 395.

113. *Richard II,* IV.i.203–13.

114. *Richard II,* IV.i.218–19.

115. *Richard II,* IV.i.222–26.

116. Forker, note to *Richard II,* 402.

117. *Richard II,* IV.i.249–52.

118. *Richard II,* IV.i.271.

119. *Richard II,* IV.i.272–73.

120. *Richard II,* IV.i.319–20.

121. *Richard II,* V.i.5.

122. *Richard II,* V.i.30.

123. *Richard II,* V.v.49–50.

124. *Richard II,* V.v.108–12.

125. *Richard II,* V.iii.130.

126. *Richard II,* V.iii.135.

127. *Richard II,* V.vi.49–50.

128. *Richard II,* IV.i.89–90.

129. *Richard II,* IV.i.94–101.

130. *Henry IV, Part 1,* I.i.3–4.

131. *Henry IV, Part 1,* I.i.18–29, 47–48, and 100–101.

132. *Henry IV, Part 2,* III.i.107–8.

133. *Henry IV, Part 2,* IV.iii.360–62.

134. *Henry IV, Part 2,* IV.iii.363–68.

135. This is widely discussed, but see, for example, David Norbrook, "A Liberal Tongue: Language and Rebellion in *Richard II,*" in J. M. Mucciolo, ed., *Shakespeare's Universe: Renaissance Ideas and Conventions—Essays in Honour of W. R. Elton* (Aldershot: Scolar Press, 1996), 37–51.

136. Again this is widely discussed, but see Naomi Conn Liebler, *Shakespeare's Festive Tragedy: The Ritual Foundations of Genre* (London: Routledge, 1995), 57–58.

137. Ure, "Introduction," in *Richard II,* lvii-iii.

138. Gurr, "Introduction," *King Richard II,* 53.

139. *Richard II,* IV.I; see Forker, "Introduction," *Richard II,* 19.

140. Forker, "Introduction," *Richard II,* 10, 10 n. 2, 165.

141. *Richard II,* IV.i.108–13. See III.iii.204–5.

142. *Richard II,* IV.i.155–318.

143. For a discussion see Dawson and Yachnin, "Introduction," *Richard II,* 9–16. There is a related debate about the cuts in the Quarto edition of *Henry IV, Part 2.* As Humphreys suggests, "These lacunae may result from political censorship. Passages

innocent in 1597 would sound dangerous in 1600. By then, two dangers were at their climaxes—the Irish insurrection which Essex failed to crush in 1599, and Essex's own disaster." A. R. Humphreys, "Introduction," *in The Second Part of Henry IV,* ed. A. R. Humphries (London: Routledge, 1967), lxxi.

144. *Henry IV, Part 2,* III.i.38–40.

145. *Henry IV, Part 2,* III.i.41–43.

CHAPTER FIVE

1. Ben Jonson, *Every Man in his Humour,* Prologue, 9–11; ed. Robert S. Miola (Manchester: Manchester University Press, 2000), 244–45.

2. See Kenneth Branagh, *Beginning* (London: Chatto & Windus, 1990), chaps. 6 (theater), 9 and 10 (film); and his comments in "In Performance: The RSC and Beyond," in *Henry V,* ed. Jonathan Bate and Eric Rasmussen (London: Macmillan, 2010), 184.

3. Andrew Gurr, "Introduction," in *King Henry V,* ed. Andrew Gurr, updated ed. (Cambridge: Cambridge University Press, 2005), 9.

4. See Gary Taylor, "Introduction," in *Henry V,* ed. Gary Taylor (Oxford: Oxford University Press, 1982), 12–26. For a challenge to the whole theory of memorial reconstruction, see Bourus, *Young Shakespeare's Young Hamlet,* chap. 2, esp. 58–59.

5. Taylor, "Introduction," *Henry V,* 12; see Ann Kaegi, "Introduction," in *Henry V,* ed. A. R. Humphreys, rev. Ann Kaegi (London: Penguin, 2010), xxvii. One of Shakespeare's models for this play, and some parts of the *Henry IV* plays, was *The Famous Victories of Henry V,* available in *The Oldcastle Controversy: Sir John Oldcastle, Part I and The Famous Victories of Henry V,* ed. Peter Corbin and Douglas Sedge (Manchester: Manchester University Press, 1991).

6. *Henry V,* prologue, 11–12. I have used the Arden edition for quotations from the Folio text, unless otherwise noted; and compared to *The First Quarto of King Henry V,* ed. Andrew Gurr (Cambridge: Cambridge University Press, 2000); *King Henry V: Parallel Texts of the First Quarto (1600) and First Folio (1623) Editions,* ed. B. R. Nicolson (London: The New Shakespeare Society, 1877).

7. The word appears twenty-two times in the Folio, not counting stage directions.

8. *Henry V,* II.iv.14; III.ii.9–11; IV.ii.27–28; IV.vi.18; IV.iii.85–88. Other examples could be given.

9. *Henry V,* prologue, 19–22 (Folio only).

10. *Henry V,* IV.i.15 (Folio only).

11. *Henry V,* IV.iii.98–99 (Folio). The Quarto reads: "find graves within your realm of France: / Tho buried in your dunghills."

12. Craik, "Introduction," *Henry V,* 41.

13. Taylor, "Introduction," *Henry V,* 37.

14. *Henry V,* I.i.9 (Folio only).

15. Many of these are discussed in *The Birth of Territory,* esp. chap. 6. On the British context, see R. H. Tawney, *Religion and the Rise of Capitalism: A Historical Study* (1926; London, Verso, 2015), 142–54.

16. *Henry V,* III.vi.38–40, 98–112 (slight variations in Quarto).

17. *Henry V,* I.i.1–5 (Folio only). See Craik note, 123 n. 4.

18. *Henry V,* I.i.6 (Folio only).

19. *Henry V,* I.i.7–19 (Folio only).

20. *Henry V,* I.i.20 (Folio only).

21. Shakespeare often confuses the terms. See Sokol and Sokol, *Shakespeare's Legal Language*, 367–68, 410–17.

22. *Henry V,* I.i.22–23 (Folio only).

23. *Henry V,* I.i.70–72 (Folio only).

24. *Henry V,* I.i.75–81 (Folio only).

25. Notes to *Henry V*, ed. Taylor, 98.

26. *Henry V,* I.i.86–89 (Folio only).

27. *Henry V,* I.ii.4–6 (variations in Quarto).

28. *Henry IV, Part 2,* IV.iii.342–44.

29. *Henry IV, Part 2* V.v.104–6.

30. *Henry V,* I.ii.9–12 (variations in Quarto).

31. *Henry V,* I.ii.14–16 (variations in Quarto).

32. *Henry V,* I.ii.18–23 (not all in Quarto)

33. *Henry V,* I.ii.35–45 (Folio); the Quarto does not have the Latin, and reads: "No female shall succeed in Salicke land."

34. See Craik, note to *Henry V*, 130; and compare to *Edward III,* 1.i.5–29.

35. A. R. Humphreys, "Introduction," in *Henry V*, ed. A. R. Humphreys (London: Penguin, 1968), 26.

36. Jonathan Bate, "Introduction," in *Henry V*, ed. Bate and Rasmussen, 2.

37. Derek Traversi, *Shakespeare from Richard II to Henry V* (London: Hollis & Carter, 1957), 170.

38. Craik, "Introduction," *Henry V*, 67; Taylor, "Introduction," *Henry V*, 29, 34. The relevant passages are reproduced in the Oxford edition, ed. Taylor, 306–7.

39. Taylor, "Introduction," *Henry V*, 21. See 21–22 for a longer analysis of the Quarto passage: "Hugh Capet also, that usurped the crown. / To fine his title with some show of truth, / When in pure truth it was corrupt and naught, / Conveyed himself as heir to the Lady Inger, / Daughter to Charles, the foresaid Duke of Lorraine; / So that, as clear as is the summer's sun, / King Pépin's title and Hugh Capet's claim, / King Charles his satisfaction all appear / To hold in right and title of the female." Taylor notes that "the Lady Lingard was the daughter of Charlemain, not Charles the Duke of Lorraine; 'the foresaid Duke of Lorraine' has not in fact been mentioned at all; nor has King Pépin; nor has King Charles (presumably an error for King Louis, as in F; but Q has not mentioned Louis either)." See also P. A. Daniel, "Introduction," in *King Henry V: Parallel Texts*, xi-xii.

40. *Henry V,* I.ii.96. The Quarto is missing "claim" here.

41. Saccio, *Shakespeare's English Kings*, 77–80, discusses the genuine history behind this scene.

42. *Henry V,* I.ii.65, 69, 78. See Kaegi, "Introduction," *Henry V*, lxxxi; Humphreys, "Commentary," *Henry V*, 141.

43. These possibilities are discussed by Taylor, "Introduction," *Henry V*, 34–35. See also the discussion in "In Performance: The RSC and Beyond," *Henry V*, 188–90.

44. Craik, "Introduction," *Henry V*, 42.

45. Taylor, "Introduction," *Henry V*, 35.

46. Taylor, "Introduction," *Henry V*, 37.

47. *Henry V*, I.ii.100–107. The Quarto is missing the "Look back . . ." line; has "grave" instead of "tomb"; and is missing "invoke his warlike spirit."

48. *Henry V*, ed. Taylor, 105 n. 106 and 107.

49. Kaegi, "Introduction," *Henry V*, lxviii.

50. *Henry V*, I.ii.117–19 (Folio only).

51. *Henry V*, I.ii.122–24 (Folio only).

52. *Henry V*, I.ii.129–30 (Folio only).

53. *Henry V*, I.ii.132–35 (Folio only).

54. *Henry V*, I.ii.136–39 (Folio). The Quarto is slightly different here: "We must not only arm vs against the French, / But lay down our proportion for the Scot, / Who will make road upon us / With all advantages."

55. *Henry V*, I.ii.140–42 (Folio). The Quarto reads: "The Marches gracious sovereign, shall be sufficient / To guard your England from the pilfering borderers."

56. *Henry V*, I.ii.143–54 (Folio). The Quarto reads: "We do not mean the coursing sneakers, / But fear the main intendment of the Scot, . . . / For you shall read that my great-grandfather / Unmasked his power for France . . . / That England, being empty of defences, / Hath shook and trembled at the brute hereof."

57. *Henry V*, ed. Taylor, 107 n. 143.

58. *Henry V*, I.ii.160–61 (minor differences in Quarto); see note to *Henry V*, ed. Craik, 141.

59. *Henry V*, ed. Taylor, 107 nn. 148 and 151.

60. *Henry V*, I.ii.166–73. The Quarto is almost the same for this passage.

61. *Henry V*, I.ii.174–89. "We are well resolved" is not in Quarto; "our power" reads "our land" in that text.

62. *Henry V*, I.ii.214–17 (both texts).

63. See *Henry V* III.ii (not in Quarto). On the different nations in Henry's army, see Claire McEachern, *The Poetics of English Nationhood, 1590–1612* (Cambridge: Cambridge University Press, 1996), chap. 3; and Edwards, *Threshold of a Nation*, chap. 4.

64. *Henry V*, III.i.27. In the prologue to Act II, the Chorus had already said that "they sell the pasture now to buy the horse"—exchanging landed wealth for the preparations of war. For a discussion of these themes in the play, see Charlotte Scott, *Shakespeare's Nature: From Cultivation to Culture* (Oxford: Oxford University Press, 2014), chap. 3.

65. On the ambiguous role of the Dauphin in the different Folio and Quarto texts, and the Quarto's replacement of him by the Duke of Bourbon in the Agincourt scenes, see Gurr, "Textual Analysis," *King Henry V*, 231–32; Taylor, "Introduction," *Henry V*, 24–26. Taylor's Oxford edition follows the Quarto in this regard, unlike almost all other modern texts.

66. Craik, "Introduction," 42.

67. *Henry V*, I.ii.223–26 (minor differences in the Quarto).

68. *Henry V*, I.ii.248–58. Quarto: "certain towns in France" in place of "certain dukedoms in the right"; "crave" in place of "claim."

69. *Henry V*, I.ii.259 (both).

70. *Henry V*, I.ii.260–64. The second line reads: "Your message and his present we accept" in the Quarto.

71. *Henry V,* I.ii.282 (both).

72. *Henry V,* II.o.5–6 (not in Quarto).

73. *Henry V,* V.ii.172–76. There are only minor differences in the Quarto.

74. McEachern, *The Poetics of English Nationhood,* 136.

75. *Henry V,* V.ii.288–311 (not in Quarto). See Lance Wilcox, "Katherine of France as Victim and Bride," *Shakespeare Studies,* 1985, 61–76, 72.

76. *Henry V,* III.iv. See Wilcox, "Katherine of France as Victim and Bride," 67–68.

77. *Henry V,* III.iv.46–47. There are possibilities of more than just this one part. See Humphreys, "Commentary," 170–72.

78. On this, see Kaegi, "Introduction," *Henry V,* lxxiv.

79. *Henry V,* III.iii.6–27 (not in Quarto from line 11).

80. *Henry V,* III.iii.34–41 (not in Quarto).

81. *Henry V,* III.iii.42–43 (both).

82. *Henry V,* IV.vi.35–38. There are differences in the Quarto, notably the lack of the suggestion that Henry is doing this because the French had reinforced their troops and a new attack was imminent, but enough of the sense remains. Indeed, the Quarto also has Pistol repeat a line from earlier in the play, in both versions: "Couple gorge"—bad French for "cut the throat." See II.i.73; IV.iv.38.

83. Karen Newman, *Fashioning Femininity and English Renaissance Drama* (Chicago: University of Chicago Press, 1991), 101.

84. McEachern, *The Poetics of English Nationhood,* 134.

85. *Henry V,* V.ii.96–97 (not in Quarto).

86. *Henry V,* V.ii.181–86. This is more of a dialogue in the Quarto.

87. *Henry V,* V.ii.326–28. The Quarto has a speech by the French king here, agreeing to "the articles," but not mentioning Katherine.

88. *Henry V,* III.o.28–32 (not in Quarto).

89. *Henry V,* V.ii.333–36 (both).

90. *Henry V,* V.ii.345–49. The Quarto text is quite different here: "This and what else, / Your majesty shall crave. / God that disposeth all, give you much joy." The "pale" shores may be another play on the dual sense of this word. See chapter 2 above and the coda.

91. *Henry V,* V.epilogue.7 (not in Quarto).

92. *Henry V,* V.epilogue.9–12 (not in Quarto). Henry VI became King of France on the death of his grandfather, two months after the death of Henry V.

93. *Henry V,* V.epilogue.13 (not in Quarto).

94. Gurr, "Introduction," *King Henry V,* 16.

95. Taylor, "Introduction," *Henry V,* 34.

96. Marie Axton, *The Queen's Two Bodies: Drama and the Elizabethan Succession* (London: Royal Historical Society, 1977), 112–13.

97. Gurr, "Introduction," *King Henry V,* 17.

98. John Knox, "The First Blast of the Trumpet," in *On Rebellion,* ed. Roger A. Mason (Cambridge: Cambridge University Press, 1994), 3–47.

99. Knox, "The First Blast of the Trumpet," 8.

100. Gurr, "Introduction," *King Henry V,* 17–18. *John Stubbs's Gaping Gulf with Letters and Other Relevant Documents,* ed. Lloyd E. Berry (Charlottesville: The Folger Shakespeare Library/University Press of Virginia, 1968).

101. Gurr, "Introduction," *King Henry V*, 19.

102. Gurr, "Introduction," *King Henry V*, 21: "The genealogical claim is rehearsed in the first act, the conquest is completed in the fourth, and the covenant, whereby the marriage left the French crown to Henry's heirs, is completed in the final act."

103. *Shakespeare's Edward III: An Early Play Restored to the Canon*, ed. Eric Sams (New Haven: Yale University Press, 1996); *Edward III*, ed. Giorgio Melchiori (Cambridge: Cambridge University Press, 1998); *Edward III*, ed. Richard Proudfoot and Nicola Bennett (London: Arden Shakespeare, 2017). As well as recognition, these provide much better texts to work with. Proudfoot says of the edition by George Parfitt (Nottingham: Nottingham Drama Texts, 1985): "Though level-headed in its brief but informative introduction and commentary, it presents a grossly inaccurate text." Proudfoot, *"The Reign of King Edward the Third,"* 185 n. 2.

104. Inna Koskenniemi, "Themes and Imagery in *Edward III*," *Neuphilologische Mitteilungen* 65, no. 4 (1964): 446–80, 480.

105. On linguistic evidence for Shakespeare or Marlowe as authors of parts, see Alfred Hart, *Shakespeare and the Homilies* (New York: Octagon, 1977); M. W. A. Smith, "The Authorship of *The Raigne of King Edward the Third*," *Literary and Linguistic Computing* 6, no. 3 (1991): 166–75; T. V. N. Merriam, "Influence Alone? Reflections on the Newly Canonized *Edward III*," *Notes and Queries*, June 1999, 200–206; and his "Marlowe's Hand in *Edward III*," *Literary and Linguistic Computing* 8, no. 2 (2000): 59–72. For a general discussion, based on reading rather than data, claiming that Thomas Kyd was the other author, see Brian Vickers, "The Two Authors of Edward III," *Shakespeare Survey* 67 (2014): 102–18.

106. Eliot Slater, *The Problem of the Reign of King Edward III: A Statistical Approach* (Cambridge: Cambridge University Press, 1988), 135.

107. *Edward III*, II.i.452; Sonnet 94, in *Shakespeare's Sonnets*, ed. Katherine Duncan-Jones (London: Arden Shakespeare, 1997), 14.

108. Giorgio Melchiori, "Introduction," in *Edward III*, ed. Melchiori, 38–39. For the others, see *Shakespeare's Edward III*, ed. Sams, 186.

109. See, for example, Proudfoot, *"The Reign of King Edward the Third,"* which defers the authorship question until late in the piece; or Larry S. Champion, "'Answer to this Perillous Time': Ideological Ambivalence in *The Raigne of King Edward III* and the English Chronicle Plays," *English Studies* 2 (1988): 117–29.

110. Marlowe, *Edward II*, in *Doctor Faustus and Other Plays*.

111. Proudfoot, *"The Reign of King Edward the Third,"* 165–66.

112. David Bevington and Eric Rasmussen, "Introduction," in Marlowe, *Doctor Faustus and Other Plays*, xxxiv.

113. Proudfoot, *"The Reign of King Edward the Third,"* 166.

114. On the genealogical relation, see *Edward III*, I.i. See Melchiori, "Introduction," 39–40.

115. Koskenniemi, "Themes and Imagery in *Edward III*," 477.

116. *Edward III*, I.i.79–86.

117. *King John*, I.i.3–4 (see chapter 5) and *Henry V*, II.iv.79; *Henry VI, Part 1*, III.iii.7; *Henry VI, Part 2*, III.i.75. See Sams, note to *Shakespeare's Edward III*, 82.

118. See also *Edward III*, III.iii.35.

119. *Edward III*, II.ii.93–94. "Limit" here means territory or region. See the discussion in chapter 7.

120. *Edward III*, IV.i.11; reading "dominiums" with the Arden edition.

121. *Edward III*, III.iii.46–51.

122. *Edward III*, II.ii.8–12.

123. Sams, note to *Shakespeare's Edward III*, 105; see *Edward III*, II.i.448; *Titus Andronicus*, ed. Jonathan Bate (London: Arden Shakespeare, 1995), II.i.115.

124. *Edward III*, III.i.54–57.

125. *Edward III*, III.i.125–26.

126. *Edward III*, I.i.123–30.

127. *Edward III*, I.ii.142–58.

128. Sams, notes to *Shakespeare's Edward III*, 91; see "The Case for Shakespeare", in *Shakespeare's Edward III*, 201–2.

129. *Edward III*, V.i.228–35.

130. McEachern, *The Poetics of English Nationhood*, 134.

CHAPTER SIX

1. Edwards, *Threshold of a Nation*, 107.

2. *The Tempest*, ed. Virginia Mason Vaughan and Alden T. Vaughan (London: Arden Shakespeare, 1999), I.ii.107–16.

3. *The Tempest*, I.ii.166–68.

4. See *The Tempest*, II.I, esp. 270–95.

5. *The Tempest*, I.ii.258, 263–67. For a fascinating study that puts the witch at center stage, see Silvia Federici, *Caliban and the Witch: Women, the Body, and Primitive Accumulation* (Brooklyn: Autonomedia, 2004).

6. *The Tempest*, I.ii.269–70.

7. On the parallels between Prospero and Sycorax, see Stephan Orgel, "Introduction," in *The Tempest*, ed. Stephen Orgel (Oxford: Oxford University Press, 1987), 19–20.

8. *The Tempest*, I.ii.281–84.

9. *The Tempest*, I.ii.284.

10. *The Tempest*, I.ii.320–21. "Slave," "service," and "servant" are frequently used of both Caliban and Ariel.

11. *The Tempest*, I.ii.332–33.

12. Orgel, "Introduction," *The Tempest*, 25.

13. *The Tempest*, I.ii.338–39.

14. *The Tempest*, I.ii.343–45.

15. *The Tempest*, I.ii.351–52.

16. *The Tempest*, I.ii.347–48.

17. Catherine Belsey, *Shakespeare in Theory and Practice* (Edinburgh: Edinburgh University Press, 2008), 58.

18. Simon C. Estok, *Ecocriticism and Shakespeare: Reading Ecophobia* (New York: Palgrave Macmillan, 2011), 103–4.

19. *The Tempest*, III.ii.40–42.

20. *The Tempest*, III.ii.50–51.

21. For a survey of the interpretations and debates, see Alden T. Vaughan, "Shakespeare's Indian: The Americanization of Caliban," *Shakespeare Quarterly* 39, no. 2 (1988): 137–53; and his "Caliban in the 'Third World': Shakespeare's Savage as Sociopolitical Symbol," *Massachusetts Review* 29, no. 2 (1988): 289–313. See Trevor Griffiths, "'This Island's Mine': Caliban and Colonialism," *Yearbook of English Studies* 13 (1983): 159–80; Meredith Anne Skura, "Discourse and the Individual: The Case of Colonialism in *The Tempest*," *Shakespeare Quarterly* 40, no. 1 (1989): 42–69; Jonathan Hart, *Columbus, Shakespeare, and the Interpretation of the New World* (London: Palgrave, 2003), chap. 6; Julia Reinhard Lupton, "Creature Caliban," *Shakespeare Quarterly* 51, no. 1 (2000): 1–23; and, more generally, Alden T. Vaughan and Virginia Mason Vaughan, *Shakespeare's Caliban: A Cultural History* (Cambridge, Cambridge University Press, 1991). The last is good on both literary aspects and representations on stage and screen and in art and other literature. For more general readings using the play as a spur, see Octave Mannoni, *Prospero and Caliban: The Psychology of Colonization*, trans. Pamela Powesland (Ann Arbor: University of Michigan Press, 1990); Stephen Greenblatt, *Marvelous Possessions: The Wonder of the New World* (Oxford: Clarendon Press, 1991); and Monique Allewaert, *Ariel's Ecology: Plantations, Personhood, and Colonialism in the American Tropics* (Minneapolis: University of Minnesota Press, 2013).

22. Martin Butler, "Introduction," in *The Tempest*, ed. Martin Butler (London: Penguin, 2007), xxviii; Virginia Mason Vaughan and Alden T. Vaughan, "Introduction," in *The Tempest*, ed. Virginia Mason Vaughan and Alden T. Vaughan, 44. See also Peter Hulme and William H. Sherman, eds., *"The Tempest" and Its Travels* (London: Reaktion, 2000).

23. Vaughan and Vaughan, "Introduction," *The Tempest*, 48; see 4–5. See *The Tempest*, II.i.84, where Tunis is equated with classical Carthage. For a discussion, see Jerry Brotton, "'This Tunis, sir, was Carthage': Contesting Colonialism in *The Tempest*," in Ania Loomba and Martin Orkin, eds., *Post-Colonial Shakespeares* (London: Routledge, 1998), 23–42.

24. Daniel Brayton, *Shakespeare's Ocean: An Ecocritical Exploration* (Charlottesville: University of Virginia Press, 2012), 167.

25. Annabel Patterson, *Shakespeare and the Popular Voice* (Oxford: Basil Blackwell, 1989), 143.

26. *The Tempest*, II.ii.145–49.

27. Vaughan and Vaughan, "Introduction," *The Tempest*, 45.

28. *The Tempest*, I.i.65–68.

29. *The Tempest*, IV.i.60–75.

30. Scott, *Shakespeare's Nature*, 195.

31. Oliver Rackham, *Trees and Woodland in the British Landscape*, rev. ed. (1976; London: J.M. Dent & Sons, 1990), 184.

32. *The Tempest*, IV.i.80–81.

33. *The Tempest*, IV.i.110–17.

34. Vaughan and Vaughan, "Introduction," *The Tempest*, 71.

35. *The Tempest*, II.i.37.

36. *The Tempest*, II.i.144–45.

37. *The Tempest*, II.i.148–69.

38. Jonathan Bate, "Introduction," in *The Tempest*, ed. Jonathan Bate and Eric Rasmussen (London: Palgrave, 2008), 7. Montaigne's essay in reprinted in *The Tempest*, ed. Vaughan and Vaughan, 303–14.

39. *The Tempest*, I.ii.58.

40. Butler, "Introduction," *The Tempest*, xl; "Commentary," 99–100.

41. *The Tempest*, II.i.112–13.

42. For a contemporary examination of this notion, see Gerard Toal, *Near Abroad: Putin, the West, and the Contest over Ukraine and the Caucasus* (Oxford: Oxford University Press, 2017).

43. *Othello*, I.i.148.

44. The text can be found in Bullough, *Narrative and Dramatic Sources of Shakespeare*, 7:241–52, trans. Bullough, and reprinted in *Othello*, ed. Honigmann, 370–86.

45. Jonathan Bate, "Introduction," in *Othello*, ed. Jonathan Bate and Eric Rasmussen (London: Macmillan, 2009), 3; see Norman Sanders, note to *Othello*, ed. Norman Sanders (Cambridge: Cambridge University Press, 2003), 64.

46. The literature on the other themes is extensive. For a useful discussion of Venice, Venetians, and the Spanish context of the time of writing, see Eric Griffin, "Un-sainting James: or, *Othello* and the 'Spanish Spirits' of Shakespeare's Globe," *Representations* 62 (1998): 58–99. For a discussion of the corporeal, see Peter Stallybrass, "Patriarchial Territories: The Body Enclosed," in Margaret W. Ferguson, Maureen Quilligan, and Nancy J. Vickers, eds., *Rewriting the Renaissance: The Discourses of Sexual Difference in Early Modern Europe* (Chicago: University of Chicago Press, 1986), 123–42.

47. Tom McAlindon, "Introduction," in *Othello*, ed. Kenneth Muir (London: Penguin, 2005), lvii.

48. Richard Knolles, *The General Historie of the Turkes, from The first beginning of that Nation to the rising of the Othoman Familie with all the notable expeditions of the Christian Princes against them*, 3rd ed. (1603; London: Adam Islip, 1621). Relevant passages are printed in Bullough, *Narrative and Dramatic Sources of Shakespeare*, 7:262–65.

49. *Othello*, I.ii.39–44.

50. Mark Rose, *Shakespearean Design* (Cambridge, MA: Harvard University Press, 1972), 92.

51. *Othello*, I.iii.1–12.

52. W. Watkiss Lloyd, *Critical Essays on the Plays of Shakespeare* (London: George Bell and Sons, 1892), 463, part cited in E. A. J. Honigmann, "Introduction," in *Othello*, ed. E. A. J. Honigmann, 63.

53. Michael Neill, note to *Othello, the Moor of Venice*, ed. Michael Neill (Oxford: Oxford University Press, 2006), 215.

54. *Othello*, II.i.291.

55. *Othello*, I.iii.15.

56. *Othello*, I.iii.18–31.The First Quarto text of this speech ends with "facile question bear it." For that text, see *The First Quarto of Othello*, ed. Scott McMillan (Cambridge: Cambridge University Press, 2001). For discussions, see McMillan's introduction; Neill, "Appendix B: The Texts of the Play," *Othello*, 405–33; and E. A. J. Honigmann, *The Texts of "Othello" and Shakespearian Revision* (London: Routledge, 1996).

57. *Othello*, I.iii.32–40.

58. Harold C. Goddard, *The Meaning of Shakespeare, Volume II* (1951; Chicago: University of Chicago Press, 1960), 81.

59. *Othello,* I.iii.49–50.

60. See Neill, "Introduction," *Othello,* 25.

61. *Othello,* I.iii.211–12. Neill, note to *Othello*, 227, observes that "lose" appears as "loose" in the Folio, suggesting the term from animal husbandry, meaning to "release the female to the male." See also Polonius: "At such a time I'll loose my daughter to him" (*Hamlet,* II.ii.159).

62. *Othello,* I.iii.135–37.

63. Wolfgang Clemen, *The Development of Shakespeare's Imagery,* 2nd ed. (1951; London: Methuen and Co., 1977), 96.

64. *Othello,* I.iii.141–42.

65. One source may be John Leo, *A Geographical Historie of Africa*, trans. John Pory (London: Georg Bishop, 1600). See Ayanna Thompson, "Introduction," in *Othello: Revised Edition*, ed. E. A. J. Honigmann (London: Arden Shakespeare, 2016), 15; Alden T. Vaughan and Virginia Mason Vaughan, "Before Othello: Elizabethan Representations of Sub-Saharan Africans," *William and Mary Quarterly* 54, no. 1 (1997): 19–44.

66. *Othello,* I.iii.456–65. This speech is one of the key cuts from the 1622 Quarto text.

67. McAlindon, "Introduction," *Othello*, lvii.

68. *Othello,* I.iii.222–24.

69. *Othello,* II.i.1–19.

70. Neill, note to *Othello*, 241.

71. McMillan, note to *The First Quarto of Othello*, 74.

72. Kenneth Muir, "Commentary," *Othello*, ed. Muir, 167.

73. *Othello,* II.i.20–24.

74. *Othello,* II.i.68–69.

75. Neill, "Introduction," *Othello*, 25–26.

76. Honigmann, "Introduction," *Othello,* 11.

77. McAlindon, "Introduction," *Othello*, lviii-lix.

78. Emrys James, "'Othello,' 'Lepanto' and the Cyprus Wars," *Shakespeare Survey* 21 (1969): 47–52; see Neill, "Appendix A: The Date of the Play," *Othello*, 399; Vaughan, *Othello*, 27–28. More generally, see Michael Neill, "Changing Places in *Othello*," *Shakespeare Survey* 37 (1984): 115–32.

79. Honigmann, "Introduction," *Othello*, 8.

80. *Othello,* II.i.201.

81. Goddard, *The Meaning of Shakespeare, Volume II*, 80. Iago's line is in *Othello,* II.i.114.

82. Bate, "Introduction," *Othello*, 7.

83. Neill, "Introduction," *Othello*, 148.

84. Sanders, "Introduction," *Othello*, 18–19.

85. *Othello,* II.iii.166–68. See Daniel Vitkus, *Turning Turk: English Theater and the Multicultural Mediterranean, 1570–1630* (London: Palgrave Macmillan, 2003), esp. chap. 4; Virginia Mason Vaughan, *Othello: A Contextual History* (Cambridge: Cambridge University Press, 1994), chap. 1; Thompson, "Introduction," *Othello*, 34. See also *Richard III,* III.v.41: "What? Think you we are Turks or infidels?"

86. *Othello,* I.i.110; I.iii.356.

87. *Othello,* IV.iii.24, 31. For a discussion of these passages, see Thompson, "Introduction," *Othello,* 24–25; and John W. Draper, "Shakespeare and Barbary," *Études Anglaises* 14, no. 4 (1961): 306–13.

88. William Shakespeare and George Wilkins, *Pericles,* ed. Suzanne Gossett (London: Arden Shakespeare, 2004).

89. Suzanne Gossett, "Introduction," in *Pericles,* ed. Gossett, 1, 36.

90. For a hugely detailed discussion, see MacDonald Jackson, *Defining Shakespeare: Pericles as Test Case* (Oxford: Oxford University Press, 2003); and Vickers, *Shakespeare, Co-Author,* chap. 5. While Gossett's Arden Third Series edition presents both Shakespeare and Wilkins as its authors, other recent editions assert Shakespeare was its sole author. See Doreen DelVecchio and Antony Hammond, "Introduction," in *Pericles, Prince of Tyre,* ed. Doreen DelVecchio and Antony Hammond (Cambridge: Cambridge University Press, 1998), 8–15. See also Roger Warren, "Introduction," in William Shakespeare and George Wilkins, *A Reconstructed Text of Pericles, Prince of Tyre,* ed. Roger Warren, on the basis of a text prepared by Gary Taylor and MacD. Jackson (Oxford: Oxford University Press, 2003), 60–71.

91. There is an extensive literature on these questions. See, for example, Philip Edwards, "An Approach to the Problem of *Pericles,*" *Shakespeare Survey* 5 (1952): 25–49; Gary Taylor, "The Transmission of *Pericles,*" *Papers of the Bibliographical Society of America* 80, no. 2 (1986): 193–217.

92. The most extensive reconstruction is Shakespeare and Wilkins, *A Reconstructed Text of Pericles, Prince of Tyre,* ed. Warren, which reprints the text of the Quarto edition in an appendix, 231–86. Wilkins's text is reprinted in Bullough, *Narrative and Dramatic Sources of Shakespeare,* vol. 6. The opposite approach to the Oxford edition is taken in the DelVecchio and Hammond Cambridge edition.

93. On the background, see Jonathan Bate and Lucy Munro, "Introduction," in *Pericles,* ed. Jonathan Bate, Eric Rasmussen, and Lucy Munro (Basingstoke: Macmillan, 2012), 1; DelVecchio and Hammond, "Introduction," *Pericles, Prince of Tyre,* 1–8. Gower's text is reprinted in Bullough, *Narrative and Dramatic Sources of Shakespeare,* vol. 6.

94. Gossett, "Introduction," *Pericles,* 127.

95. *Pericles,* I.iii.28.

96. *Pericles,* I.o.19.

97. Gossett, "Introduction," *Pericles,* 127.

98. Richard Halpern, *Shakespeare among the Moderns* (Ithaca: Cornell University Press, 1997), 150; Linda McJannet, "Genre and Geography: The Eastern Mediterranean in *Pericles* and *The Comedy of Errors,*" in John Gillies and Virginia Mason Vaughan, eds., *Playing the Globe: Genre and Geography in English Renaissance Drama* (London: Associated University Presses, 1988), 86–106; see 88.

99. Gossett, "Introduction," *Pericles,* 127. For a general discussion, see Constance C. Relihan, "Liminal Geography: *Pericles* and the Politics of Place," *Philological Quarterly* 71, no. 3 (1992): 281–99; Lisa Hopkins, "'The Shores of My Mortality': Pericles' Greece of the Mind," in David Skeele, ed., *Pericles: Critical Essays* (New York and London: Garland, 2000), 228–37; Bernhard Klein, "The Sea in *Pericles,*" in Haberman and Witen, *Shakespeare and Space,* 121–40.

100. For a discussion, see DelVecchio and Hammond, "Introduction," *Pericles, Prince of Tyre*, 1; Gossett, "Introduction," *Pericles*, 54–56. Eric Sams, *The Real Shakespeare II: Retrieving the Later Years, 1594–1616* (New York: Centro Studi Eric Sams, 2008), 173–77, suggests that the first two acts date from the 1580s, and are by Shakespeare's hand; the others were revised in the early seventeenth century.

101. Gossett, "Introduction," *Pericles*, 129. See http://dcc.newberry.org/system/artifacts/390/original/Ortelius_Theatrum_Plate_122.jpg.

102. Gossett, "Introduction," *Pericles*, 129; see McJannet, "Genre and Geography," 91; Mentz, *At the Bottom of Shakespeare's Ocean*, 72.

103. Gossett, "Introduction," *Pericles*, 132. The first reference is to Lesley B. Cormack, "Britannia Rules the Waves? Images of Empire in Elizabethan England," in Gordon and Klein, *Literature, Mapping, and the Politics of Space*, 45–68; the Gillies reference is to *Shakespeare and the Geography of Difference*, 101. See also Bradin Cormack, "Marginal Waters: *Pericles* and the Idea of Jurisdiction," in Gordon and Klein, *Literature, Mapping, and the Politics of Space*, 155–80.

104. Mentz, *At the Bottom of Shakespeare's Ocean*, 71.

105. Mentz, *At the Bottom of Shakespeare's Ocean*, 71.

106. *Pericles*, I.i.24–25.

107. Estok, *Ecocriticism and Shakespeare*, 79.

108. *Pericles*, II.i.49.

109. Mentz, *At the Bottom of Shakespeare's Ocean*, 72. See Sarah Hanna, "Shakespeare's Greek World: The Temptations of the Sea," in Gillies and Vaughan, *Playing the Globe*, 107–28.

110. *Pericles*, IV.iv.1–7. On Gower, see Steven Mullaney, *The Place of the Stage: License, Play, and Power in Renaissance England* (Chicago: University of Chicago Press, 1988), 148: "Shakespeare had never before felt compelled to bring his authors or authorities on stage; that he does so here, in a romance structured around the genealogical entanglements and contaminations of authority, should at least give us pause . . . [yet] Gower in fact conceals as much as he reveals, especially where the genealogical entanglements of Shakespeare's sources are concerned."

111. See Eugene Giddens, "Commentary," in *Pericles*, ed. Eugene Giddens (London: Penguin, 2008), 142.

112. *Pericles*, III.i.1–6.

113. *Pericles*, III.iii.12–13; V.i.146–47.

114. *Pericles*, III.i.46–53.

115. *Pericles*, III.i.57–64. "Ooze" is an emendation from the Quarto's "oare."

116. *Pericles*, III.i.72–75.

117. *Pericles*, III.iii.12–21; and see I.iv.83–94.

118. This story is told by Gower, *Pericles*, IV.0.11–52.

119. *Pericles*, IV.iii.5–6.

120. See *Pericles*, IV.i.89–98; IV.ii.37–50.

121. Gossett, "Introduction," *Pericles*, 131.

122. On piracy, see Daniel Heller-Roazen, *The Enemy of All: Piracy and the Law of Nations* (New York: Zone Books, 2009); Marcus Rediker, *Outlaws of the Atlantic: Sailors, Pirates, and Motley Crews in the Age of Sail* (London: Verso, 2014); and Mark Neocleous,

The Universal Adversary: Security, Capital, and the "Enemies of All Mankind" (London: Routledge, 2016).

123. Gossett, "Introduction," *Pericles*, 133.

124. *Pericles*, II.iii.85–86.

125. *Pericles*, V.i.180–84.

126. *Pericles*, IV.iv.10.

127. See Gossett, note to *Pericles*, 342; Warren, note to *Pericles*, 190.

128. See David Skeele, *Thwarting the Wayward Seas: Critical and Theatrical History of Shakespeare's "Pericles" in the Nineteenth and Twentieth Centuries* (Newark: University of Delaware Press, 1998); and more generally John H. Astington, "Fortune, the Sea, and Shakespeare's Late Plays," in Leslie Thomson, ed., *Fortune: "All is But Fortune"* (Washington, D.C.: The Folger Shakespeare Library, 2000), 23–29; Brayton, *Shakespeare's Ocean*.

129. *Antony and Cleopatra*, ed. John Wilders (London: Arden Shakespeare, 1995), II.ii.120–21.

130. *Antony and Cleopatra*, II.ii.121–23.

131. *Antony and Cleopatra*, II.ii.160.

132. John Wilders, "Introduction," *Antony and Cleopatra*, ed. Wilders, 2.

133. Wilders, "Introduction," *Antony and Cleopatra*, 2–3.

134. *Antony and Cleopatra*, I.i.1–2.

135. *Antony and Cleopatra*, I.i.12–13.

136. See *Julius Caesar*, ed. David Daniell (London: Arden Shakespeare, 1998), IV.i.12–15.

137. *Antony and Cleopatra*, II.vii.71.

138. *Antony and Cleopatra*, I.v.24–25.

139. John Dover Wilson, note to *Antony and Cleopatra*, ed. John Dover Wilson (Cambridge: Cambridge University Press, 1950), 157.

140. *Antony and Cleopatra*, II.vii.91.

141. *Antony and Cleopatra*, II.vii.92.

142. *Antony and Cleopatra*, II.vii.61–62.

143. Maurice Charney, *Shakespeare's Roman Plays: The Function of Imagery in the Drama* (Cambridge, MA: Harvard University Press, 1961), 93; see 80–93 for a general discussion.

144. *Antony and Cleopatra*, I.i.14. See Wilders, note to *Antony and Cleopatra*, 91.

145. *Antony and Cleopatra*, I.i.15–16.

146. *Antony and Cleopatra*, I.i.17.

147. *Antony and Cleopatra*, I.i.34–41.

148. Wilders, note to *Antony and Cleopatra*, 93.

149. *Antony and Cleopatra*, IV.xiv.58.

150. *Antony and Cleopatra*, II.iii.1–2.

151. *Antony and Cleopatra*, II.iii.5–7.

152. *Antony and Cleopatra*, I.ii.105–9.

153. Emrys Jones, "Commentary," *Antony and Cleopatra*, ed. Emrys Jones (London: Penguin, 2015), 157.

154. *Antony and Cleopatra*, III.vi.8–16. See David Bevington, "Introduction," in *Antony and Cleopatra*, ed. David Bevington, updated ed. (1990; Cambridge: Cambridge University Press, 2005), 72.

155. *Antony and Cleopatra,* II.vi.8–10.

156. *Antony and Cleopatra,* II.vi.20–21.

157. *Antony and Cleopatra,* II.vi.24–26.

158. *Antony and Cleopatra,* III.v.11–12.

159. *Antony and Cleopatra,* III.v.13–15.

160. Bevington, note to *Antony and Cleopatra,* 173.

161. *Antony and Cleopatra,* I.ii.191–92.

162. *Antony and Cleopatra,* I.iii.51, 47.

163. *Antony and Cleopatra,* I.iv.36.

164. *Antony and Cleopatra,* I.iv.43–47

165. *Antony and Cleopatra,* I.iv.49.

166. *Antony and Cleopatra,* I.iv.51–53.

167. *Antony and Cleopatra,* I.iv.79–80. See *Julius Caesar,* I.iii.87–88: "And he shall wear his crown by sea and land / In every place save here in Italy."

168. *Antony and Cleopatra,* II.i.8–9.

169. *Antony and Cleopatra,* II.i.16–17.

170. *Antony and Cleopatra,* II.ii.171–73.

171. *Antony and Cleopatra,* II.vi.34–39.

172. *Antony and Cleopatra,* II.vi.84–95.

173. Wilders, note to *Antony and Cleopatra,* 160.

174. Augustine, *Concerning the City of God against the Pagans,* trans. and ed. R. W. Dyson (Cambridge: Cambridge University Press, 1998), 147–48 (Book IV, Chapter 4). Compare the story in Cicero, *De re publica,* in *On the Commonwealth and On the Laws,* ed. and trans. James E. G. Zetzel (Cambridge: Cambridge University Press, 1999), 67 (Book III, 24a).

175. *Antony and Cleopatra,* II.vii.61.

176. *Antony and Cleopatra,* II.vii.69–70.

177. *Antony and Cleopatra,* II.vii.71–72.

178. *Antony and Cleopatra,* I.i.34.

179. *Antony and Cleopatra,* II.v.78; II.v.94–95; III.vii.15. See Bevington, "Introduction," *Antony and Cleopatra,* 33. For a more general discussion, see Clemen, *The Development of Shakespeare's Imagery,* 159–67.

180. *Antony and Cleopatra,* III.vii.27–28.

181. *Antony and Cleopatra,* III.vii.31–34.

182. *Antony and Cleopatra,* III.vii.34–53.

183. Drawing on notes to *Antony and Cleopatra,* ed. Wilders, 196–97.

184. *Antony and Cleopatra,* III.vii.61–66.

185. *Antony and Cleopatra,* III.vii.3–4.

186. *Antony and Cleopatra,* III.x.6–8.

187. See *OED* and Wilders, note to *Antony and Cleopatra,* 201.

188. *Antony and Cleopatra,* III.x.18–24.

189. Wilders, notes to *Antony and Cleopatra,* 202.

190. *Antony and Cleopatra,* III.xi.55–56.

191. *Antony and Cleopatra,* III.xi.56–58.

192. *Antony and Cleopatra,* V.ii.114.

193. *Antony and Cleopatra,* III.xi.64.

194. *Antony and Cleopatra,* IV.ii.4–5.

195. *Antony and Cleopatra,* IV.x.1–9.

196. *Antony and Cleopatra,* IV.xi.1–4.

197. *Antony and Cleopatra,* V.ii.78–91.

198. Wilders, notes to *Antony and Cleopatra,* 282–83.

199. *Antony and Cleopatra,* V.ii.119.

200. *Antony and Cleopatra,* III.xiii.76.

201. *Antony and Cleopatra,* V.i.17–19.

202. See Donald K. Anderson Jr., "A New Gloss for the 'Three-Nook'd World' of *Antony and Cleopatra," English Language Notes* 17, no. 2 (1979): 103–6; and Alfred Hiatt, "Mapping the Ends of Empire," in Ananya Jahanara Kabir and Deanne Williams, eds., *Postcolonial Approaches to the European Middle Ages* (Cambridge: Cambridge University Press, 2010), 48–76; and on these maps generally, see Leo Bagrow, *History of Cartography,* rev. and enl. R. A. Skelton (London: CA Watts, 1964), chap. 3.

203. On this, see Suzanne Conklin Akbari, "From Due East to True North: Orientalism and Orientation," in Jeffrey Jerome Cohen, ed., *The Postcolonial Middle Ages* (New York: St Martin's Press, 2000), 19–34.

204. For a longer discussion, see Elden, *The Birth of Territory,* 114–15.

205. Wilders, "Introduction," *Antony and Cleopatra,* 52–53.

CHAPTER SEVEN

1. On this period generally, though with no reference to Shakespeare, see E. G. R. Taylor, *Late Tudor and Early Stuart Geography, 1583–1650: A Sequel to Tudor Geography, 1485–1583* (London: Methuen & Co., 1934); and Taylor, *The Mathematical Practitioners of Tudor and Stuart England* (Cambridge: Cambridge University Press, 1954).

2. Cohen, *Shakespeare and Technology,* 1. See also Fletcher, *Time, Space, and Motion in the Age of Shakespeare*; and Paula Blank, *Shakespeare and the Mismeasure of Renaissance Man* (Ithaca: Cornell University Press, 2006). My thinking on calculation and its relation to territory more generally has been shaped, above all, by the work of Martin Heidegger and Foucault. See Stuart Elden, *Speaking against Number: Heidegger, Language, and the Politics of Calculation* (Edinburgh: Edinburgh University Press, 2006); and Elden, *Foucault's Last Decade,* esp. chap. 2.

3. J. M. Nosworthy, "Introduction," in *Cymbeline* (London: Arden Shakespeare, 1955), lxxi.

4. *A Midsummer Night's Dream,* ed. Sukanta Chaudhuri (London: Arden Shakespeare, 2017), III.ii.9.

5. *Julius Caesar* I.i.3, 7; see also the "poor mechanic porters" in *Henry V,* I.ii.200; and the "mechanics" in *Coriolanus,* V.iii.83.

6. *Henry VI, Part 2,* I.iii.194; *Henry IV, Part 2,* V.v.34.

7. *Love's Labour's Lost,* I.ii.40–41.

8. *Troilus and Cressida,* I.ii.109–10.

9. *Troilus and Cressida,* III.iii.254–55.

10. Bevington, note to *Troilus and Cressida,* ed. Bevington, 282.

11. The phrase "measure still for measure" is said by the Duke in the final scene: *Measure for Measure*, ed. J. W. Lever (London: Routledge, 1965), V.i.409. For a contextual reading of the play, see Debora Kuller Shuger, *Political Theologies in Shakespeare's England: The Sacred and the State in "Measure for Measure"* (Basingstoke: Palgrave, 2001). For political readings of *Measure for Measure*, see Leonard Tennenhouse, "Representing Power: *Measure for Measure* in Its Time," in Stephen Greenblatt, ed., *The Powers of Forms in the English Renaissance* (Norman, OK: Pilgrim Books, 1982), 139–56; Kiernan Ryan, "*Measure for Measure*: Marxism before Marx," in Howard and Shershow, *Marxist Shakespeares*, 227–44; Conal Conden, "Unfolding 'the Properties of Government': The Case of *Measure for Measure* and the History of Political Thought," in Armitage, Condren, and Fitzmaurice, *Shakespeare and Early Modern Political Thought*, 157–75. For a reading inspired by Foucault's work on governmentality, see Richard Wilson, *Shakespeare in French Theory: King of Shadows* (London: Routledge, 2007), chap. 2.

12. Jacques Lezra, *Unspeakable Subjects: The Genealogy of the Event in Early Modern Europe* (Stanford: Stanford University Press, 1997), 259. The reference is to *Measure for Measure*, I.i.18: "Elected him our absence to supply."

13. *Measure for Measure*, I.i.9–11.

14. J. W. Lever, note to *Measure for Measure*, ed. Lever, 4.

15. *Measure for Measure*, I.i.45–47.

16. *Measure for Measure*, I.iii.11–13.

17. *Measure for Measure*, I.iii.50–61.

18. Lever, note to *Measure for Measure*, 25.

19. Jonathan Bate, "Introduction," in *Measure for Measure*, ed. Jonathan Bate and Eric Rasmussen (London: RSC Shakespeare/Macmillan, 2010), 2; see Julia Briggs, "Introduction," in *Measure for Measure*, ed. J. M. Nosworthy (London: Penguin, 2005), liii-iv.

20. *Measure for Measure*, I.iii.50.

21. *Measure for Measure*, III.i.93.

22. The first two instances in this play may be down to the revision work of Thomas Middleton. See the edition of the text in Middleton, *The Collected Works*, 1550, 1553.

23. *Love's Labour's Lost*, V.ii.406.

24. *The Merry Wives of Windsor*, II.ii.20–22.

25. *Henry IV, Part 2*, II.iii.39–41.

26. Nosworthy, note to *Measure for Measure*, 117.

27. *Measure for Measure*, I.ii.69–70.

28. N. W. Bawcutt, note to *Measure for Measure*, ed. N. W. Bawcutt (Oxford: Oxford University Press, 1991), 95.

29. Bate, "Introduction," in *Measure for Measure*, 2.

30. *Measure for Measure*, II.i.54.

31. *Measure for Measure*, III.i.94–96.

32. Lever, note to *Measure for Measure*, 72.

33. *Measure for Measure*, II.i.50–56.

34. Bawcutt, note to *Measure for Measure*, 112; *Othello*, V.ii.233.

35. Gibbons, note to *Measure for Measure*, 102.

36. Nosworthy, note to *Measure for Measure*, 125–26. See Lever, note to *Measure for Measure*, 72; Br. Nicholson, "Measure for Measure, II. i. 95, 98," *Notes and Queries*, September 10, 1892, 203–4.

37. Bawcutt, "Appendix A: Longer Textual Notes," in *Measure for Measure*, 234–35; see also "Editorial Procedures," 77: "Even when explanation is not possible, as with 'prenzie' at 3.I.95 and 98, F is retained because no emendation made so far seems convincing."

38. Brian Gibbons, note to *Measure for Measure*, ed. Brian Gibbons (Cambridge: Cambridge University Press, 1991), 132.

39. Bate and Rasmussen, note to *Measure for Measure*, ed. Bate and Rasmussen, 60.

40. *Measure for Measure*, II.iv, 86.

41. *Measure for Measure*, V.i.47–48.

42. *Measure for Measure*, III.i.255–56.

43. *Measure for Measure*, V.i.218–21.

44. *Measure for Measure*, III.i.213–23.

45. *Measure for Measure*, IV.i.144.

46. On this theme generally, see Steve Mentz, *Shipwreck Modernity: Ecologies of Globalization, 1550–1719* (Minneapolis: University of Minnesota Press, 2015).

47. M. M. Mahood, "Introduction," in *The Merchant of Venice*, ed. M. M. Mahood (Cambridge: Cambridge University Press, 2003), 1.

48. *The Merchant of Venice*, ed. John Drakakis (London: Arden Shakespeare, 2010), I.i.7–13.

49. *The Merchant of Venice*, I.i.16–18.

50. Drakakis, note to *The Merchant of Venice*, 171.

51. *The Merchant of Venice*, I.i.25–26.

52. *The Merchant of Venice*, I.i.41–44.

53. John Drakakis, "Introduction," in *The Merchant of Venice*, ed. Drakakis, 12. On these themes in the play generally, see Mark Netzloff, "The Lead Casket: Capital, Mercantalism, and *The Merchant of Venice*," in Linda Woodbridge, ed., *Money in the Age of Shakespeare* (New York: Palgrave Macmillan, 2003), 159–76; Steven R. Mentz, "The Fiend Gives Friendly Counsel: Launcelot Gobbo and Polyglot Economics in *The Merchant of Venice*," in Woodbridge, *Money in the Age of Shakespeare*, 177–87; and more generally, Halpern, *The Poetics of Primitive Accumulation: English Renaissance Culture and the Genealogy of Capital.*

54. *The Merchant of Venice*, I.i.177–79.

55. *The Merchant of Venice*, I.III.49–52.

56. *The Merchant of Venice*, I.III.99–100

57. *The Merchant of Venice*, I.III.140–47.

58. W. Moelwyn Merchant, "Commentary," in *The Merchant of Venice*, ed. W. Moelwyn Merchant (London: Penguin, 2005), 118.

59. *The Merchant of Venice*, I.iii.153–55.

60. *The Merchant of Venice*, I.iii.14–25.

61. Mahood, "Introduction," *The Merchant of Venice*, 13.

62. Mahood, "Introduction," *The Merchant of Venice*, 27.

63. *The Merchant of Venice,* I.iii.176–77.

64. *The Merchant of Venice,* II.viii.27–30.

65. *The Merchant of Venice,* III.i.2–7.

66. *The Merchant of Venice,* III.i.89–96.

67. *The Merchant of Venice,* III.Ii.266–70.

68. *The Merchant of Venice,* III.ii.278–82.

69. *The Merchant of Venice,* III.ii.283–89.

70. *The Merchant of Venice,* III.ii.314–18.

71. *The Merchant of Venice,* III.iii.4, 5, 11, 12, 17, etc.

72. *The Merchant of Venice,* IV.i.254–58.

73. *The Merchant of Venice,* IV.i.34–38.

74. See also *The Merchant of Venice,* III.ii.277.

75. James Shapiro, *Shakespeare and the Jews* (New York: Columbia University Press, 1996), 122.

76. *The Merchant of Venice,* IV.i.98–99.

77. Jay L. Halio, "Introduction," in *The Merchant of Venice*, ed. Jay L. Halio (Oxford: Oxford University Press, 1993), 52. See O. Hood Phillips, *Shakespeare and the Lawyers* (London: Methuen & Co., 1972), chap. 8; E. F. J. Tucker, "The Letter of the Law in *The Merchant of Venice*," *Shakespeare Survey* 29 (1976): 93–102.

78. *The Merchant of Venice,* IV.i.302–8, 320–28. In line 322 "tak'st" is an emendation of the Folio's "cut'st," though both make sense.

79. Halio, notes to *The Merchant of Venice*, ed. Halio, 204.

80. *The Merchant of Venice,* V.i.276–77.

81. *The Merchant of Venice,* II.vi.36.

82. *The Merchant of Venice,* III.ii.149–59. The Quarto has "sum of something"; most editions follow the Folio's correction to "sum of nothing."

83. Generally, see Sullivan, *The Drama of Landscape*, esp. chap. 3. As Blank notes, *Shakespeare and the Mismeasure of Renaissance Man*, 121, even some of Shakespeare's uses of the term "place" should be understood in the arithmetic sense.

84. *Coriolanus,* II.iii.20–23.

85. *The Two Noble Kinsmen,* IV.i.141–44.

86. *Hamlet* ,V.i.129–30 (also in Folio); see Thompson and Taylor, note to *Hamlet*, 419. Osric's description of Laertes as "the card or calendar of gentry" (*Hamlet* V.ii.95; not in Folio) seems to be in the metaphorical sense of the image or exemplar.

87. Cohen, *Shakespeare and Technology*, 40.

88. *Macbeth,* I.iii.14–17.

89. Braunmuller, "Introduction," *Macbeth*, 39, provides a reproduction of one from the mid-sixteenth century—a map of Europe, with sea routes in the Mediterranean, a compass, and landmarks.

90. *Macbeth,* I.iii.11–12.

91. *Macbeth,* I.iii.28–29.

92. See Clark and Mason, notes to *Macbeth*, 138–39; Gillies, "Introduction: Elizabethan Drama and the Cartographizations of Space," 32. See Cohen, *Shakespeare and Technology*, 38, 39; and more generally Mentz, *At the Bottom of Shakespeare's Ocean*.

93. *Macbeth,* IV.i.51–52.

94. Brayton, *Shakespeare's Ocean*, 176.

95. *The Tempest*, I.i.65–66.

96. Poole, *Supernatural Environments in Shakespeare's England*, 202.

97. Poole, *Supernatural Environments in Shakespeare's England*, 203.

98. *The Tempest*, IV.i.237; see also *Henry VIII*, V.iii.40–41.

99. Munday, Chettle, et al., *Sir Thomas More*, 13, 24–26.

100. Jowett, notes to *Sir Thomas More*, ed. Jowett, 286–87.

101. Edward Wright, *Certaine Errors in Navigation, Arising either of the ordinarie erroneous making or using of the sea Chart, Compass, Cross staffe, and Tables of declination of the Sunne, and fixed Starres detected and corrected* (London: Valentine Sims, 1599).

102. See Paul Glennie and Nigel Thrift, *Shaping the Day: A History of Timekeeping in England and Wales, 1300–1800* (Oxford: Oxford University Press, 2009), esp. chap. 8.

103. Leonarde Digges, *A Boke Named Tectonicon* (1556; London: Thomas Marshe, 1566); John Norden, *The Surveyor's Dialogue (1618): A Critical Edition*, ed. Mark Netzloff (Aldershot: Ashgate, 2010).

104. William Cuningham, *The Cosmographical Glasse, conteinyng the pleasant Geographie, Hydrographie, or Navigation* (London, 1559).

105. Georg Braun and Frans Hogenberg, *Civitates Orbis Terrarum*, 6 vols. (Cologne, 1572–1617); John Norden, *Speculum Britanniae: Description of Middlesex* (1593; Amsterdam: Da Capo, 1971).

106. Cohen, *Shakespeare and Technology*, 2.

107. Cohen, *Shakespeare and Technology*, 42.

108. *The Merchant of Venice*, II.v.35–36.

109. Drakakis, note to *The Merchant of Venice*, 253.

110. Genesis 32.10.

111. Merchant, "Commentary," *The Merchant of Venice*, ed. Merchant, 124.

112. David G. Krehbiel, "Jacob's Staff," *Ontario Land Surveyor*, Spring 1990, http://www.surveyhistory.org/jacob's_staff1.htm; Poole, *Supernatural Environments in Shakespeare's England*, 177–78, 273 n. 82.

113. *Henry IV, Part 1*, II.ii.12–13.

114. See Cohen, *Shakespeare and Technology*, 173.

115. *Love's Labour's Lost*, V.ii.474.

116. Woudhuysen, note to *Love's Labour's Lost*, 269; Hibbard, note to *Love's Labour's Lost*, 213.

117. *Antony and Cleopatra*, II.iii.5–7.

118. *The Two Noble Kinsmen*, IV.iii.94–95.

119. Cohen, *Shakespeare and Technology*, 174. See D.A. Harvey, *Maps in Tudor England* (Chicago: University of Chicago Press, 1993), chap. 5; Klein, *Maps and the Writing of Space*, chap. 2.

120. *Henry IV, Part 2*, I.iii.41–62. The Quarto text omits lines 35–55 of this scene.

121. Luke 14.28–32. See Davison, "Commentary," in *Henry IV, Part Two*, ed. Peter Davison (London: Penguin, 2005), 171.

122. Poole, *Supernatural Environments in Shakespeare's England*, 200.

123. Davison, "Commentary," *Henry IV, Part Two*, 172.

124. Bulman, note to *Henry IV Part 2*, ed. Bulman, 207. For a discussion, see Martin Brückner and Kristen Poole, "The Plot Thickens: Surveying Manuals, Drama, and the Materiality of Narrative Form in Early Modern England," *English Literary History* 69, no. 3 (2002): 617–48; and more generally Lorna Hutson, "Fortunate Travelers: Reading for the Plot in Sixteenth-Century England," *Representations* 41 (1993): 83–103.

125. Bulman, note to *Henry IV Part 2*, ed. Bulman, 208.

126. Jay L. Halio, note to *Henry VIII or, All is True*, ed. Jay L. Halio (Oxford: Oxfrod University Press, 1999), 81.

127. *Henry VIII*, I.i.172–73.

128. *Henry VIII*, I.i.223–24.

129. McMullan, note to *Henry VIII*, ed. McMullan, 230; McMullan, "Introduction," *Henry VIII*, 100–101.

130. D. R. Hainsworth, *Stewards, Lords, and People: The Estate Steward and His World in Later Stuart England* (Cambridge: Cambridge University Press, 1992), 109, part cited by McMullan, note to *Henry VIII*, 222. See Brückner and Poole, "The Plot Thickens."

131. McMullan, note to *Henry VIII*, 222. For a discussion of these lines, see Poole, *Supernatural Environments in Shakespeare's England*, 201. On the play generally, see William M. Baillie, "*Henry VIII*: A Jacobean History," *Shakespeare Studies* 12 (1979): 247–66.

132. *Henry IV, Part 2*, II.iv.97, 102–3, 140.

133. Davison, "Commentary," *Henry IV Part Two*, 192. There is a similar play in *Titus Andronicus*, V.iii.111, where Aaron treats Titus's hand as forfeit: "I played the cheater for thy father's hand." The only other instance of the word "cheater" is in Sonnet 151, line 3, though it there just means deceiver.

134. For a discussion of these, see Virginia Mason Vaughan, "Preface: The Mental Maps of English Renaissance Drama," in Gillies and Vaughan, *Playing the Globe*, 7–16, 12.

135. *Titus Andronicus*, III.ii.11.

136. *Henry VI, Part 2*, I.i.203.

137. Sonnet 68, lines 1 and 13.

138. *The Rape of Lucrece* 1712–13. Lucrece's sleep is also described as "the map of death," in *The Rape of Lucrece* 402.

139. *The Rape of Lucrece* 755. See Duncan-Jones and Woudhysen, notes to *Shakespeare's Poems*, 300, 372. Other instances are "the map of my microcosm" (*Coriolanus*, II.i.60); "I see, as in a map, the end of all" (*Richard III*, II.iv.55); and "I may remain the map of infamy" (*Edward III*, III.iii.218). The first refers to a face (and is Shakespeare's only use of the word "microcosm"); the second to a representation; the third to an emblem. See Cohen, *Shakespeare and Technology*, 44–46.

140. *Macbeth*, I.v.62–63.

141. *Henry IV, Part 2*, I.i.60–63. See also *Romeo and Juliet*, ed. René Weis (London: Arden Shakespeare, 2012), I.iii.82: "Read o'er the volume of young Paris' face."

142. Davison, "Commentary," *Henry IV, Part Two*, ed. Davison, 153; Bulman, note to *Henry IV, Part 2*, 171.

143. *Richard II*, V.i.11–12.

144. Forker, note to *Richard II*, 416.

145. Ure, note to *Richard II*, 146. Contrast, however, Dawson and Yachnin, note to *Richard II*, 254, who see this use again in the sense of "embodiment, representation."

146. John Gillies, "Introduction: Elizabethan Drama and the Cartographizations of Space," in Gillies and Vaughan, *Playing the Globe*, 19–45, 29. My count is eighteen, though this includes one in *Edward III*.

147. *Henry V*, IV.vii.22–23.

148. *Cymbeline*, ed. Valerie Wayne (London: Arden Shakespeare, 2017), IV.i.1–2.

149. *The Merchant of Venice*, I.i.16–18.

150. *Twelfth Night, or, What You Will*, ed. Keir Elam (London: Arden Shakespeare, 2008), III.ii.74–76.

151. *Twelfth Night*, III.ii.24–26. See Keir Elam, notes to *Twelfth Night*, 264, 268. On this map, see C. H. Coote, "Shakspere's 'New Map in Twelfth Night'" (London: Dulau & Co., 1878); reprinted from *New Shakspere Society's Transactions, 1877–79* (1878), 88–100; discussed in John Gillies, "Introduction: Elizabethan Drama and the Cartographizations of Space," 20–21; and Hopkins, *Shakespeare on the Edge*, 11 n. 34. See Nicholas Crane, *Mercator: The Man who Mapped the Planet* (London: Weidenfeld & Nicholson, 2002), 325; and Hoenselaars, "Mapping Shakespeare's Europe," 238. For a general discussion of maps in Shakespeare's time, see J. D. Rogers, "Voyages and Exploration: Geography; Maps," in *Shakespeare's England: An Account of the Life & Manners of His Age* (Oxford: Clarendon Press, 1932), 1:170–98.

152. *Richard III*, V.iii.23–24. This is partly echoed by Richard III's own order to "Give me some ink and paper" a few lines later (V.iii.49).

153. Gillies, "Introduction: Elizabethan Drama and the Cartographizations of Space," 30–31.

154. *Troilus and Cressida*, I.iii.205.

155. Bevington, note to *Troilus and Cressida*, 193.

156. *Othello*, I.i.19–32.

157. The Cambridge edition has "devision" rather than "division," though the meaning is the same. See Sanders, note to *Othello*, 66.

158. Honigmann, note to *Othello*, 116–17; Neill, notes to *Othello*, 197–98; Sanders, note to *Othello*, 67.

159. Sanders, note to *Othello*, 66.

160. Honigmann, note to *Othello*, 116.

161. *Henry IV, Part 1*, II.iii.46–54.

162. I am indebted to Kastan, notes to *Henry IV, Part 1*, 201; and Peter Davison, "Commentary," in *Henry IV, Part One*, ed. Peter Davison (London: Penguin, 2005), 155.

163. Kastan, note to *Henry IV, Part 1*, 201; Sir John Davies, "In Gallum. 24," in *The Poems*, ed. Richard Krueger (Oxford: Clarendon Press, 1975), 139.

164. Cohen, *Shakespeare and Technology*, 173; see Harvey, *Maps in Tudor England*, 7.

165. J. B. Harley, "Meaning and Ambiguity in Tudor Cartography," in Sarah Tyacke, ed., *English Map-Making 1500–1650* (London: The British Library, 1983), 22–45, 22.

166. *Henry IV, Part 1*, I.i.3–4. This pilgrimage was promised in *Richard II*, V.vi.49–50. See also the discussion of a crusade in *Henry IV, Part 1*, I.i.18–29; and the repeated excuse for delay at 47–48 and 100–101. For a discussion, see James Black, "Henry IV's Pilgrimage," *Shakespeare Quarterly* 34, no. 1 (1983): 18–26. On the play generally, see Stephen Greenblatt, *Shakespearean Negotiations: The Circulation of Social Energy in Renaissance England* (Oxford: Clarendon Press, 1988), chap. 2.

167. *Henry IV, Part 1,* I.i.5–6. This image was discussed in chapter 3.

168. *Henry IV, Part 1,* I.i.14–16; see Hopkins, *Shakespeare on the Edge,* 15.

169. *Henry IV, Part 1,* I.iii.144–56.

170. *Henry IV, Part 1,* I.i.34–46. On this battle, see Terence Hawkes, *Shakespeare in the Present* (London: Routledge, 2002), 26–34. Hotspur later recounts a story of them clashing in single combat on the banks of the Severn, though the veracity of this tale is disputed (I.iii.93–118).

171. See Graham Cattle, "'The Detested Blot': The Representation of the Northern English in Shakespeare's *Henry IV Part One,*" *Parergon* 13, no. 1 (1995): 25–32.

172. *Richard II,* V.i.55–62.

173. *Richard II,* V.i.63–65.

174. For example, *Henry IV, Part 1,* III.i.8.

175. See *Henry IV, Part 1,* IV.iii.54–105; and V.i.39–71.

176. See *Henry IV, Part 1,* I.i.85–89.

177. *Henry IV, Part 1,* III.ii.93–99.

178. *King John,* V.ii.89.

179. *Henry IV, Part 1,* III.ii.100–105.

180. *Henry IV, Part 1,* III.ii.109–10.

181. David Scott Kastan, "Introduction," in *Henry IV, Part 1,* ed. Kastan, 13.

182. See Davison, "Commentary," *Henry IV, Part One,* 139, 141; and Kastan's notes to *Henry IV, Part 1,* 137, 143, 172–73. The "Lord Mortimer of Scotland" mentioned in III.ii.164 is another figure entirely, the Earl of March (i.e., the Scottish marches or borders), George Dunbar. Shakespeare gives him the name Mortimer, assuming he was part of the same family. See Davison, "Commentary," *Henry IV, Part One,* 154, 180–81; Kastan, note to *Henry IV, Part 1,* 266.

183. *Henry IV, Part 1,* I.iii.15–21.

184. *Henry IV, Part 1,* II.iv.349–51.

185. *Henry IV, Part 1,* II.iv.352–54; see Kastan, note to *Henry IV, Part 1,* 227.

186. *Henry IV, Part 1,* II.iv.355–56.

187. Kastan, "Introduction," *Henry IV, Part 1,* 64–65; for a discussion of the wider themes, see 62–69; Sandra K. Fischer, "'He means to pay': Value and Metaphor in the Lancastrian Tetralogy," *Shakespeare Quarterly* 40, no. 2 (1989): 149–64; Nina Levine, "Extending Credit in the Henry IV Plays," *Shakespeare Quarterly* 51, no. 4 (2000): 403–31; Jesse M. Lander, "'Crack'd Crowns' and Counterfeit Sovereigns: The Crisis of Value in *1 Henry IV,*" *Shakespeare Studies* 30 (2002): 137–61.

188. *Henry IV, Part 1,* II.i.12–13; see Davison, "Commentary," *Henry IV, Part One,* 145.

189. *Henry IV, Part 1,* III.i.1–2.

190. *Henry IV, Part 1,* III.i.4–6.

191. Jonathan Bate, "Introduction," in *Henry IV, Part I,* ed. Jonathan Bate and Eric Rasmussen (London: Macmillan, 2009), 5.

192. *Henry IV, Part 1,* III.i.62–67.

193. The king himself had previously used it in that sense (*Henry IV, Part 1,* I.i.29).

194. Some readings suggest the scene is primarily comic. See, for instance, Gillies, *Shakespeare and the Geography of Difference,* 46–47.

195. *Henry IV, Part 1*, III.i.68–69.

196. *Henry IV, Part 1*, III.i.70–80.

197. *Holinshed's Chronicles*, Vol. III, 521.

198. *Edward III*, II.ii.93–94.

199. Sonnet 44, lines 4–5.

200. *Venus and Adonis*, in *Shakespeare's Poems*, 235.

201. Onions, *A Shakespeare Glossary*, 157.

202. *Holinshed's Chronicles*, Vol. III, 521.

203. On these rivers generally in the geographical imagination, see Julie Sanders, *The Cultural Geography of Early Modern Drama, 1620–1650* (Cambridge: Cambridge University Press, 2011), 45–52.

204. See Cattle, "'The Detested Blot,'" 27.

205. *Henry IV, Part 1*, III.i.81–88.

206. *Henry IV, Part 1*, III.i.89–93. This is not to be, and Glendower is unable to "draw his power this fourteen days." His failure to arrive is one element in the rebels' defeat—he was "a rated sinew," a strong support. See IV.i.124–25; IV.iv.15–16.

207. *Henry IV, Part 1*, III.i.94–103.

208. Bruce Avery, "Gelded Continents and Plenteous Rivers: Cartography as Rhetoric in Shakespeare," in Gillies and Vaughan, *Playing the Globe*, 46–62, 55–56.

209. Peter Davison, "An Account of the Text," in *Henry IV, Part One*, ed. Davison, 115. Compare *Antony and Cleopatra*, III.x.6: "The greater cantle of the world is lost."

210. Davison, "Commentary," *Henry IV, Part One*, 173.

211. *The First Part of King Henry IV*, ed. Herbert Weil and Judith Weil (Cambridge: Cambridge University Press, 1997), 135.

212. *Henry IV, Part 1*, III.i.104–8.

213. David Bevington, notes to *Henry IV, Part One*, ed. David Bevington (Oxford: Oxford University Press, 1987), 183, 212, 213.

214. *Henry IV, Part 1*, III.i.109–13.

215. See Bevington, note to *Henry IV, Part One*, 213.

216. *Henry IV, Part 1*, III.i.133–37.

217. *Henry IV, Part 1*, III.i.138–39.

218. *Henry IV, Part 1*, III.i.219; 256 and 260.

219. Kastan, "Introduction," *Henry IV, Part 1*, 31.

220. Stephen Greenblatt, *Shakespeare's Freedom* (Chicago: University of Chicago Press, 2010), 3.

221. Sullivan, *The Drama of Landscape*, 94–95.

222. Avery, "Gelded Continents and Plenteous Rivers," 59–60.

223. *Henry V*, IV.vii.22–31. The Quarto text has several small differences.

224. Gillies, *Shakespeare and the Geography of Difference*, 46–47.

225. This is supported by Bevington, note to *Henry IV, Part One*, 95.

226. Bevington, note to *Henry IV, Part One*, 213, suggests that this pertains to the original course (from source to Burton). Nevertheless, he is correct that Glendower's "share is not even threatened by this dispute."

227. For a contemporary discussion of the relation, see Stuart Elden, "Legal Terrain: The Political Materiality of Territory," *London Review of International Law* 5, no. 2 (2017): 199–224.

228. *Henry IV, Part 1*, V.iv.82–100.

229. *Henry IV, Part 1*, III.i.136.

230. *Henry IV, Part 1*, V.v.34–44.

231. Poole, *Supernatural Environments in Shakespeare's England*, 222–23.

232. Jonathan Gil Harris, *Untimely Matter in the Time of Shakespeare* (Philadelphia: University of Pennsylvania Press, 2009), 24–25.

CHAPTER EIGHT

1. On the play generally, see Adrian Poole, *Coriolanus* (New York: Harvester Wheatsheaf, 1988); Jan H. Blits, *Spirit, Soul, and City: Shakespeare's "Coriolanus"* (Lanham, MD: Lexington Books, 2006); and Andreas Höfele, *Stage, Stake, and Scaffold: Humans and Animals in Shakespeare's Theatre* (Oxford: Oxford University Press, 2011), chap. 3.

2. Robin Headlam Wells, *Shakespeare's Politics: A Contextual Introduction* (London: Continuum, 2009), 58; Frank Kermode, *Shakespeare's Language* (London: Penguin, 2000), 243.

3. Zvi Jagendorf, "*Coriolanus*: Body Politic and Private Parts," *Shakespeare Quarterly* 41, no. 4 (1990): 455–69, 457.

4. William Hazlitt, from *Characters of Shakespear's Plays*, excerpted in B. A. Brockman, ed., *Shakespeare: Coriolanus—A Casebook* (Houndmills: Macmillan, 1977), 26. See D. J. Enright, "*Coriolanus:* Tragedy or Debate?," *Essays in Criticism* 4, no. 1 (1954): 1–19.

5. Patterson, *Shakespeare and the Popular Voice*, 121.

6. T. S. Eliot, "Coriolan," in *Collected Poems, 1909–1962* (London: Faber & Faber), 137–43; Bertolt Brecht, "Coriolan," in *Collected Plays: Nine*, ed. and trans. Ralph Manheim and John Willett (New York: Vintage, 1972). See also Brecht's "Study of the First Scene of Shakespeare's *Coriolanus*," in *Brecht on Theatre*, ed. and trans. John Willett (London: Methuen, 1964), 252–65.

7. See Lee Bliss, "Introduction," in *Coriolanus*, ed. Lee Bliss (Cambridge: Cambridge University Press, 2000), 17; Arthur Riss, "The Belly Politic: *Coriolanus* and the Revolt of Language," *English Literary History* 59, no. 1 (1992): 53–75.

8. Wells, *Shakespeare's Politics*, 58. On this, see E. C. Pettet, "*Coriolanus* and the Midlands Insurrection of 1607," *Shakespeare Survey* 3 (1950): 34–42.

9. Plutarch's text, in the North translation, appears in *Coriolanus*, ed. Philip Brockbank (London: Arden Shakespeare, 1976), 313–68. On the relation, see Kwang-Ho Kim, "Shakespeare's Treatment of the Source in *Coriolanus*," *English Studies* 7 (1983): 49–62; and David George, "Plutarch, Insurrection, and Dearth in *Coriolanus*," in Catherine M. S. Alexander, ed., *Shakespeare and Politics* (Cambridge: Cambridge University Press, 2004), 110–29; more generally, see Kenneth Muir, "The Background of *Coriolanus*," *Shakespeare Quarterly* 10, no. 2, 1959, 137–45; and Anne Barton, "Livy, Machiavelli, and Shakespeare's *Coriolanus*," in Alexander, *Shakespeare and Politics*, 67–90.

10. Philip Brockbank, "Introduction," in *Coriolanus*, ed. Brockbank, 26.

11. Brockbank, "Introduction," *Coriolanus*, 36.

12. Kermode, *Shakespeare's Language*, 243.

13. Jay L. Halio, "Introduction," in *King Henry VIII, or, All is True*, ed. Jay L. Halio (Oxford: Oxford University Press, 1999), 42.

14. For a discussion of these themes in that play, see David Bevington, "Introduction," in *Troilus and Cressida*, ed. Bevington, 85–87; Colin Burrow, "Introduction," in *Troilus and Cressida*, ed. R. A. Foakes (London: Penguin, 2006), xxxvi-viii.

15. J. C. Maxwell, "Animal Imagery in *Coriolanus*," *Modern Language Review* 42, no. 4 (1947): 417–21.

16. Patterson, *Shakespeare and the Popular Voice*, 141.

17. Kermode, *Shakespeare's Language*, 254.

18. Agnes Heller, *The Time Is Out of Joint: Shakespeare as a Philosopher of History* (Lanham, MD: Rowman & Littlefield, 2002), 291. There is actually, *pace* Heller, quite a bit going in the opening part of this scene that has political connotations. Nonetheless, chapter 12 of her study is helpful on *Coriolanus* generally.

19. Stanley Cavell, "Who Does the Wolf Love? Reading *Coriolanus*," *Representations* 3 (1983): 1–20, 2.

20. On others, see G. Thomas Tanselle and Florence W. Dunbar, "Legal Language in *Coriolanus*," *Shakespeare Quarterly* 13, no. 2 (1962): 231–38; Paul A. Cantor, *Shakespeare's Rome: Republic and Empire* (Ithaca: Cornell University Press, 1976), pt. 1; W. Hutchings, "Beast or God: The Coriolanus Controversy," *Critical Quarterly* 24, no. 2 (1982): 35–50; Charles Mitchell, "Coriolanus: Power as Honor," *Shakespeare Studies* 1 (1965): 199–226; Alexander Leggatt, *Shakespeare's Political Drama: The History Plays and the Roman Plays* (London: Routledge, 1989), chap. 3; Alan Hager, *Shakespeare's Political Animal: Schema and Schemata in the Canon* (Newark: University of Delaware Press, 1990), chap. 4; Cathy Shrank, "Civility and the City in *Coriolanus*," *Shakespeare Quarterly* 54, no. 1 (2003): 406–23; Brayton Polka, "*Coriolanus* and the Roman World of Contradiction: A Paradoxical World Elsewhere," *The European Legacy* 15, no. 2 (2010): 171–94.

21. Michael D. Bristol, "Lenten Butchery: Legitimation Crisis in *Coriolanus*," in Jean E. Howard and Marion F. O'Connor, eds., *Shakespeare Reproduced: The Text in History and Ideology* (New York: Methuen, 1987), 207–24, 213.

22. *Coriolanus*, I.i.91–114.

23. There is no complete English translation, but the best single edition is John of Salisbury, *Policraticus: On the Frivolities of Courtiers and the Footprints of Philosophers*, ed. and trans. Cary J. Nederman (Cambridge: Cambridge University Press, 1990); the body politic is discussed in Book V, Chapter 2. Bliss, note to *Coriolanus*, 109, suggests Shakespeare knew the translation of a passage from Livy in William Camden, *Remaines of a Greater Worke, Concerning Britaine* (London: S. Waterson, 1605), 199.

24. Edward Forset, *Comparative Discourse of the Bodies Natural and Politique* (London: Iohn Bill, 1606). On this, see Riss, "The Belly Politic," 64–67; and Muir, "The Background of *Coriolanus*," who notes that there is no direct evidence Shakespeare knew Forset's work (141). See also Bliss, "Introduction," *Coriolanus*, 13, who notes that the story of Coriolanus features in Forset's book, as well as in Bodin's *Six Books* (translated in 1606), William Fulbecke's *The Pandectes of the Law of Nations* (London: Thomas Wight, 1602), and Thomas and Dudley Digges, *Foure paradoxes, or politique Discourses* (London: H. Lownes, 1604). Bliss suggests that the most likely text of these to have been known by Shakespeare is *Foure Paradoxes*; see especially *Foure paradoxes*, 104–5.

25. The literature on this is quite extensive. See, for example David G. Hale, "*Coriolanus*: The Death of a Political Metaphor," *Shakespeare Quarterly* 22, no. 3 (1971):

197–202; Andrew Gurr, "'Coriolanus' and the Body Politic," *Shakespeare Survey* 28 (1975): 63–69; James Holstun, "Tragic Superfluity in *Coriolanus*," *English Literary History* 50, no. 3 (1983): 485–507; Jagendorf, "*Coriolanus*"; Riss, "The Belly Politic"; and John Jowett, "Coriolanus and the Hypostatized Body," *IRCL*, 2007, http://www.ircl.cnrs.fr/pdf/2007/Coriolanpdf/JohnJowett.pdf. For a contextual reading that puts it in conversation with other works and the situation in the first years of James I's reign, see Alex Garganigo, "'Coriolanus,' the Union Controversy, and Access to the Royal Person," *Studies in English Literature, 1500–1900* 42, no. 2 (2002): 335–59. The classic work on the idea is Kantorowicz, *The King's Two Bodies*, discussed in chapter 4. See Claire Rasmussen and Michael Brown, "The Body Politic as Spatial Metaphor," *Citizenship Studies* 9, no. 5 (2005): 469–84.

26. Peter Holland, "Introduction," in *Coriolanus*, ed. Holland, 77. See Clark Lunberry, "In the Name of *Coriolanus*: The Prompter (Prompted)," *Comparative Literature* 54, no. 3 (2002): 229–41.

27. There are other potential sources for this scene. For a discussion, see Holland, "Introduction," *Coriolanus*, 35–36 and 66–67.

28. *Coriolanus*, I.i.116.

29. *Coriolanus*, I.i.123–24, 128–29.

30. *Coriolanus*, I.i.130–35.

31. *Coriolanus*, I.i.143–50.

32. *Coriolanus*, III.i.124, 150.

33. *Coriolanus*, V.vi.93.

34. *Coriolanus*, III.i.118–19.

35. R. B. Parker, "Introduction," in *Coriolanus*, ed. R. B. Parker (Oxford: Oxford University Press, 1994), 77–78.

36. Cavell, "Who Does the Wolf Love?"

37. Jagendorf, "*Coriolanus*," 457–58.

38. *Coriolanus*, V.iii.122–25, discussed at the end of this chapter. On the male aspects, see Phyllis Rackin, "*Coriolanus*: Shakespeare's Anatomy of *Virtus*," *Modern Language Studies* 13, no. 2 (1983): 68–79; Cynthia Marshall, "Wound-Man: *Coriolanus*, Gender, and the Theatrical Construction of Interiority," in Valerie Traub, M. Lindsay Kaplan, and Dympna Callaghan, eds., *Feminist Readings of Early Modern Culture* (New York: Cambridge University Press, 1996), 93–118; Coppélia Kahn, *Man's Estate: Masculine Identity in Shakespeare* (Berkeley: University of California Press, 1981), chap. 6.

39. See also Coppélia Kahn, *Roman Shakespeare: Warriors, Wounds, and Women* (London: Routledge, 1997), 152–54.

40. *Coriolanus*, I.ix.28–29.

41. *Coriolanus*, I.iv.50.

42. *Coriolanus*, II.i.111–16.

43. Brockbank, note to *Coriolanus*, 158; Holland, note to *Coriolanus*, 223.

44. *Coriolanus*, II.i.118–20.

45. *Coriolanus*, II.i.139–52.

46. See Jagendorf, "*Coriolanus*," 464.

47. *Coriolanus*, V.iii.36; Marshall, "Wound-Man," 103.

48. *Coriolanus*, II.i.225–29.

49. *Coriolanus,* II.ii.67–69.

50. *Coriolanus,* II.ii.134–38.

51. *Coriolanus,* II.iii.4–12. The first citizen's reply invokes "the many-headed multitude" (15–16). See also *Henry IV, Part 2,* Induction, 18–19: "the blunt monster with uncounted heads / The still discordant wav'ring multitude." The crowd is unsubtle, unruly, and numerous.

52. *Coriolanus,* II.iii.48–53.

53. Jarrett Walker, "Voiceless Bodies and Bodiless Voices: The Drama of Human Perception in *Coriolanus,*" *Shakespeare Quarterly* 43, no. 2 (1992): 170–85, 177.

54. *Coriolanus,* II.iii.75–76.

55. *Coriolanus,* II.iii.104–5, 106–7.

56. *Coriolanus,* II.iii.126.

57. *Coriolanus,* II.iii.160–61.

58. *Coriolanus,* I.iv.59. This point is indebted to Rebecca Lemon, "Arms and the Law in *Coriolanus,*" in Jordan and Cunningham, *The Law and Shakespeare,* 233–48, 239. See also Riss, "The Belly Politic," 55–56, although this overstretches the spatial resonances of Coriolanus's separation from the people.

59. *Coriolanus,* III.iii.48–51.

60. More generally, see R. R. Simpson, *Shakespeare and Medicine* (Edinburgh: E. & S. Livingstone, 1959); Greg W. Bentley, *Shakespeare and the New Disease: The Dramatic Function of Syphilis in "Troilus and Cressida," "Measure for Measure," and "Timon of Athens"* (New York: Peter Lang, 1989).

61. *Coriolanus,* I.i.160–61.

62. *Coriolanus,* I.iv.31–39.

63. *Coriolanus,* III.i.78–82.

64. Brockbank, note in *Coriolanus,* 199.

65. *Coriolanus,* I.v.3. See Parker, "Introduction," *Coriolanus,* 81, for a longer list of diseases invoked in the play. "Murrain" is also used as a curse in *The Tempest,* III.ii.78 (and "red plague" at I.ii.365); and *Troilus and Cressida,* II.i.18.

66. *Coriolanus,* III.i.155–58.

67. *Coriolanus,* III.i.89.

68. *Coriolanus,* III.i.296–309. There are parallels with *Henry IV, Part 2,* III.i.38–43 here. See Hale, "*Coriolanus,*" 201.

69. *Coriolanus,* IV.ii.18–20.

70. On this theme, see Maxwell, "Animal Imagery in *Coriolanus*"; and Ineke Murakami, "The 'Bond and Privilege of Nature' in *Coriolanus,*" *Religion & Literature* 38, no. 3 (2006): 121–36. There are some useful thoughts in the latter, but the application of Giorgio Agamben seems a stretch. On animals in Shakespeare generally, though with no reference to *Coriolanus,* see Bruce Boehrer, *Shakespeare among the Animals: Nature and Society in the Drama of Early Modern England* (New York: Palgrave, 2002); and more generally, Laurie Shannon, *The Accommodated Animal: Cosmopolity in Shakespearean Locales* (Chicago: University of Chicago Press, 2013).

71. *Coriolanus,* III.i.34.

72. *Coriolanus,* II.i.92; see also III.ii.33.

73. *Coriolanus,* II.ii.76.

74. *Coriolanus*, III.i.90.

75. *Coriolanus*, I.i.217. See also "shreds" (I.i.203).

76. See Patterson, *Shakespeare and the Popular Voice*, chap. 6.

77. *Coriolanus*, I.i.163–67.

78. *Coriolanus*, I.i.157.

79. *Coriolanus*, I.i.201.

80. *Coriolanus*, I.i.243–44.

81. *Coriolanus*, I.vi.38.

82. *Coriolanus*, I.vi.43.

83. *Coriolanus*, II.i.241, 245, 251.

84. *Coriolanus*, I.iii.86.

85. *Coriolanus*, IV.v.215.

86. *Coriolanus*, I.viii.3.

87. *Coriolanus*, III.i.177.

88. *Coriolanus*, V.iii.162. For a near-complete list, see Maxwell, "Animal Imagery in *Coriolanus*," 420.

89. *Coriolanus*, I.i.154; I.vi.44; and, less obviously, IV.v.176. See Holland, note to *Coriolanus*, 163.

90. *Coriolanus*, I.iv.32–37.

91. *Coriolanus*, V.iii.35.

92. *Macbeth*, II.iv.12–13. See Maxwell, "Animal Imagery in *Coriolanus*," 420.

93. See Clark and Mason, notes to *Macbeth*, 198.

94. See Gail Kern Pastor, "To Starve with Feeding: The City in *Coriolanus*," *Shakespeare Studies* 11 (1978): 123–44, 135–36.

95. *Coriolanus*, II.i.6.

96. *Coriolanus*, II.i.7–12.

97. *Coriolanus*, II.iii.114. The Folio reads "this Wooluiſh tongue." Some editors retain "tongue," though this is a simple misreading of "toge"; "woolish" is possible, relating to a wolf in sheep's clothing, or the gown. Brockbank notes that other editors have suggested "the garb should be the sheep's; but it is the wolf's property and symbolizes his treacherous nature." See Holland, "Longer Notes," *Coriolanus*, 425; and Höfele, *Stage, Stake, and Scaffold*, 111, who agrees with "wolvish" and notes that "toge" is itself an emendation of "tongue" from the Folio, but claims this "seems to me irrefutable" (111 n. 52).

98. *Coriolanus*, I.i.220–21. Holstun, "Tragic Superfluity in *Coriolanus*," 488–89, notes that this evokes both the "medical venting of excess humours and the commercial vending of excess stock."

99. *Coriolanus*, I.iii.33.

100. *Coriolanus*, I.i.230–31.

101. *Coriolanus*, I.viii.7.

102. Holland, note to *Coriolanus*, 204.

103. *Coriolanus*, IV.i.1–2. He had earlier (III.i.94) referred to a "Hydra," but here the use of "butts" seems to imply a rather more herd-like animal.

104. *Coriolanus*, IV.i.30–31.

105. *Coriolanus*, V.iv.11–14.

106. *Coriolanus*, III.i.313.

107. *Coriolanus,* V.iv.27–28.

108. *Coriolanus,* IV.vii.33–35.

109. *Coriolanus,* III.i.139.

110. *Coriolanus,* IV.v.44–45.

111. *Coriolanus,* IV.vi.94–96.

112. *Coriolanus,* IV.v.189–90.

113. *Coriolanus,* I.i.213.

114. *Coriolanus,* III.i.198–200, 205–8.

115. *Coriolanus,* III.i.239–41.

116. *Coriolanus,* III.i.244.

117. Cantor, *Shakespeare's Rome,* 82.

118. *Coriolanus,* I.i.193–94.

119. Pastor, "To Starve with Feeding," 131. See, for example, the claim that "if we lose the field / We cannot keep the town" (I.vii.4–5); and Robert S. Miola, *Shakespeare's Rome* (Cambridge: Cambridge University Press, 1983), 164: "the constricted and constrictive city . . . sharply defined by outlying battlefields, rival towns, and its own vividly realized topography—its walls, gates, Capitol, Tiber, Tarpeian rock, forum, private houses, and streets."

120. *Coriolanus,* III.i.249.

121. *Coriolanus,* III.i.265–67. At 288 he describes him as a "viperous traitor."

122. *Coriolanus,* III.i.273–74.

123. *Coriolanus,* IV.iv.1–6.

124. *Coriolanus,* V.vi.112–17.

125. *Coriolanus,* V.vi.121.

126. Jagendorf, "*Coriolanus,*" 467.

127. *Coriolanus,* III.iii.15.

128. *Coriolanus,* III.iii.87–92.

129. *Coriolanus,* III.iii.98–104.

130. *Coriolanus,* III.iii.105–6.

131. *Coriolanus,* III.iii.116–22, 133–35.

132. See Höfele, *Stage, Stake, and Scaffold,* 100.

133. *Coriolanus,* III.iii.136.

134. Greenblatt, *Shakespeare's Freedom,* 108.

135. Holland, "Introduction," *Coriolanus,* 109; see 91. For a fuller discussion, see Brian Parker, "A Tale of Three Cities: Staging in *Coriolanus,*" in Augusta Lynne Magnussen and C. Edward McGee, eds., *The Elizabethan Theatre XIII: Papers Given at the Thirteenth International Conference* (Toronto: D. Meany, 1994), 119–45.

136. For a discussion, see Fionnuala O'Neill, "Review of Shakespeare's *Coriolanus* (adapted for film and directed by Ralph Fiennes)," *Shakespeare* 8, no. 4 (2012): 456–60; Slavoj Žižek, *The Year of Dreaming Dangerously* (London: Verso, 2012), 118–25; Holland, "Introduction," *Coriolanus,* 133–40.

137. This is an explicit reference to John Osborne's 1973 play, *A Place Calling Itself Rome,* included in *Four Plays* (London: Oberon, 2000), 145–216.

138. John Locke, *Second Treatise,* in *Two Treatises of Government,* ed. Peter Laslett (1960; Cambridge: Cambridge University Press, 1988), V.26.

139. Locke, *Second Treatise*, V.49.

140. *Coriolanus*, IV.v.79–80.

141. Brockbank, note to *Coriolanus*, 254; Holland, note to *Coriolanus*, 338.

142. *Coriolanus*, V.vi.29–31.

143. *Coriolanus*, IV.vi.131.

144. *Coriolanus*, IV.vi.133–35.

145. *Coriolanus*, IV.vi.147–48.

146. *Coriolanus*, V.i.4–6.

147. *Coriolanus*, V.ii.40–50.

148. *Coriolanus*, V.v.4–5.

149. *Coriolanus*, IV.v.128–37. There are echoes here of the Earl of Richmond's speech in *Richard III*, V.ii.1–12, where he motivates his supporters who have marched "thus far into the bowels of the land" to defeat the "wretched, bloody and usurping boar."

150. *Coriolanus*, IV.vi.38–42.

151. *Coriolanus*, IV.vi.75–80.

152. See Elden, *The Birth of Territory*, chap. 2.

153. *Coriolanus*, IV.v.132.

154. *Coriolanus*, V.iii.33–34.

155. Isidore of Seville, *The Etymologies of Isidore of Seville*, trans. Steven A. Barney, W. J. Lewis, J. A. Beach, and Oliver Berghof (Cambridge: Cambridge University Press, 2006), 16.5.22.

156. *Coriolanus*, IV.v.88–89; Holland, note to *Coriolanus*, 339.

157. *Coriolanus*, IV.v.93–94.

158. *Coriolanus*, IV.v.130–33.

159. *Coriolanus*, IV.ii.50–51. For a discussion, see Janet Adelman, "'Anger's my Meat': Feeding, Dependency, and Aggression in *Coriolanus*," in Murray M. Schwartz and Coppélia Kahn, eds., *Representing Shakespeare: New Psychoanalytic Essays* (Baltimore: Johns Hopkins University Press, 1980), 129–49; Adelman, *Suffocating Mothers: Fantasies of Maternal Origin in Shakespeare's Plays from Hamlet to The Tempest* (London: Routledge, 1992), chap. 6.

160. *Coriolanus*, V.iii.101–3.

161. *Coriolanus*, V.iii.122–25.

162. See, in particular, Stuart Elden, "Foucault and Geometrics," in Philippe Bonditti, Didier Bigo, and Frédéric Gros, eds., *Foucault and the Modern International: Silences and Legacies for the Study of World Politics* (London: Palgrave Macmillan, 2017), 295–311; and Elden, *The Birth of Territory*.

CHAPTER NINE

1. See Elden, *The Birth of Territory*, chap. 2 on Rome, and pt. 2 on the Middle Ages.

2. Sophocles, *Oedipus at Colonus*, in *The Theban Plays*, trans. Robert Fagles (London: Penguin, 1984), line 406.

3. For a reading of *Antigone* in this regard, see Stuart Elden, "The Place of the *Polis*: Political Blindness in Judith Butler's *Antigone's Claim*," *Theory and Event* 8 (2005).

4. *Beowulf*, ed. Michael Swanton, rev. ed. (Manchester: Manchester University Press, 1997), line 103. At lines 1348–52, both Grendel and his mother are described this way, as those who trod the paths of exile. On these themes, see Manish Sharma, "Metalepsis and Monstrosity: The Boundaries of Narrative Structure in *Beowulf*," *Studies in Philology* 102, no. 3 (2005): 247–75, 265–66; and S. L. Higley, "Aldor on Ofre, or the Reluctant Hart: A Study of Liminality in *Beowulf*," *Neuphilologische Mitteilungen* 87 (1986): 342–53. For a fuller discussion, see Stuart Elden, "Place Symbolism and Land Politics in *Beowulf*," *Cultural Geographies* 16, no. 4 (2009): 447–63.

5. Snorri Sturlson, *The Prose Edda*, trans. Jesse L. Byock (London: Penguin, 2005); *The Poetic Edda*, trans. Carolyne Larrington (Oxford: Oxford University Press, 1996).

6. On this, see Turner, "*King Lear* Without."

7. *King Lear*, I.i.39–40. See chapter 1 above.

8. See Turner, "*King Lear* Without"; James Ogden, "Lear's Blasted Heath," in James Ogden and Arthur H. Scouten, eds., *Lear from Study to Stage: Essays in Criticism* (London: Associated University Presses, 1997), 135–45; Jones, *Shakespeare's Storms*, chap. 4.

9. *Macbeth*, I.iii.77. See chapter 7 above.

10. *Troilus and Cressida*, I.i.2.

11. *Troilus and Cressida*, I.i.109–11.

12. *Troilus and Cressida*, note to 162.

13. *The Two Noble Kinsmen*, ed. Lois Porter (London: Arden Shakespeare, 1997). See Vickers, *Shakespeare, Co-Author*, chap. 6.

14. For a general discussion, see Charlotte Scott, "Dark Matter: Shakespeare's Foul Dens and Forests," *Shakespeare Survey* 64 (2011): 276–89; and for slightly later period to Shakespeare, see Sanders, *The Cultural Geography of Early Modern Drama, 1620–1650*, chap. 2, "Into the Woods: Spatial and Social Geographies in the Forest." More generally, see Harrison, *Forests*.

15. *Two Gentlemen of Verona*, III.i.163–67.

16. *Henry VI, Part 2*, III.ii.295–97; see, for example, *Romeo and Juliet*, III.i.188–99.

17. *Two Gentlemen of Verona*, III.i.170–87.

18. *Romeo and Juliet*, III.iii.17–23, 29–51; *Henry VI, Part 2*, III.ii.388–402.

19. On the locations, see William C. Carroll, "Introduction," in *Two Gentlemen of Verona*, ed. Carroll, 78.

20. *Two Gentlemen of Verona*, IV.i.62

21. *Two Gentlemen of Verona*, V.ii.44.

22. *Two Gentlemen of Verona*, V.i.11.

23. *Two Gentlemen of Verona*, IV.i.

24. Jonathan Bate, "Introduction," in *Two Gentlemen of Verona*, ed. Jonathan Bate and Eric Rasmussen (London: RSC/Macmillan, 2011), 9.

25. *Two Gentlemen of Verona*, V.iv.1–10.

26. The play appears in Shakespeare and Others, *Collaborative Plays*, 7–69. For the Shakespearean attribution, see MacDonald Jackson, "Shakespeare and the Quarrel Scene in *Arden of Faversham*," *Shakespeare Quarterly* 57 (2006): 249–93; and his *Determining the Shakespeare Canon: "Arden of Faversham" and "A Lover's Complaint"* (Cambridge: Cambridge University Press, 2015); along with Craig and Kinney, *Shakespeare,*

Computers, and the Mystery of Authorship, chap. 4. This is disputed by Brian Vickers, "Thomas Kyd, Secret Sharer," *Times Literary Supplement*, April 18, 2008, 13–15.

27. *Arden of Faversham*, ed. Martin White, rev. ed. (London: AC & Black, 2007), I.2–8.

28. See *Arden of Faversham*, XIII.12, 32, and Epilogue, 10. See Tom Lockwood, "Introduction," in *Arden of Faversham*, ed. White, xii-xiv. For a general discussion, see Sullivan, *The Drama of Landscape*, chap. 1.

29. *Mucedorus* is included in Shakespeare and Others, *Collaborative Plays*, 503–50; see Richard Finkelstein, "Censorship and Forgiven Violence in *Mucedorus*," *Parergon* 17, no. 1 (1999): 89–108.

30. On this theme generally, see Jeanne Addison Roberts, *The Shakespearean Wild: Geography, Genus, and Gender* (Lincoln: University of Nebraska Press, 1991).

31. *A Midsummer Night's Dream*, I.i.157–68.

32. See Peter Holland, "Introduction," in *A Midsummer Night's Dream* (Oxford: Oxford University Press, 1994), 103–4; Mark Rose, *Shakespearean Design* (Cambridge, MA: Harvard University Press, 1972), 17–19. Rose (88–89, 182 n. 7) suggests that two sets of paired scenes (II.ii and III.i; and III.ii and IV.i) should really be seen as two scenes, giving seven in total—the fewest of any play by Shakespeare.

33. On the play generally, Karl Klein, "Introduction," *Timon of Athens* (Cambridge: Cambridge University Press, 2001), is a useful guide.

34. Nicholas Walton, "Introduction," in *Timon of Athens*, ed. G. R. Hibbard (London: Penguin, 2005), xlv.

35. *Timon of Athens*, ed. Anthony B. Dawson and Gretchen E. Minton (London: Arden Shakespeare, 2008), I.ii.225–30. References are to this edition. On Middleton's role, see Middleton, *The Collected Works*, 467–70; and *Thomas Middleton and Early Modern Textual Culture: A Companion to the Collected Works*, ed. Gary Taylor and John Lavagnino (Oxford: Oxford University Press, 2007), 704–11.

36. Anthony B. Dawson and Gretchen E. Minton, note to *Timon of Athens*, ed. Dawson and Minton, 199.

37. *Timon of Athens*, I.ii.231.

38. *Timon of Athens*, II.ii.145.

39. *Timon of Athens*, II.ii.146–50.

40. *Timon of Athens*, II.ii.151.

41. Cicero, *De re publica* 3.9.15; Plutarch, *Moralia* 210e, trans. Frank Cole Babbitt, Greek-Latin edition (London: William Heinemann, 1931).

42. *Timon of Athens*, II.ii.155–57.

43. G. R. Hibbard, "Commentary," in *Timon of Athens*, ed. Hibbard, 183.

44. On the parallels and contrast with *King Lear*, see John Jowett, "Introduction," in *Timon of Athens*, ed. John Jowett (Oxford: Oxford University Press, 2004), 26–28.

45. *Timon of Athens*, IV.i.1–3.

46. Dawson and Minton, note to *Timon of Athens*, 264 n. 2.

47. *Timon of Athens*, IV.iii.346–47. See Philip Brockbank, "Jesus, Shakespeare, and Karl Marx—*Timon of Athens* and *The Merchant of Venice*: Parables for the City," in *On Shakespeare: Jesus, Shakespeare, and Karl Marx, and Other Essays* (Oxford: Blackwell, 1989), 3–29.

48. *Titus Andronicus,* III.i.54. The scene divisions differ in other editions, because Bate's Arden edition reincorporates II.I into I.i.

49. For general readings, see Richard Marienstras, *New Perspectives on the Shakespearean World,* trans. Janet Lloyd (Cambridge: Cambridge University Press, 1985), chap. 2, 40–47; Francis Barker, *The Culture of Violence: Essays on Tragedy and History* (Manchester: Manchester University Press, 1993), 143–206; and Naomi Conn Liebler, *Shakespeare's Festive Tragedy: The Ritual Foundations of Genre* (London: Routledge, 1995), 131–48.

50. For a comprehensive discussion, see Hereward Price, "The Authorship of *Titus Andronicus,*" *Journal of English and Germanic Philology* 42, no. 1 (1943): 55–81, 81: "We must conclude, however regretfully, that Shakespeare was the author of *Titus Andronicus.*" Most editions include discussions of staging; see also Mariangela Tempera, *Feasting with Centaurs: "Titus Andronicus" from Stage to Text* (Bologna: CLUEB, 1999).

51. Eugene M. Waith, "Introduction," in *Titus Andronicus,* ed. Eugene M. Waith (Oxford: Oxford University Press, 1984), 10–11, 20. Accordingly, modern editions are based on the 1594 First Quarto, with the exception of III.ii, which uses the Folio text. Jonathan Bate, "Introduction," *Titus Andronicus,* ed. Bate, esp. 72–83, broadly agrees with this account. In contrast, see Vickers, *Shakespeare, Co-Author,* chap. 3, and Jonathan Bate, "Reconsiderations and Reinventions," in *Titus Andronicus,* ed. Jonathan Bate, rev. ed. (London: Bloomsbury, 2018), 121–62.

52. For the texts, see *Titus Andronicus,* ed. Waith, Appendices A and B, 195–207. Some other possible source material is provided in "Appendix: Patterns and Precedents," *Titus Andronicus,* ed. Bate, 279–89. The best discussion is G. K. Hunter, "Sources and Meanings in *Titus Andronicus,*" in J. C. Gray, ed., *Mirror Up to Shakespeare: Essays in Honour of G. R. Hibbard* (Toronto: University of Toronto Press, 1984), 171–88.

53. On the play's structure, see Ruth Nevo, "Tragic Form in *Titus Andronicus,*" in A. A. Mendilow, ed., *Scripta Hierosolymitana,* vol. 25, *Further Studies in English Language and Literature* (Jerusalem: Magnes Press, 1975), 1–18; and Hereward T. Price, *Construction in Shakespeare,* University of Michigan Contributions in Modern Philology 17 (Ann Arbor: University of Michigan Press, 1951), 37–41.

54. *Titus Andronicus,* I.i.28.

55. *Titus Andronicus,* I.i.73.

56. *Titus Andronicus,* I.i.189.

57. *Titus Andronicus,* I.i.201–2.

58. *Titus Andronicus,* I.i.324–25.

59. On this theme, see Ronald Broude, "Roman and Goth in Titus Andronicus," *Shakespeare Survey* 6 (1970): 27–34; and generally Samuel Kliger, *The Goths in England: A Study in Seventeenth- and Eighteenth-Century Thought* (Cambridge, MA: Harvard University Press, 1952).

60. *Titus Andronicus,* I.i.496–98.

61. *Titus Andronicus,* II.i.1–2.

62. Jonathan Bate, "Introduction," *Titus Andronicus,* ed. Bate, 7.

63. *Titus Andronicus,* II.ii.12–15.

64. *Titus Andronicus,* II.ii.30–31.

65. *Titus Andronicus,* II.ii.93–97.

66. On this switch, which takes place in a single scene, see Jacques Berthoud, "Introduction," in *Titus Andronicus* (London: Penguin, 2005), lii-liii.

67. *Titus Andronicus,* I.i.593–94.

68. Edward Berry, *Shakespeare and the Hunt: A Cultural and Social Study* (Cambridge: Cambridge University Press, 2001), 85, 25. See Roger B. Manning, *Hunters and Poachers: A Social and Cultural History of Unlawful Hunting in England, 1485–1640* (Oxford: Clarendon Press, 1993), 182–83.

69. *Titus Andronicus,* I.i.628.

70. *Titus Andronicus,* I.i.612–18.

71. *Titus Andronicus,* II.i.25–26.

72. *Titus Andronicus,* III.i.89–91.

73. *Titus Andronicus,* III.i.92, 103. For an excellent discussion of these lines, see Heather James, "Cultural Disintegration in *Titus Andronicus*: Mutilating Titus, Vergil, and Rome," in *Violence in Drama,* Themes in Drama 13, ed. James Redmond (Cambridge: Cambridge University Press, 1991), 123–40, 127. The argument generally is developed in *Shakespeare's Troy: Drama, Politics, and the Translation of Empire* (Cambridge: Cambridge University Press, 1997).

74. Berry, *Shakespeare and the Hunt,* 35; on the play generally, see Berry's chap. 3.

75. *Titus Andronicus,* II.iii.16–19.

76. Marienstras, *New Perspectives on the Shakespearean World,* 42. On Lavinia, see Kahn, *Roman Shakespeare,* chap. 3.

77. *Titus Andronicus,* II.ii.193, 198–201.

78. *Titus Andronicus,* II.ii.202, 210, 224, 230, 235–36. Cocytus is a river in hell, standing here for hell generally. See Bate, note to *Titus Andronicus,* 182.

79. Alan Sommers, "'Wilderness of Tigers': Structure and Symbolism in *Titus Andronicus,*" *Essays in Criticism* 10, no. 2 (1960): 275–89, 276.

80. *Titus Andronicus,* III.i.54.

81. *Titus Andronicus,* III.i.63.

82. *Titus Andronicus,* III.i.66–72.

83. *Titus Andronicus,* III.i.94–98

84. *Titus Andronicus,* III.i.285–86.

85. *Titus Andronicus,* III.i.297–301.

86. Ovid is a likely inspiration for much of the play. See Eugene M. Waith, "The Metamorphosis of Violence in *Titus Andronicus,*" *Shakespeare Survey* 10 (1957): 39–49; and Jonathan Bate, *Shakespeare and Ovid* (Oxford: Clarendon Press, 1993), 101–17.

87. *Titus Andronicus,* IV.i.51–53.

88. *Titus Andronicus,* IV.i.55–58.

89. *Titus Andronicus,* IV.i.78.

90. Sommers, "*'Wilderness of Tigers,'*" 285.

91. *Titus Andronicus,* V.iii.197, 194.

92. *Titus Andronicus,* V.iii.5.

93. *Titus Andronicus,* V.iii.178–79.

94. Broude, "Roman and Goth in *Titus Andronicus,*" 30–31. For the plea and the rites, see *Titus Andronicus,* I.i.107–23; 146.

95. *Titus Andronicus,* IV.iii.33–35.

96. *Titus Andronicus,* IV.iv.61–67.

97. *Titus Andronicus,* IV.iv.68–76.

98. See Broude, "Roman and Goth in *Titus Andronicus,*" 29.

99. Bate, "Introduction," *Titus Andronicus,* 19.

100. See Elden, *The Birth of Territory,* chap. 9.

101. *As You Like It,* I.i.94–99. On the play generally, see Paul A. Kottman, *Tragic Conditions in Shakespeare: Disinheriting the Globe* (Baltimore: Johns Hopkins University Press, 2009), chap. 1; Robert N. Watson, *Back to Nature: The Green and the Real in the Late Renaissance* (Philadelphia: University of Pennsylvania Press, 2006), chap. 3.

102. *As You Like It,* I.i.109–13.

103. *As You Like It,* I.iii.39–42.

104. *As You Like It,* I.iii.134–35.

105. Dusinberre, note to *As You Like It,* 188. See *As You Like It,* II.iii.68; and III.ii. 13–29.

106. *As You Like It,* II.i.1–6, 15–17.

107. *As You Like It,* II.i.21–25.

108. See Daley, "The Idea of Hunting in *As You Like It,*" 84–85.

109. On the term at this time, see John Manwood, *A Treatise of the Laws of the Forest,* 3rd ed. (London, 1665), 110, 113.

110. On desert, see also *As You Like It,* II.vi.17; and III.ii.122–23.

111. Dusinberre, note to *As You Like It,* 191.

112. *As You Like It,* II.i.27–28.

113. *As You Like It,* II.i.58–63.

114. Dusinberre, "Introduction," *As You Like It,* 53.

115. Jonathan Bate, "Introduction," in *As You Like It,* ed. Jonathan Bate and Eric Rasmussen (London: Palgrave Macmillan, 2008), 4, 9.

116. *As You Like It,* IV.i.19–20.

117. Juliet Dusinberre, "Introduction," *As You Like It,* ed. Dusinberre, 2–3.

118. *As You Like It,* III.ii.291–93.

119. See Katherine Duncan-Jones, "Introduction," in *As You Like It,* ed. H. J. Oliver (London: Penguin, 2005), xlvi-ii; Alan Brissenden, "Introduction," in *As You Like It,* ed. Alan Brissenden (Oxford: Oxford University Press, 1993), 39–41. For a general discussion of these themes, see Berry, *Shakespeare and the Hunt,* chap. 6.

120. Addison Roberts, *The Shakespearean Wild,* 25, sees this as a general shift in Shakespeare's work, where "later forests tend to become tamer."

121. *As You Like It,* II.iv.89.

122. *As You Like It,* III.v.108.

123. James Shapiro, *1599: A Year in the Life of William Shakespeare* (London: Faber & Faber, 2005), 271–72, 279; Bate, "Introduction," *As You Like It,* 7–8.

124. See "*As You Like It* in Performance: The RSC and Beyond," in *As You Like It,* ed. Bate and Rasmussen, 147–48.

125. A. Stuart Daley, "The Dispraise of the Country in *As You Like It,*" *Shakespeare Quarterly* 36, no. 3 (1985): 300–314.

126. See Richard Wilson, *Will Power: Essays on Shakespearean Authority* (Hemel Hempstead: Harvester Wheatsheaf, 1993), chap. 3; and more textually, Andrew Barnaby,

"The Political Consciousness of Shakespeare's *As You Like It*," *Studies in English Literature: 1500–1900* 36, no. 2 (1996): 373–95. On the complications, see A. Stuart Daley, "The Dispraise of the Country in "*As You Like It*," *Shakespeare Quarterly* 36, no. 3 (1985): 300–314. For a general discussion, see Berry, *Shakespeare and the Hunt*, chap. 6; Anne Barton, "Parks and Ardens," *Proceedings of the British Academy* 80 (1991): 49–71; A. Stuart Daley, "The Idea of Hunting in *As You Like It*," *Shakespeare Survey* 21 (1993): 72–95.

127. Dusinberre, "Introduction," *As You Like It*, 113.

128. Berry, *Shakespeare and the Hunt*, 167.

129. *As You Like It*, II.iv.15.

130. Daley, "The Dispraise of the Country," 301. See also his "Where Are the Woods in *As You Like It*?," *Shakespeare Quarterly* 34, no. 2 (1983): 172–80.

131. Manwood, *A Treatise of the Laws of the Forest*, 40–41. For a general discussion of Manwood, see Marienstras, *New Perspectives on the Shakespearean World*, chap. 1.

132. Manwood, *A Treatise of the Laws of the Forest*, 42.

133. Manwood, *A Treatise of the Laws of the Forest*, 43–44, 91–96.

134. Rackham, *Trees and Woodland in the British Landscape*, 164–65.

135. Daley, "The Dispraise of the Country in *As You Like It*," 305; drawing upon Caroline F. E. Spurgeon, *Shakespeare's Imagery* (Cambridge: Cambridge University Press, 1935), Chart VII.

136. Daley, "The Dispraise of the Country in *As You Like It*," 308.

137. *As You Like It*, III.i.5–12.

138. Luke 15.8. See H. J. Oliver, "Commentary," in *As You Like It*, ed. Oliver, 130; Dusinberre, note to *As You Like It*, 234.

139. *As You Like It*, III.i.15–18.

140. *As You Like It*, IV.iii.74–76.

141. For a discussion, see Manwood, *A Treatise of the Laws of the Forest*, 318.

142. *As You Like It*, III.ii.324–25. See V.iv.157: "skirts of this wild wood." "Skirts" has the same sense as "outskirts": "1. the border, boundary, or outlying part *of* a territory, country, kingdom, etc. . . . 2. "The edge, margin, verge *of* a wood, lake, cloud, etc." (*OED*). See also "the circle of this forest" (V.iv.34).

143. Marienstras, *New Perspectives on the Shakespearean World*, 206 n. 32; see Dusinberre, note to *As You Like It*, 307.

144. Marienstras, *New Perspectives on the Shakespearean World*, 19; see Manwood, *A Treatise of the Laws of the Forest*, chap. 20; Manning, *Hunters and Poachers*, chap. 4 and 238.

145. Berry, *Shakespeare and the Hunt*, 25; Dusinberre, "Introduction," *As You Like It*, 53–55.

146. Berry, *Shakespeare and the Hunt*, 165.

147. See Berry, *Shakespeare and the Hunt*, esp. 188–89, 209–10; and J. Frankis, "The Testament of the Deer in Shakespeare," *Neuphilologische Mitteilungen* 59, no. 2 (1958): 65–68.

148. Harrison, *Forests*, 100–101.

149. Edward Bond, *Plays Three: Bingo, The Fool, The Woman, Stone* (London: Methuen, 1987). For a contemporary updating of Shakespeare to land struggles, see Orkin, *Shakespeare against Apartheid*. An excellent study is Álvaro Sevilla-Buitrago, "Territory

and the Governmentalisation of Social Reproduction: Parliamentary Enclosure and Spatial Rationalities in the Transition from Feudalism to Capitalism," *Journal of Historical Geography* 38, no. 3 (2012): 209–19.

150. Bond bases his account on the papers in E. K. Chambers, *William Shakespeare: A Study of Facts and Problems* (Oxford: Clarendon Press, 1930), vol. 2.

151. Bond, "Bingo: Introduction," in *Plays Three*, 4.

152. Bond, "Bingo: Introduction," 6.

153. On the wild, see Michael Steffes, "The Ancient Greek Wild in 'The Winter's Tale,'" *Renaissance and Reformation* 27, no. 4 (2003): 31–51.

154. Gillies, *Shakespeare and the Geography of Difference*, 40.

155. John Pitcher, "Introduction," in *The Winter's Tale*, ed. John Pitcher (London: Arden Shakespeare, 2010), 30; see 100.

156. Pitcher, "Introduction," *The Winter's Tale*, 100 (and the reproduction of the Ortelius map on 101); Gillies, *Shakespeare and the Geography of Difference*, 34–36; Vaughan, "Preface: The Mental Maps of English Renaissance Drama," 8.

157. See Bate, "Shakespeare's Islands"; Daryl W. Palmer, "Jacobean Muscovites: Winter, Tyranny, and Knowledge in *The Winter's Tale*," *Shakespeare Quarterly* 46, no. 3 (1995): 323–39; Desai, "'What means Sicilia?'" On the pastoral, see Craig Horton, "'. . . the Country must diminish': Jacobean London and the Production of Pastoral Space in *The Winter's Tale*," *Parergon* 20, no. 1 (2003): 85–107; and more generally Michele Marrapodi, "'Of that Fatal Country': Sicily and the Rhetoric of Topography in *The Winter's Tale*," in Michele Marrapodi, A. J. Hoenselaars, Marcello Cappuzzo, and L. Falzon Santucci, eds., *Shakespeare's Italy: Functions of Italian Locations in Renaissance Drama* (Manchester: Manchester University, 1993), 213–28.

158. *The Winter's Tale*, II.iii.171–77.

159. *The Winter's Tale*, III.iii.2.

160. *The Winter's Tale*, III.iii.30, 32.

CODA

1. See also Simon Barker, *War and Nation in the Theatre of Shakespeare and His Contemporaries* (Edinburgh: Edinburgh University Press, 2007).

2. Shapiro, *1606*, 48–49.

3. Valerie Wayne, "Introduction," in *Cymbeline*, ed. Valerie Wayne, 66–67.

4. Wayne, "Introduction," *Cymbeline*, 39.

5. *Cymbeline*, II.iv.15–20.

6. See John Pitcher, "Commentary," in *Cymbeline*, ed. John Pitcher (London: Penguin, 2005), 214.

7. *Cymbeline*, III.i.12–13.

8. *Cymbeline*, III.i.16–29.

9. King, *Cymbeline*, 76 and 90 n. 73.

10. See the discussion in *Cymbeline*, ed. Roger Warren (Oxford: Oxford University Press, 2008), 154.

11. See Pitcher, "Commentary," in *Cymbeline*, 224–25.

12. *Richard II*, II.i.48, 62–63.

13. King, *Cymbeline*, 77. See David Bevington, "Foundational Myth in *Cymbeline*," in Christa Janson, Lena Cowen Orlin, and Stanley Wells, eds., *Shakespeare without Boundaries: Essays in Honor of Dieter Mehl* (Newark: University of Delaware Press, 2011), 219–40.

14. *Cymbeline*, III.iv.135–38.

15. See Pitcher, "Commentary," *Cymbeline*, 241.

16. See Jodi Mikalachki, "The Masculine Romance of Roman Britain: *Cymbeline* and Early Modern English Nationalism," *Shakespeare Quarterly* 46, no. 3 (1995): 301–22.

17. Glenn Clark, "The 'Strange' Geographies of *Cymbeline*," in Gillies and Vaughan, *Playing the Globe*, 230–59, 236; see Rhonda Lemke Sanford, "A Room Not One's Own: Feminine Geography in *Cymbeline*," in Gillies and Vaughan, *Playing the Globe*, 63–85.

18. *Cymbeline*, II.ii.12–14.

19. *Rape of Lucrece* 402, 407–8. See Clark, "The 'Strange' Geographies of *Cymbeline*," 243–44; and more generally, Georgianna Ziegler, "My Lady's Chamber: Female Space, Female Chastity in Shakespeare," *Textual Practice* 4, no. 1 (1990): 73–90.

20. Stallybrass, "Patriarchial Territories," 127. Stallybrass's main example is *Othello* and the status of Desdemona.

21. *Cymbeline*, III.i.72–76.

22. J. M. Nosworthy, note to *Cymbeline*, Arden Second Series, ed. J. M. Nosworthy (London: Methuen, 1955), 78.

23. *Cymbeline*, III.vii.1–7.

24. See also *Cymbeline*, II.iv.18; III.vii.12; IV.ii.331–33; IV.iii.24.

25. Wayne, "Introduction," *Cymbeline*, 65.

26. Maley, "Postcolonial Shakespeare," 146.

27. *Cymbeline*, IV.ii.35; Clark, "The 'Strange' Geographies of *Cymbeline*," 251.

28. Brian Levack, *Formation of the British State: England, Scotland, and the Union, 1603–1707* (Oxford: Clarendon Press, 1987), 2. See Maley, "Postcolonial Shakespeare," 146, though Maley drops "usually," "the word," and the specific date in his quotation. See also Maley, "'This Sceptred Isle'"; and Eric Heinze, "Imperialism and Nationalism in Early Modernity: The 'Cosmopolitan' and the 'Provincial' in Shakespeare's *Cymbeline*," *Social and Legal Studies* 18, no. 3 (2009): 373–96.

29. Maley, "Postcolonial Shakespeare," 146.

30. Maley, "Postcolonial Shakespeare," 146; McEachern, *The Poetics of English Nationhood*, 1. For a discussion, see Philip Edwards, *Nation and Empire: A Study in English and Irish Drama* (Cambridge: Cambridge University Press, 1979), chap. 4.

31. Sullivan, *The Drama of Landscape*, 140; see also 143: "Wales is constituted here in terms of a landscape of sovereignty, characterized as it is by its conceptual annexation into England."

32. See J. M. Nosworthy, "Introduction," *Cymbeline*, ed. Nosworthy, xv-xvii, who suggests 1606–11, probably 1608 or 1609; Pitcher, "Introduction," *The Winter's Tale*, 88; and Wayne, "Introduction," *Cymbeline*, 32–33.

33. Emrys Jones, "Stuart Cymbeline," *Essays in Criticism* 11, no. 1 (1961): 84–99, 90; see Marcus, *Puzzling Shakespeare*.

34. For a brief discussion, see Marcus, *Puzzling Shakespeare*, 124–25.

35. *Cymbeline*, V.v.436–39.

36. *Cymbeline*, V.v.452–54.

37. Maley, "Postcolonial Shakespeare," 151.

38. Maley, "Postcolonial Shakespeare," 151–52.

39. *Richard III,* IV.iv.532–33.

40. Gillies, *Shakespeare and the Geography of Difference,* 48; see Jones, "Stuart Cymbeline," 93; Heinze, "Imperialism and Nationalism in Early Modernity," 389.

41. Sullivan, *The Drama of Landscape,* 135–37; see Hawkes, *Shakespeare in the Present,* chap. 4.

42. *Cymbeline,* III.ii.59–61.

43. *Cymbeline,* III.v.16–17.

44. Pitcher, "Commentary," *Cymbeline,* 245–46.

45. *Cymbeline,* III.iii.68–70.

46. *Cymbeline,* III.iii.102–3.

47. *Cymbeline,* V.iii.52; see 57–58.

48. *Cymbeline,* V.iii.7, 11, 14.

49. *Cymbeline,* V.ii.11–13.

50. See *Cymbeline,* V.v.457–64.

51. Pitcher, "Commentary," *Cymbeline,* 222.

52. *Cymbeline,* V.v.468–74.

53. *Cymbeline,* V.v.474–83.

54. Jones, "Stuart Cymbeline," 96.

55. See King, *Cymbeline,* 49; and more generally Elden, *The Birth of Territory,* 290.

56. Maley, "Postcolonial Shakespeare," 148.

57. *Cymbeline,* III.i.19.

58. *A Midsummer Night's Dream,* II.i.2–4.

59. *The Comedy of Errors,* II.i.99–100.

60. *Hamlet,* I.iv.28; *Macbeth,* III.ii.51. See chapter 2 above.

61. *Troilus and Cressida,* II.iii.243–44.

62. *Venus and Adonis,* 229–40.

63. See Duncan-Jones and Woudhuysen, notes to *Shakespeare's Poems,* 153–54. For a discussion of this passage, see also Berry, *Shakespeare and the Hunt,* 51.

64. Rackham, *Trees and Woodland in the British Landscape,* 151.

65. Rackham, *Trees and Woodland in the British Landscape,* 153.

66. *The Winter's Tale,* IV.iii.4.

67. Pitcher, note to *The Winter's Tale,* ed. Pitcher, 250.

68. Ernest Schanzer, "Commentary," in *The Winter's Tale,* ed. Ernest Schanzer (London: Penguin, 2005), 160.

69. *Richard II,* III.iv.40; *King John,* II.i.23.

70. *Henry VI, Part 1,* IV.ii.45–56.

71. *The First Part of the Contention 1594,* folio E1; Marlowe, *Edward II,* II.ii.163–64.

INDEX

Act of Enclosure, 25, 239
Act of Supremacy, 77
Act of Union, 58
agriculture: agricultural language in *Richard II*, 105–6; in *As You Like It*, 234, 239; and colonization, 140–41; land enclosure, 25, 140, 141, 198, 213–14, 233–34, 237, 239; in *The Tempest*, 139–41. *See also* fields; land
All's Well That Ends Well (Shakespeare), 6
Althusius, Johannes, 4, 5, 86, 231
Altick, Richard, 92
America, 2, 141, 151–52
Ammianus Marcellinus, 218
Anglo-Saxon, 20, 27, 219, 248
animal imagery: in *Coriolanus*, 198–99, 206, 207–11, 212; in *Henry V*, 123; in *Henry VI, Part 1*, 249; in *Macbeth*, 208; in *The Tempest*, 137, 138; in *Titus Andronicus*, 227, 228, 229; in *Troilus and Cressida*, 198–99; in *Venus and Adonis*, 247–48
Antigone (Sophocles), 219
Antony and Cleopatra (Shakespeare): battle of Actium, 160–62; calculation, 156, 158; colonialism, 6, 7, 154–64; date, 164; interrelation of public and private worlds, 155–56, 163–64; land, 155, 158–64; and maps, 10, 163; marriage alliances, 155; measurement, 155, 156–57, 164, 178–79; pales, 160; pirates, 158–60; Roman Empire, 154–64; sea, 155, 158–64; surveying metaphors, 157, 178–79; triumvirate, 156, 157–58, 163; warfare, 157–63; world imagery, 156–64
Arden of Faversham (anonymous), 9, 222
Arthur, King, 245
As You Like It (Shakespeare): agriculture, 234, 239; banishment, 70, 231–33, 236; contrast of city and country, 232, 233–34; degrees of being outside, 233–34, 236–37; desert, use of word, 232, 235; disguise, 231–32, 233; and enclosure, 233–34; forest, 231–37, 239; hunger, 232–33; hunting, 232–33, 234, 237; infection and disease, 235; outside of territory, 6, 8, 231–37, 239; purlieu, use of word, 236; temporality, 233; territory, use of word, 4, 14, 70, 73, 240; usurpation, 231, 233
authority: and colonization, 152; in *King John*, 68–69, 76–78; in *Measure for Measure*, 1, 166–67; and the papacy, 68–69, 76–78; in *Pericles*, 152
authorship: *Arden of Faversham*, 9, 222; collaborative, 6, 8–9; *Double Falsehood*, 8–9; *Edmund Ironside*, 9; *Edward III*, 9, 129–30; *Henry VIII*, 8; *Macbeth*, 8, 56; *Measure for Measure*, 8; *Mucedorus*, 222; *Pericles*, 8, 150, 153; *Sir Thomas More*, 6, 8, 177; *The Spanish Tragedy*, 9; *Timon of Athens*, 8, 223–24; *Titus Andronicus*, 8, 225; *The Troublesome Reign of King John*, 71; *The Two Noble Kinsmen*, 8
Avery, Bruce, 189–90, 192–93

banishment: in *As You Like It*, 70, 231–33, 236; in *Coriolanus*, 211–16, 220, 221; in *Cymbeline*, 246; in *Henry VI, Part 2*, 12, 221; and pilgrimage, 91; in *Richard II*, 5, 67, 88–91, 101, 108–10, 220; in *Romeo*

banishment (*cont.*)
and *Juliet*, 221; in *Sir Thomas More*, 6; in *The Tempest*, 137, 138, 220; in *Titus Andronicus*, 228–29; in *The Two Gentlemen of Verona*, 221–22; in *The Two Noble Kinsmen*, 220

Barker, Francis, 20

Bate, Jonathan: on *As You Like It*, 233; on *Henry IV, Part 1*, 187; on *King John*, 81; on *Macbeth*, 50; on *Measure for Measure*, 167, 168; on *Othello*, 143, 149; on *Tamburlaine*, 9–10; on *Titus Andronicus*, 225, 226, 230; on *The Two Gentlemen of Verona*, 221

Bawcutt, N. W., 167, 168

Belleforest, François de, 30

Beowulf (anonymous), 27, 219

Berry, Edward, 227, 237

Berthoud, Jacques, 225

Betterton, Thomas, 9

Bevington, David, 130, 182

Bible, 27, 178, 179

Bingo (Bond), 237–38

biopolitics, 91, 216, 217

Birth of Territory, The (Elden), 2, 4, 249

bodies. *See* body politic; corporeality; female body

Bodin, Jean, *Six Books of the Commonwealth*, 4, 86

body politic: in *Comparative Discourse of the Bodies Natural and Politique*, 201; in *Coriolanus*, 198, 200–202, 217; in *Hamlet*, 6, 46; in *Policraticus*, 201; in *Richard II*, 98–100

Boke Named Tectonicon, A (Digges), 177

Bond, Edward, 237–38

book metaphors, 181

borders. *See* boundaries

boundaries: in *Coriolanus*, 213–14; in *Cymbeline*, 246; of estates, 180; in *King Lear*, 17; and land enclosure, 140, 141; in *Macbeth*, 55, 56; and the map division in *Henry IV, Part 1*, 187–93; as political technology, 3; rivers as boundaries, 187–93, 246; and surveying, 180; in *The Tempest*, 140, 141–42; in *Titus Andronicus*, 228; Welsh border, 187–88, 246

Bourus, Terri, 30

Branagh, Kenneth, 48, 113

Braunmuller, A. R., 65, 68, 71–72, 79, 82

Brayton, Daniel, 138

Brecht, Bertolt, 198

Bristol, Michael D., 200

Britain: Act of Union, 58; all nations represented in the army in *Henry V*, 123; Britain/British, use of word, 240, 241; colonialism, 140–41, 152, 244; and national identity, 5, 240–47; physical geography, 241–42; relation to Rome in *Cymbeline*, 241–47; relation to the Catholic church, 241, 244; Roman invasion, 241, 244, 245–46; union of English and Scottish crowns, 5, 25, 27–28, 58–59, 244–45. *See also* England; Scotland; Wales

Brockbank, Philip, 198

Brontë, Charlotte, 197

Caird, John, 48

Calais, Pale of, 61, 74, 135, 249

calculation: in *Antony and Cleopatra*, 156, 158; in *Coriolanus*, 204, 205; in *King Lear*, 17, 28; in *Love's Labour's Lost*, 165; in *Measure for Measure*, 165–69; in *The Merchant of Venice*, 170–72, 174–75; in *Othello*, 144, 182–83; as political technology, 3; in *Troilus and Cressida*, 165. *See also* measurement

cannibalism, 202, 216, 229

Cantor, Paul, 210

Capitano Moro, Un (Cinthio), 142, 147

Cardenio (Shakespeare and Fletcher), 8

cartography. *See* maps

"Case of the Post Nati," 245

Catholicism, 43, 45, 83, 128. *See also* church; papacy

Cavell, Stanley, 199

Cawdrey, Robert, 5, 27

censorship, 25, 77, 111

Certaine Errors in Navigation (Wright), 177, 182

Charnes, Linda, 38, 48

Charney, Maurice, 156

Chemnitz, Bogislaw Philipp von, 4

Chronicles (Holinshed), 50, 56, 96, 115, 119, 186, 188–89

church: Britain's relation to the Catholic church, 241, 244; in *Henry V*, 82, 114–17, 134; in *King John*, 82–83, 114–15; lands and property, 82–83, 114–17, 134; Luther's theses against, 45; in *The Troublesome Reign of King John*, 83. *See also* papacy; spiritual power

Cicero, 216, 218
Cinthio, Giovanni Battista Giraldi, 142, 147
cities: contrast of city and country, 221–22, 232, 233–34, 238; maps of, 177; relation to the people in *Coriolanus*, 209–10; *territorium* as lands surrounding, 5, 27, 216, 218, 231, 235
Civitates Orbis Terrarum, 177
Clark, Glenn, 243, 244
Clark, Sandra, 59
class: in *Coriolanus*, 197, 198, 209–10; in *Henry VI, Part 2*, 12
Clemen, Wolfgang, 146
clothing metaphors, 53
Cohen, Adam Max, 165, 176, 178, 179, 184
collaborative authorship, 6, 8–9
colonialism: and agriculture, 140–41; in *Antony and Cleopatra*, 6, 7, 154–64; and authority, 152; British, 140–41, 152, 244; colonization of adjoining lands, 152; in *Coriolanus*, 216; in *Cymbeline*, 244; and the female body, 138; and jurisdiction, 152; noncontiguous colonization, 152; in *Othello*, 6, 7, 142–50; in *Pericles*, 6, 7, 150–54; plantations, 140–41, 151–52; Roman, 152, 216, 244; and the sea, 152; in *The Tempest*, 1, 6, 7, 136–42
combinate, use of word, 169
Comedy of Errors, The (Shakespeare), 1–2, 247
commodity, 80–81
Comparative Discourse of the Bodies Natural and Politique (Forset), 201
Confessio Amantis (Gower), 150
contemporary political resonances: in *Coriolanus*, 198; in *Cymbeline*, 245; in *King Lear*, 18, 25, 27–28, 110; in *Pericles*, 151; in *Richard II*, 110–11
Coriolanus (Shakespeare): animal imagery, 198–99, 206, 207–11, 212; banishment, 211–16, 220, 221; body politic, 198, 200–202, 217; boundaries, 213–14; calculation, 204, 205; cannibalism, 202, 216; class, 197, 198, 209–10; colonialism, 216; contemporary political resonances, 198; corn riots, 198, 199, 200–202; corporeality, 6, 7–8, 197–217; disguise, 213; fable of the belly, 200–202; female body, 203, 216; Fiennes's film version, 213–14; hunger, 199, 200–202, 207–8; individual vs. collective bodies, 200, 205, 207, 208–9, 211; infection and disease, 205–7, 216, 235; navigation metaphors, 175; outside of territory, 213–14, 238; political nature of, 197–98, 199; power, 199, 204, 209; relation of people and city, 209–10; sexual puns, 205; sources, 198, 201–2; spatial disruption, 6, 213–14; synopsis, 199; territories, use of word, 69, 73, 215–17, 240; used by nineteenth- and twentieth-century authors, 197–98; warfare, 6, 199, 203–5; wounds, 202–5
corn riots, 198, 199, 200–202
coronations: in *King John*, 67–69; in *Richard II*, 108; use of orbs and sceptres, 58–59
corporeality: animals (*see* animal imagery); body politic, 6, 46, 98–100, 198, 200–202, 217; in *The Comedy of Errors*, 1–2; in *Coriolanus*, 6, 7–8, 197–217; *corpus mysticum*, 98–100, 104; female body (*see* female body); in *Henry IV, Part 2*, 111; in *Henry VIII*, 198; individual vs. collective bodies, 200, 205, 207, 208–9, 211; infection and disease, 205–7, 216; in *King John*, 83–85; the king's two bodies, 98–100, 111; in *Richard II*, 98–100; territory as body, 1–2, 217; in *Troilus and Cressida*, 198–99, 203; wounds, 202–5
corpus mysticum, 98–100, 104
Cosmographica (Munster), 2
Cosmographical Glasse, The (Cuningham), 177
Craik, T. W., 119, 123
Crécy, battle of, 61
Critchley, Simon, 42–43
Cuningham, William, *The Cosmographical Glasse*, 177
Cymbeline (Shakespeare): banishment, 246; boundaries, 246; Britain/British, use of word, 241; Britain's relation to Rome, 241–47; colonialism, 244; conquest and the female body, 243; contemporary political resonances, 245; contrast of city and country, 238; date, 247; Folio text, 242; maps, 181; measurement, 165; national identity, 5, 8, 28, 241–47; pales, 242, 247; patriotism, 241–42; property, 243; Roman invasion, 241, 244, 245–46; Wales, 238, 242, 244, 245–46; women as property, 243
Cyprus, 142–50, 182–83

Daley, Stuart, 234, 235
Davies, Sir John, 184
Davison, Peter, 190

Day, John, 37
de Grazia, Margreta, 38, 45, 48
De sublimi et regio territorii iure (Knichen), 4
Denmark, 7, 29–50, 59–60
desert, use of word, 221–22, 232, 235, 238
Devereux, Robert (Earl of Essex). *See* Essex, Earl of (Robert Devereux)
Dictionary (Johnson), 169
Diet of Worms, 45
Digges, Leonarde, 177
Discipline and Punish (Foucault), 3
Discoverie of a Gaping Gulf (Stubbs), 128
disease. *See* infection
disguise: in *As You Like It*, 231–32, 233; in *Coriolanus*, 213; in *King Lear*, 19, 219; in *Measure for Measure*, 169, 220; in *The Merchant of Venice*, 174, 175; in *Mucedorus*, 222; in *Twelfth Night*, 220; women disguised as men, 174, 175, 220, 231–32, 233
dispossession: in *Hamlet*, 34–39; in *King Lear*, 25; and land enclosure, 25; in *Richard II*, 96–97; in *The Tempest*, 136–39. *See also* usurpation
division of territories: in *Gorboduc*, 24–25; in *Henry IV, Part 1*, 7, 184–95; in *Henry VI, Part 1*, 11–12; in *Henry VI, Part 2*, 11, 12; in *Henry VI, Part 3*, 11, 12; in *King Lear*, 14–20; in *Richard III*, 11, 12–14; in *Thomas of Woodstock*, 94
Dodd, William, 19, 25
Double Falsehood (disputed authorship), 8–9
Dover Wilson, John, 31, 43, 70–71, 72
dowries: in *King John*, 64, 76; in *King Lear*, 16–20; in *Measure for Measure*, 169. *See also* marriage alliances
Drakakis, John, 170
Drayton, Michael, 10
Dusinberre, Juliet, 232, 233, 234

earth: in *Beowulf*, 27; in *Hamlet*, 43; in *Henry IV, Part 1*, 194; in *King Lear*, 7, 25–26, 27, 28; in *Richard II*, 98–100, 103, 108–9, 112; in *Richard III*, 26
economic language: in *Henry IV, Part 1*, 187; in *King Lear*, 17, 28; in *Othello*, 144
economics: in *Henry V*, 114–17, 134; in *King John*, 68, 79–83; of land, 5, 7, 28, 79, 91–98, 112, 114–17; in *Richard II*, 5, 7, 79, 91–98, 108, 110; in *Thomas of Woodstock*, 94
Edda (Norse myths), 219
Edmund Ironside (anonymous), 9
Edward, the Black Prince, 120, 130, 134
Edward II (Marlowe), 61, 130, 249
Edward III: conquests in France, 131; frequency of mentions in the history plays, 130; lineage, 64, 118, 120, 130–31
Edward III (disputed authorship): Arden edition, 129; authorship, 9, 129–30; Cambridge edition, 129; conquest and the female body, 132–33; conquests in France, 131–32; date, 129; defense against Scotland, 132; entirely in verse, 79; genealogies, 130; influence of *Tamburlaine*, 130; law, 7, 129–34; limit, use of word, 188; linguistic analysis, 129; phrases from the Sonnets used in, 129; pursuit of Countess Salisbury, 132–33, 134; relation to *Edward II*, 130; spacious, use of word, 132; territories, use of word, 70, 73, 129–30, 134; thematic links with *Henry V*, 9, 130–32; usurpation, 131; warfare, 131–32; Yale edition, 129
Edwards, Philip, 31, 34, 36, 48, 136
Eliot, George, 197–98
Eliot, T. S., 198
Elizabeth I: allusions to in *King Lear*, 25; attempts to depose, 110; depicted as Europe, 2; and personalized politics, 25; as Queen of France, 74; and succession, 48, 110, 127–28
empire, contemporary usage of word, 244. *See also* colonialism; Ottoman Empire; Roman Empire
enclosure, 25, 140, 141, 198, 213–14, 233–34, 237, 239
England: corn riots, 198; division of in *Henry IV, Part 1*, 187–93; division of in *King Lear*, 14–20; England/English, use of word, 240; Gaunt's speech on in *Richard II*, 93–95, 99–100, 112; land enclosure, 25, 198, 213–14, 233–34, 237, 239; Midland Revolt, 198; and national identity, 5, 93–94, 240–41; physical geography, 94; and regional relations in *Hamlet*, 7, 29, 44, 46–48, 60; and regional relations in *Macbeth*, 7, 29, 50, 55–58, 60, 219, 230; represented in the army in *Henry V*, 123; union of crowns with Scotland, 5, 25, 27–28, 58–59, 244–45. *See also* Britain
Essex, Earl of (Robert Devereux), 110
estate maps, 177, 184

Estok, Simon, 138, 152
exchange, 104, 175. *See also* trade
excommunication, 64, 78, 82–83
exile. *See* banishment

farming (agriculture). *See* agriculture
farming (leasing) of land, 42, 91–98, 105–6
Fawkes, Guy, 201
Felix Holt (Eliot), 197–98
female body: and colonization, 138; and conquest, 124–26, 132–33, 134, 217, 243; in *Coriolanus*, 203, 216; in *Cymbeline*, 243; in *Edward III*, 132–33, 134; geography represented as, 1–2, 243; in *Henry V*, 124–26, 134, 217; as property, 243; and rape, 124–26, 138, 221, 226–28, 229; in *The Rape of Lucrece*, 243; sexualization of, 124; in *The Tempest*, 138
female succession, 118–21, 127–28
feudalism, 28, 81, 87, 99–100
fields, 113–14. *See also* land
Fiennes, Ralph, 213–14
First Part of the Contention (Shakespeare), 61, 249. See also *Henry VI, Part 2* (Shakespeare)
Flahiff, Frederick, 27
Fletcher, John, 8, 220
Florio, John, 168
Foakes, R. A., 25
forests: in *As You Like It*, 231–37, 239; contemporary usage of word, 234–35; etymology of word, 235; and lawlessness, 221, 225–28, 237; in *Macbeth*, 57, 219; Manwood's definition, 234–35, 236; in *A Midsummer Night's Dream*, 222–23; in *Mucedorus*, 222; rapes occurring in, 221, 226–28, 229; as spaces outside territory, 220–39; in *Timon of Athens*, 223–25; in *Titus Andronicus*, 221, 225–31; in *The Two Gentlemen of Verona*, 221–22, 233; in *The Two Noble Kinsmen*, 220
Forker, Charles, 89, 92, 95, 97, 100, 102, 103, 105, 108
Forset, Edward, 201
Foucault, Michel, 3, 90–91
France: Angevin territories in *King John*, 63, 74–79; attempted conquest of in *Edward III*, 131–32; civil war (1589–94), 2; conquest of in *Henry V*, 113–14, 116–27; invasion of England in *King John*, 63, 64, 78–79, 81–82; invasion of England in *King Lear*, 21–24, 230; and regional relations in *Hamlet*, 7, 29, 45–46, 48, 60; and succession disputes in *King John*, 63–64, 65–66, 75–76; and succession rights in *Henry V*, 117–21
Furness, H. H., 70

genealogies, 27, 118–20, 128, 130. *See also* inheritance; succession
Generall Historie of the Turkes, The (Knolles), 86, 143
Geoffrey of Monmouth, 24, 25
Gibbons, Brian, 168
Gillies, John, 2, 181, 182, 193
Goddard, Harold, 145, 149
golden age, 141, 231
Gorboduc (Norton and Sackville), 24–25
Gossett, Suzanne, 151, 154
Gower, John, 150
Greenblatt, Stephen, 12, 192, 212
Greg, W. W., 22–23
Gunter, Edmund, 177–78
Gurr, Andrew, 1, 89, 110, 127–28

Hainsworth, D. R., 180
Halio, Jay, 174
Halpern, Richard, 151
Hamel, Guy, 71, 73–74
Hamlet (Shakespeare): Arden editions, 31, 33, 34, 42; backstory of King Hamlet and King Fortinbras, 30, 32–36, 38–40, 59–60; body politic, 6, 46; Branagh's film version, 48–49; dispossession, 34–39; earth, 43; final scene, 44–45; First Quarto text, 30–31, 32–34, 36, 41–42, 60; Folio text, 31, 32–34, 36, 37–38, 41–42, 47, 49, 60; Fortinbras subplot, 34–36, 38–45, 46, 48–49, 59–60; graveyard scene, 36–39, 43, 49, 62; Hamlet and England, 42, 43, 46–48; Hamlet's encounter with Fortinbras's army, 42–43, 46, 60; Horatio's role, 48–50; infection and disease, 235; inheritance, 35–39; Laertes and France, 45–46; land, 34–39, 42–43; law, 36–39, 62; madness, 46–47; navigation metaphors, 175–76; network analysis, 49–50; opening scene, 31–36; pales, 49, 60–61, 247; Penguin edition, 34, 42; place/name connections, 20; Polacks/Pollax, use of word, 32–34; property, 36–39, 62; regional relations, 7, 29–50, 59–62; revenge, 29, 30, 46; Second Quarto text, 30–31, 32–34,

Hamlet (Shakespeare) (*cont.*)
36–37, 41–43, 49; sources, 30; spatial disruption, 6; spatial reductions, 43, 103; and the spatiality of the stage, 31, 43; strategy, 7, 39–43, 60–62; succession, 39–40, 44–45, 48–49, 60; temporality, 29–30; *Ur-Hamlet*, 30; usurpation, 29, 32; variant texts, 9, 30–31, 32–34, 36–38, 41–43, 47, 49, 60; warfare, 42–43
Harley, J. B., 184
Harris, Jonathan Gil, 195
Harrison, Robert Pogue, 237
Hazlitt, William, 197
Heller, Agnes, 199
Henry IV, Part 1 (Shakespeare): battle of Shrewsbury, 186, 194; boundaries, 187–93; comparison of Hal and Hotspur, 185–86; death of Hotspur, 43, 194; division of territory, 7, 184–95; earth, 194; economic language, 187; Folio text, 9, 190; limit, use of word, 188; map division scene, 7, 184–93; measurement, 178; military technologies, 183–84; pilgrimage, 109; Quarto text, 9, 190; sexual puns, 186–87; sources, 186, 188–89; spatial reduction, 43, 194; surveying metaphors, 178
Henry IV, Part 2 (Shakespeare): book metaphors, 181; corporeality, 111; infection and disease, 235; the king's two bodies, 111; pilgrimage, 110; precise, use of word, 167; surveying, 179–80
Henry V (Shakespeare): army constituted of all British nations, 123; Branagh's film version, 113; Canterbury's Salic law speech, 118–20; and the church, 82, 114–17, 134; conquest and the female body, 124–26, 134, 217; defense against Scotland, 121–24; economics, 114–17, 134; fields, 113–14; Folio text, 113, 119, 122; genealogies, 118–19, 128; inheritance, 117–21, 127–28, 134; Katherine scenes, 113, 124–27; land, 82, 114–17, 134; law, 7, 113–28, 134; maps, 181; marriage alliances, 126–27; national identity, 5, 241; Olivier's film version, 113; opening scene, 114–17; patriotism, 113; prologue, 113–14; property, 82, 114–17, 134; Quarto text, 113, 119, 122, 125; Salic law, 117–21, 127–28, 134; sexuality, 124–26; sources, 115, 119; and the spatiality of the stage, 113–14; succession, 117–21, 127–28, 134; taxation, 115–16, 134; thematic links with *Edward III*, 9, 130–32; threatened rape and murder of civilians, 124–26; usurpation, 119; warfare, 9, 113–14, 116–27
Henry VI, Part 1 (Shakespeare), 11–12, 69, 73, 79, 249
Henry VI, Part 2 (Shakespeare): banishment, 12, 221; class, 12; division of territory, 11, 12; and *The First Part of the Contention*, 61; Jack Cade's rebellion, 12, 198; map metaphors, 180; property, 12; territories, use of word, 12, 14, 69, 73; warfare, 12
Henry VI, Part 3 (Shakespeare), 11, 12, 79
Henry VII, 245
Henry VIII, 77
Henry VIII (Shakespeare), 8, 180, 198, 240
Henry Frederick, Prince of Wales, 18
Hibbard, G. R., 29, 32, 34
Histoires Tragiques, Les (Belleforest), 30
Historiae Danicae (Saxo Grammaticus), 30
History of the Kings of Britain (Geoffrey of Monmouth), 24, 25
Hobbes, Thomas, 4
Holinshed, Raphael, 50, 56, 96, 115, 119, 186, 188–89
Holland, Peter, 212–13
Holy Roman Empire, 4, 86, 231
Homer, 219
Hondius, Jodocus, 238
Honigmann, E. A. J., 71, 72, 80, 148
Hooker, Richard, 4
Howard, Jean, 5
Hughes, Alan, 225
Hundred Years' War, 61
hunger: in *As You Like It*, 232–33; in *Coriolanus*, 199, 200–202, 207–8
hunting: in *As You Like It*, 232–33, 234, 237; in *Coriolanus*, 208–9; in *Henry VI, Part 1*, 249; in *Titus Andronicus*, 226–28, 229; in *Venus and Adonis*, 247–48
Hydrographiae Descriptio (Wright), 182

Iliad (Homer), 219
imperialism. *See* colonialism; empire, contemporary usage of word
infection: in *As You Like It*, 235; in *Coriolanus*, 205–7, 216, 235; in *Hamlet*, 235; in *Henry IV, Part 2*, 235; in *Troilus and Cressida*, 235
inheritance: female inheritance, 118–21, 127–28; in *Hamlet*, 35–39; in *Henry V*, 117–

21, 127–28, 134; in *King John*, 65, 79–80; in *King Lear*, 17, 20–21, 27, 28; and law, 96–97; in *Macbeth*, 54–55; and primogeniture, 127–28; in *Richard II*, 96–97; in *The Tempest*, 142. *See also* succession
Innocent III (pope), 64, 77. *See also* papacy
invasions: in *Cymbeline*, 241, 244, 245–46; in *Hamlet*, 40, 45, 48; in *King John*, 63, 64, 78–79, 81–82; in *King Lear*, 21–24, 230; in *Macbeth*, 57–58, 219, 230; in *Richard III*, 230, 245; in *Titus Andronicus*, 230. *See also* warfare
Irace, Kathleen, 33, 34
Ireland: in *Edward II*, 61, 249; English Pale, 61, 249; in *The First Part of the Contention*, 61, 249; land enclosure, 214; plantations, 141; rebellion in *Richard II*, 91–93, 97–98; and regional relations in *Macbeth*, 7, 29, 50, 51, 55, 58, 60; represented in the army in *Henry V*, 123
Isidore of Seville, 163, 216

Jack Cade's rebellion, 12, 198
Jaffa, Harry, 19
Jagendorf, Zvi, 202–3, 211
James I: allusions to in *Cymbeline*, 245; allusions to in *King Lear*, 18, 25, 27–28, 110; King James Bible, 27; poem celebrating naval victory over the Turks, 149; and the Union of the Crowns, 5, 25, 27–28, 58–59, 244–45
Jamestown, 151–52
Jew of Malta, The (Marlowe), 10
John, Ivor, 70, 71, 72
John of Salisbury, 201
Johnson, Samuel, 169
Jones, Emrys, 149
Jonson, Ben, 113, 177–78
Jowett, John, 177
Julius Caesar (Shakespeare), 1, 165, 198
jurisdiction: and colonization, 152; and pales, 8, 61, 75, 249; in *Pericles*, 152; and piracy, 154

Kahn, Paul, 21
Kantorowicz, Ernst, 98–99, 100
Kastan, David Scott, 184, 186, 192
Kermode, Frank, 198, 199
King John (Shakespeare): Angevin territories in France, 63, 74–79; Arden editions, 70, 71, 72, 75, 79; authority, 68–69, 76–78; barons' rebellion, 63, 64; bipartite structure, 65; Cambridge edition, 70–71, 72; and the church, 82–83, 114–15; Clarendon edition, 71, 72; commodity, 80–81; coronations, 67–69; corporeality, 83–85; date, 63, 71–72; disguise, 19, 219; dowries, 64, 76; economics, 68, 79–83; entirely in verse, 79; excommunication, 64, 78, 82–83; Faulconbridge/Bastard subplot, 64–65, 79–80; French invasion, 63, 64, 78–79, 81–82; inheritance, 65, 79–80; land, 64, 65, 79–85, 87, 114–15; legitimacy, 65, 79–80; majesty, 65–69, 84–85, 86–87; marriage alliances, 64, 76; national identity, 5, 241; Oxford edition, 71, 72, 75, 79; pales, 74–75, 249; papacy, 63, 64, 68–69, 76–78, 82; Penguin edition, 71, 72, 75; possession, 80–81; property, 68, 79–83; RSC edition, 71, 72; similarities to *Richard II*, 72, 79; sources, 71–72; spiritual power, 64, 68–69, 76–78; succession, 63–64, 65–66, 79, 85; temporal power, 64, 68–69, 76–78; territory/territories, use of word, 7, 69–74, 86–87, 240; textual problems, 75; and *The Troublesome Reign of King John*, 9, 70, 71–74, 82, 83; usurpation, 66; weather, 78–79, 82
King Lear (Shakespeare): boundaries, 17; contemporary political resonances, 18, 25, 27–28, 110; dispossession, 25; division of territory, 14–20; dowries, 16–20; earth, 7, 25–26, 27, 28; economic language, 17, 28; Edgar/Edmund subplot, 20–21, 25, 27, 79; Folio text, 9, 14–17, 19, 22–24, 27, 69, 219, 240; genealogies, 27; heath scenes, 219, 239; inheritance, 17, 20–21, 27, 28; invasion by France and Cordelia, 21–24, 230; land, 7, 20–25, 27, 28; legitimacy, 20–21, 28; love test, 17–19, 25; madness, 18, 26; maps, 14–17; national identity, 5, 28, 241; nothing, use of word, 17–18, 21, 27; outside of territory, 219, 239; place/name connections, 20; property, 21, 28; Quarto text, 9, 15–17, 21–24, 27, 219; sources, 24–25; spatial disruption, 6; and the spatiality of the stage, 26–27; strategy, 19, 28; succession, 19, 27, 28; temporality, 22–23; territory, use of word, 4, 9, 14–16, 27–28, 70, 73, 84, 240
king's two bodies, 98–100, 111. *See also* body politic

Kingsley-Smith, Jane, 5–6
Klinck, Dennis R., 100
Knichen, Andreas, 4, 5, 86, 231
Knolles, Richard, 86, 143
Knox, John, 127–28
Knyvet, Charles, 180
Koskenniemi, Inna, 129
Kurosawa, Akira, 20
Kyd, Thomas, 9

Lambarde, William, 110
land: in *Antony and Cleopatra*, 155; in *Arden of Faversham*, 222; in *Beowulf*, 27; church lands, 82–83, 114–17, 134; dispossession from, 25, 34–39, 96–97, 136–39; division of (*see* division of territories); economics of, 5, 7, 28, 79, 91–98, 112, 114–17; enclosure of, 25, 140, 141, 198, 213–14, 233–34, 237, 239; farming (leasing) of, 42, 91–98, 105–6; in *Gorboduc*, 24–25; in *Hamlet*, 34–39, 42–43; in *Henry V*, 82, 114–17, 134; in *King John*, 64, 65, 79–85, 87, 114–15; in *King Lear*, 7, 20–25, 27, 28; legal aspects of, 36–39, 62, 80, 96–97, 114–17; names connected with, 20, 100–101; in *Othello*, 146; and political rule, 111–12; relation to cartographic representation, 189–93; in *Richard II*, 5, 7, 79, 91–98, 100–102, 105–6, 111–12, 115; Spartans' attitude to, 224; surveying of, 177–80; in *Thomas of Woodstock*, 94, 95; in *Timon of Athens*, 223–24. *See also* agriculture; earth; fields
landlord, use of word, 95, 100, 115
law: in *Edward III*, 7, 129–34; geographic applicability, 118–19, 134; in *Hamlet*, 36–39, 62; in *Henry V*, 7, 113–28, 134; and inheritance, 96–97; and land, 36–39, 62, 80, 96–97, 114–17; in *The Merchant of Venice*, 171–74; and property, 36–39, 62, 80, 96–97; in *Richard II*, 95, 96–97; Roman law, 80, 231; Salic law, 117–21, 127–28, 134; and succession, 117–21, 127–28
Law Tricks (Day), 37
legitimacy, 20–21, 28, 65, 79–80
Leibniz, Gottfried, 4–5, 86
Leider, Emily, 27
Levack, Brian, 244
Levellers, 61
Lever, J. W., 167
Lezra, Jacques, 13–14, 165–66
Life of Coriolanus (Plutarch), 198, 201–2
limit, use of word, 188
linguistic analysis, 129
Lloyd, W. Watkiss, 144
Locke, John, 213–14
Love's Labour's Lost (Shakespeare), 33, 165, 167, 178
Luther, Martin, 45
Lyle, E. B., 58–59

Macbeth (Shakespeare): animal imagery, 208; authorship, 8, 56; Banquo's line, 58–59; Birnam Wood moves, 57, 219; blasted heath, 219, 239; book metaphors, 181; boundaries, 55, 56; clothing metaphors, 53; Folio text, 56; forest, 57, 219; inheritance, 54–55; invasion of English army, 57–58, 219, 230; murder of Duncan, 29, 55, 59; navigation metaphors, 176; outside of territory, 219, 239; pales, 55, 60–61, 247; possible lost original text, 9; regional relations, 7, 29, 50–62; revenge, 29; sources, 50, 56; spatial disruption, 6, 52–53, 57; strategy, 7, 60–62; succession, 53–55, 58–59; temporality, 52–53, 59; textual problems, 56; usurpation, 29; warfare, 7, 50–54, 56–58; weather, 176; witches scenes, 52, 53, 55, 56, 57, 58, 219
madness, 18, 26, 46–47
Mahood, M. M., 172
majesty: defining concept of, 86; in *King John*, 65–69, 84–85, 86–87; in *Richard II*, 106–8; and sovereignty, 65, 69, 86
Maley, Willy, 244, 245
Manwood, John, 234–35, 236
maps: and *Antony and Cleopatra*, 10, 163; city plans, 177; *Civitates Orbis Terrarum*, 177; in *Cymbeline*, 181; estate maps, 177, 184; in *Henry IV, Part 1*, 7, 184–95; in *Henry V*, 181; in *Henry VI, Part 2*, 180; historical maps, 151, 152; *Hydrographiae Descriptio*, 182; Isidore of Seville's, 163; in *King Lear*, 14–17; Mercator's, 1, 238; in *The Merchant of Venice*, 181–82; metaphorical references, 180–81; Molyneux's, 1; in *Pericles*, 151, 152; popularity of, 1, 177; in *The Rape of Lucrece*, 180–81; relation to the physical landscape, 189–93; in *Richard II*, 181; in

Richard III, 182; of the Roman Empire, 151, 152; in the Sonnets, 180; *Speculum Britanniae*, 177; in *Tamburlaine the Great*, 10; *Theatrum Orbis Terrarum*, 9, 151, 152, 177, 238; in *Thomas of Woodstock*, 94; "T-in-O" maps, 10, 163; in *Titus Andronicus*, 180; in *Troilus and Cressida*, 182; in *Twelfth Night*, 182
Marienstras, Richard, 236
Marlowe, Christopher: *Edward II*, 61, 130, 249; *The Jew of Malta*, 10; *Tamburlaine the Great*, 9–10, 130; territory/territories, use of word, 10
marriage alliances: in *Antony and Cleopatra*, 155; in *Henry V*, 126–27; in *King John*, 64, 75; in *Pericles*, 152; in *The Tempest*, 142. *See also* dowries
Marshall, Cynthia, 204
Mason, Pamela, 59
Massai, Sonia, 225
Maxwell, J. C., 208
McAlindon, Tom, 143, 147, 148–49
McEachern, Claire, 124, 134, 244
McJannet, Linda, 151
Measure for Measure (Shakespeare): authority, 1, 166–67; authorship, 8; calculation, 165–69; Cambridge edition, 168; combinate, use of word, 169; disguise, 169, 220; dowries, 169; Folio text, 168; measurement, 165–66, 169; outside of territory, 220; precise, use of word, 167–68; Second Folio text, 168; shipwreck, 169
measurement: in *Antony and Cleopatra*, 155, 156–57, 164, 178–79; in *Cymbeline*, 165; in *Henry IV, Part 1*, 178; in *Love's Labour's Lost*, 178; in *Measure for Measure*, 165–66, 169; in *The Merchant of Venice*, 171, 174; in *The Tempest*, 176. *See also* calculation; surveying
mechanicals, 165, 223
Melchiori, Giorgi, 129
Mentz, Steve, 2, 152
Mercator, Gerardus, 1, 238
Merchant, W. Moelwyn, 178
Merchant of Venice, The (Shakespeare): Antonio and Shylock's contract, 171–74; calculation, 170–72, 174–75; disguise, 174, 175; exchange, 175; law, 171–74; maps, 181–82; measurement, 171, 174; navigation, 170; opening scene, 169–70; pirates, 171, 172; shipwrecks, 164, 169, 172; surveying metaphors, 178; trade, 169–73
Merry Wives of Windsor, The (Shakespeare), 167
Metamorphoses (Ovid), 229
Middleton, Thomas, 56, 223–24
Midland Revolt, 198
Midsummer Night's Dream, A (Shakespeare), 165, 222–23, 247
Milford Haven, 245
military technologies, 182–84. *See also* warfare
Mirror for Magistrates, The, 103–4
Molyneux, Emory, 1
"Monstrous Regiment of Women" (Knox), 127–28
Montaigne, Michel de, 141
Moretti, Franco, 24, 49–50
Morris, Brian, 52
Mucedorus (anonymous), 222
Much Ado About Nothing (Shakespeare), 6, 155
Muir, Kenneth, 22, 23, 24, 52
Munster, Sebastian, 2

national identity: and Britain, 5, 240–47; in *Cymbeline*, 5, 8, 28, 241–47; and England, 5, 93–94, 240–41; in *Henry V*, 5, 241; in *King John*, 5, 241; in *King Lear*, 5, 28, 241; in *Richard II*, 5, 93–94, 99–100, 241. *See also* patriotism
Natural History (Pliny), 146
navigation: in *Coriolanus*, 175; developments in Shakespeare's time, 177; in *Hamlet*, 175–76; in *Macbeth*, 176; in *The Merchant of Venice*, 170; in *Pericles*, 152; reconciling magnetic and geographic north, 177; in *Sir Thomas More*, 176–77; in *The Tempest*, 175; in *The Two Noble Kinsmen*, 175
Neill, Michael, 144, 148, 149
network analysis, 49–50
Nevo, Ruth, 99
Newman, Karen, 125–26
Norden, John, 177
Normandy, 46, 48, 60
Norton, Thomas, 24–25
Norway: and regional relations in *Hamlet*, 7, 29, 31–45, 48–49, 59–60; and regional relations in *Macbeth*, 7, 29, 50–54, 60
Nosworthy, J. M., 165, 168, 243

Odyssey (Homer), 219
Oedipus at Colonus (Sophocles), 219
"Of the Cannibals" (Montaigne), 141
Olivier, Laurence, 113
On the Laws of Ecclesiastical Polity (Hooker), 4
Onions, C. T., 71, 188
Orgel, Stephen, 137
Orkin, Martin, 25
Ortelius, Abraham, 9, 151, 152, 177, 238
Othello (Shakespeare): calculation, 144, 182–83; choice of name for Othello, 142–43; colonialism, 6, 7, 142–50; composition, use of word, 144; date, 164; deception, 144–45; economic language, 144; Folio text, 147, 183; Iago as Turk, 145, 149; land, 146; military technologies, 182–83; Quarto text, 147; sea, 146, 147–48; sources, 86, 142, 146, 147; warfare, 142–50; weather, 147–48
Ottoman Empire, 142–50, 151
outside of territory: in *All's Well That Ends Well*, 6; in *Antigone*, 219; in *Arden of Faversham*, 222; in *As You Like It*, 6, 8, 231–37, 239; banishment to (*see* banishment); in *Beowulf*, 219; characters arriving from the outside, 220; in *Coriolanus*, 213–14, 238; degrees of being outside, 233–34, 236–37; described as desert, 221–22, 232, 235; in the *Edda*, 219; forests, 220–39; historically, 218–19; invasion from (*see* invasions); in *King Lear*, 219, 239; and lawlessness, 221, 225–28, 237; in *Macbeth*, 219, 239; in *Measure for Measure*, 220; in *A Midsummer Night's Dream*, 222–23; in *Mucedorus*, 222; in *Much Ado About Nothing*, 6; in the *Odyssey*, 219; in *Oedipus at Colonus*, 219; in *The Tempest*, 139; in *Timon of Athens*, 223, 238; in *Titus Andronicus*, 6, 8, 221, 225–31, 238–39; in *Troilus and Cressida*, 6, 219–20; in *Twelfth Night*, 220; in *The Two Gentlemen of Verona*, 221–22, 233; in *The Two Noble Kinsmen*, 220; in *The Winter's Tale*, 238
Ovid, 141, 229

Painful Adventures of Pericles, The (Wilkins), 150
pales: in *Antony and Cleopatra*, 160; in *The Comedy of Errors*, 247; in *Cymbeline*, 242, 247; derivation of term, 8, 61, 249; in *Edward II*, 61, 249; in *The First Part of the Contention*, 61, 249; in *Hamlet*, 49, 60–61, 247; in *Henry VI, Part 1*, 249; and jurisdiction, 8, 61, 75, 249; in *King John*, 74–75, 249; in *Macbeth*, 55, 60–61, 247; in *A Midsummer Night's Dream*, 247; Pale of Calais, 61, 74, 135, 249; Pale of Ireland, 61, 249; in *Richard II*, 106, 249; in *Troilus and Cressida*, 247; in *Venus and Adonis*, 247–48; in *The Winter's Tale*, 248–49
papacy: and authority, 68–69, 76–78; in *King John*, 63, 64, 68–69, 76–78, 82; two swords doctrine, 69. *See also* church; spiritual power
Parker, R. B., 202
patriotism: in *Cymbeline*, 241–42; in *Henry V*, 113; in *Richard II*, 92–94, 99–100, 112. *See also* national identity
Patterson, Annabel, 139, 197–98, 199
Peace of Westphalia, 4
Peele, George, 8, 71, 225
Per venerabilem (papal bull), 77
performative speech, 67, 89, 90
Pericles (Shakespeare): abduction of Marina, 153–54; authority, 152; authorship, 8, 150, 153; colonialism, 6, 7, 150–54; contemporary political resonances, 151; date, 151, 164; dual ancient/contemporary significance of locations, 151; episodic nature of, 150; jurisdiction, 152; and maps, 151, 152; marriage alliances, 152; navigation imagery, 152; pirates, 153–54; possible lost original text, 9; Quarto text, 150; sea, 150–51, 152–54; shipwreck, 150, 154, 169; sources, 150; weather, 153
pilgrimage, 91, 109, 110
pirates: in *Antony and Cleopatra*, 158–60; in *Hamlet*, 47–48; in *The Merchant of Venice*, 171, 172; operate outside of jurisdiction, 154; in *Pericles*, 153–54; in Roman times, 154, 158; in Shakespeare's time, 154
Pitcher, John, 238, 246, 248
place/name connections, 20, 100–101
plantations, 140–41
Pliny the Elder, 146, 218
Plutarch, 198, 201–2
Polacks/Pollax, use of word, 32–34
Poland, 7, 29, 32–34, 40–43, 60
Policraticus (John of Salisbury), 201

Politica Methodice Digesta (Althusius), 4
political technologies: developments in Shakespeare's time, 177–78, 195–96; Foucault's concept of, 3. *See also* calculation; maps; measurement; military technologies; navigation; surveying
Poly-Olbion (Drayton), 10
Pompey the Great, 154
Poole, Kristen, 2, 50, 52, 179, 195
possession, 80–81. *See also* property
power: in *Coriolanus*, 199, 204, 209; in *King John*, 64, 68–69, 76–78; spiritual, 64, 68–69, 76–78; temporal, 64, 68–69, 76–78. *See also* majesty; sovereignty
precise, use of word, 167–68
primogeniture, 127–28
property: church property, 82–83, 114–17, 134; in *Cymbeline*, 243; in *Hamlet*, 36–39, 62; in *Henry V*, 82, 114–17, 134; in *Henry VI, Part 2*, 12; in *King John*, 68, 79–83; in *King Lear*, 21, 28; and law, 36–39, 62, 80; movable and immovable, 80; seizure of, 82–83, 92–93, 96–97; in *Timon of Athens*, 223–24; women as, 243
Protestantism, 32, 43, 45, 77, 128
Proudfoot, Richard, 8–9, 130
Pufendorf, Samuel, 4
purlieu, use of word, 236

Rackham, Oliver, 140, 235, 248
Ran (Kurosawa), 20
rape: in forest locations, 221, 226–28, 229; in *Henry V* (threatened), 124–26; in *The Rape of Lucrece*, 138; in *The Tempest* (attempted), 138; in *Titus Andronicus*, 221, 226–28, 229; in *The Two Gentlemen of Verona* (attempted), 221
Rape of Lucrece, The (Shakespeare), 138, 180–81, 243
Rasmussen, Eric, 130, 168
rebellions: corn riots in *Coriolanus*, 198, 199; in Ireland in *Richard II*, 91–93, 97–98; Jack Cade's in *Henry VI, Part 2*, 12, 198; Midland Revolt, 198; of Pannonia and Dalmatia in *Cymbeline*, 243–44
regional relations: in *Hamlet*, 7, 29–50, 59–62; in *Macbeth*, 7, 29, 50–62
Reinking, Theodor, 4
revenge, 29, 30, 46, 225, 229–30
Richard II (Shakespeare): agricultural language, 105–6; Aumerle's plot, 109; banishment, 5, 67, 88–91, 101, 108–10, 220; body politic, 98–100; Bolingbroke's return, 97, 100–105; contemporary political resonances, 110–11; coronation, 108; corporeality, 98–100; deposition scene, 106–8, 109, 110–11; dispossession, 96–97; earth, 98–100, 103, 108–9, 112; economics, 5, 7, 79, 91–98, 108, 110; entirely in verse, 79; farming of the realm, 91–98, 105–6; Folio text, 108, 110–11; Fourth Quarto text, 107–8, 110–11; garden scene, 90, 97, 249; Gaunt's speech on England, 93–95, 99–100, 112; inheritance, 96–97; Irish rebellion, 91–93, 97–98; *King John*'s similarities to, 72, 79; the king's two bodies, 98–100; land, 5, 7, 79, 91–98, 100–102, 111–12, 115; landlord, use of word, 95, 100, 115; law, 95, 96–97; majesty, 106–8; map metaphors, 181; national identity, 5, 93–94, 99–100, 241; pales, 106, 249; Parliament scene, 111; patriotism, 92–94, 99–100, 112; performative speech, 67, 89, 90; pilgrimage, 91, 109; place/name connections, 100–101; seizure of money and property, 92–93, 96–97; sources, 95; spatial reduction, 103, 104; succession, 79, 111; territories, use of word, 69, 73, 240; and *Thomas of Woodstock*, 9, 94; waste, 95, 100, 105–6, 109
Richard III (Shakespeare), 11, 12–14, 26, 182, 230, 245
Roman Empire: in *Antony and Cleopatra*, 154–64; colonialism, 152, 216, 244; in *Coriolanus*, 197–217; in *Cymbeline*, 241–47; map of in *Theatrum Orbis Terrarum*, 151, 152; and piracy, 154, 158; in *Titus Andronicus*, 225–31
Roman law, 80, 231
Romeo and Juliet (Shakespeare), 6, 221
Rose, Mark, 143

Sackville, Thomas, 24–25
Salic law, 117–21, 127–28, 134
Sams, Eric, 9, 30, 129
Sanders, Norman, 149
Saxo Grammaticus, 30
Saxony, 45, 48, 60
Scotland: "Case of the Post Nati," 245; defense against in *Edward III*, 132; defense against in *Henry V*, 121–24; land enclosure, 214; and regional relations in *Macbeth*, 7, 29,

Scotland (*cont.*)
50–62, 219, 230; represented in the army in *Henry V*, 123; union of crowns with England, 5, 25, 27–28, 58–59, 244–45. *See also* Britain
sea: in *Antony and Cleopatra*, 155; and colonization, 152; in *Othello*, 146, 147–48; in *Pericles*, 150–51, 152–54; in *The Tempest*, 139. *See also* navigation; shipwrecks
Seneca, 218
Serres, Michel, 195
sexual puns, 124, 178, 186–87, 205
sexuality, 124–26, 203, 226–28, 247–48
Shakespeare, William. *See individual titles of works*
Shakespeare Glossary, A (Onions), 71, 188
Shapiro, James, 18, 27, 174, 240
shipwrecks: in *Measure for Measure*, 169; in *The Merchant of Venice*, 164, 169, 172; in *Pericles*, 150, 154, 169; in *The Tempest*, 138, 139, 142, 169; in *Twelfth Night*, 169, 220
Shirley (Brontë), 197
Sir Thomas More (collaborative authorship), 6, 8, 176–77, 198
Six Books of the Commonwealth (Bodin), 4, 86
Smallwood, Robert, 75
Sommers, Alan, 228
Sonnets (Shakespeare), 129, 180, 188
Sophocles: *Antigone*, 219; *Oedipus at Colonus*, 219
sources: *Coriolanus*, 198, 200–201; *Hamlet*, 30; *Henry IV, Part 1*, 186, 188–89; *Henry V*, 115, 119; *King John*, 71–72; *King Lear*, 24–25; *Macbeth*, 50, 56; *Othello*, 86, 142, 146, 147; *Pericles*, 150; *Richard II*, 95; *Titus Andronicus*, 225
sovereignty: defining concept of, 86; limits of, 86, 90–91, 112; and majesty, 65, 69, 86; relation to territory, 5, 86
spacious, use of word, 132
Spanish Tragedy, The (Kyd), 9
spatial disruption, 6, 52–53, 57, 213–14
spatial reductions, 43, 103, 104, 194
Speculum Britanniae (Norden), 177
Spencer, T. J. B., 34, 40
spiritual power, 64, 68–69, 76–78. *See also* church; papacy
Spurgeon, Caroline, 235
stage, spatiality of, 2, 26–27, 31, 43, 113–14
Stallybrass, Peter, 243
storms. *See* weather
strategy: in *Hamlet*, 7, 39–43, 60–62; in *King Lear*, 19, 28; in *Macbeth*, 7, 60–62
Stubbs, John, 128
succession: to the English throne after Elizabeth, 48, 110, 127–28; female succession, 118–21, 127–28; in *Hamlet*, 39–40, 44–45, 48–49, 60; in *Henry V*, 117–21, 127–28, 134; in *King John*, 63–64, 65–66, 79, 85; in *King Lear*, 19, 27, 28; and law, 117–21, 127–28; in *Macbeth*, 53–55, 58–59; and primogeniture, 127–28; in *Richard II*, 79, 111; in *The Tempest*, 142. *See also* inheritance
Sullivan, Garrett A., Jr., 2, 94, 192, 244, 245
surveying: in *Antony and Cleopatra*, 157, 178–79; and boundaries, 180; developments in Shakespeare's time, 177–78; and dishonesty, 180; in *Henry IV, Part 1*, 178; in *Henry IV, Part 2*, 179–80; in *Henry VIII*, 180; in *Love's Labour's Lost*, 178; in *The Merchant of Venice*, 178; as political technology, 3; in *Thomas of Woodstock*, 94; in *The Two Noble Kinsmen*, 179. *See also* maps; measurement
Surveyor's Dialogue, The (Norden), 177

Table Alphabeticall (Cawdrey), 5, 27
Tacitus, 216
Tamburlaine the Great (Marlowe), 9–10, 130
taxation, 77, 97–98, 115–16, 134
Taylor, Gary, 113, 119–20, 127
Taylor, Michael, 24
Taylor, Neil, 31, 37, 38, 50
technologies. *See* calculation; maps; measurement; military technologies; navigation; political technologies; surveying
Tempest, The (Shakespeare): agriculture, 139–41; animal imagery, 137, 138; attempted rape of Miranda, 138; banishment, 137, 138, 220; boundaries, 140, 141–42; Caliban's identity, 138–39; colonialism, 1, 6, 7, 136–42; colonization and the female body, 138; date, 164; dispossession, 136–39; Gonzalo's utopian commonwealth, 141–42; inheritance, 142; location, 138–39; marriage alliances, 142; measurement, 176; navigation, 176; outside of territory, 139; Prospero's masque, 139–40; sea, 139; shipwreck, 138, 139, 142, 169; succession, 142; weather, 139, 176

temporal power, 64, 68–69, 76–78
temporality: in *As You Like It*, 233; in *Hamlet*, 29–30; Harris's analysis of, 195–96; in *King Lear*, 22–23; in *Macbeth*, 52, 59
territorium, 4, 5, 27, 216, 218, 231, 235, 237
territory, defining concept of, 2–5, 86–87, 218–19
territory/territories, use of word: in *As You Like It*, 4, 14, 70, 73, 240; contemporary usage, 4–5, 27; in *Coriolanus*, 69, 73, 215–17, 240; definite article uses, 70–74, 86–87, 240; in *Edward III*, 70, 73, 129–30, 134; in *Henry VI, Part 1*, 11–12, 69, 73; in *Henry VI, Part 2*, 12, 14, 69, 73; in *King John*, 7, 69–74, 86–87, 240; in *King Lear*, 4, 9, 14–16, 27–28, 70, 73, 84, 240; in Marlowe's plays, 10; possessive uses, 7, 72–73, 86–87, 134, 240; in *Richard II*, 69, 73, 240; in *The Troublesome Reign of King John*, 70, 71–74, 86; in *The Two Gentlemen of Verona*, 69–70, 73, 240
Terror and Territory (Elden), 2, 249
Theatrum Orbis Terrarum (Ortelius), 9, 151, 152, 177, 238
Theobald, Lewis, 8–9
Thirty Years' War, 4
Thomas of Woodstock (anonymous), 9, 94–95, 130
Thompson, Ann, 31, 37, 38, 50
time. *See* temporality
Timon of Athens (Shakespeare), 8, 223–25, 238
"T-in-O" maps, 10, 163
Titus Andronicus (Shakespeare): animal imagery, 227, 228, 229; authorship, 8, 225; banishment, 228–29; blurring of Rome and the outside, 226, 228, 229–30; boundaries, 228; cannibalism, 229; Folio text, 225; forest, 221, 225–31; hunting, 226–28, 229; Lucius invades with army of Goths, 230; map metaphors, 180; outside of territory, 6, 8, 221, 225–31, 238–39; rape and mutilation of Lavinia, 221, 226–28, 229; revenge, 225, 229–30; sexuality, 226–28; sources, 225; spacious, use of word, 132; warfare, 225
trade, 169–73, 187. *See also* exchange
Troilus and Cressida (Shakespeare): animal imagery, 198–99; calculation, 165; corporeality, 198–99, 203; infection and disease, 235; maps, 182; military technologies, 182; outside of territory, 6, 219–20; pales, 247; sexuality, 203
Troublesome Reign of King John, The (anonymous), 9, 70, 71–74, 82, 83, 86
Turner, Henry, 26
Twelfth Night (Shakespeare), 169, 182, 220
Two Gentlemen of Verona, The (Shakespeare), 69–70, 73, 221–23, 240
Two Noble Kinsmen, The (Shakespeare and Fletcher), 8, 175, 179, 220
Tyndale, William, 27

Union of the Crowns, 5, 25, 27–28, 58–59, 244–45
uprisings. *See* rebellions
Ure, Peter, 95, 181
Ur-Hamlet, 30
usurpation: in *As You Like It*, 231, 233; in *Edward III*, 131; in *Hamlet*, 29, 32; in *Henry V*, 119; in *King John*, 66; in *Macbeth*, 29

Varro, 216, 218
Vaughan, Alden, 139, 140
Vaughan, Virginia, 139, 140
vengeance. *See* revenge
Venice, 142–50, 169–73
Venus and Adonis (Shakespeare), 188, 247–48
Visser, Nicholas, 20

Waith, Eugene, 225
Wales: border, 187–88, 246; in *Cymbeline*, 238, 242, 244, 245–46; English colonization of, 244; Glendower's defense of in *Henry IV, Part 1*, 187; kingship of in *Macbeth*, 58–59; land enclosure, 214; and the map division in *Henry IV, Part 1*, 187–89; Milford Haven, 245; represented in the army in *Henry V*, 123. *See also* Britain
Walton, Nicholas, 223
warfare: in *Antony and Cleopatra*, 157–63; in *Coriolanus*, 6, 199, 203–5; in *Edward III*, 131–32; in *Hamlet*, 42–43; in *Henry V*, 9, 113–14, 116–27; in *Henry VI, Part 1*, 11–12; in *Henry VI, Part 2*, 12; in *Henry VI, Part 3*, 12; in *Macbeth*, 7, 50–54, 56–58; in *Othello*, 142–50; in *Richard III*, 12–14; technologies of (*see* military technologies); in *Titus Andronicus*, 225. *See also* Hundred Years' War; Thirty Years' War; Wars of the Roses
Wars of the Roses, 11–14, 85, 113, 127, 245
waste, 95, 100, 105–6, 109
Wayne, Valerie, 241, 244

weather: in *King John*, 78–79, 82; in *Macbeth*, 176; in *Othello*, 147–48; in *Pericles*, 153; in *The Tempest*, 139, 176
Webster, Jamieson, 42–43
Wells, Robin Headlam, 198
Wells, Stanley, 106–7
Wilders, John, 155, 159, 163–64
Wilkins, George, 150, 153
Williams, George Walton, 59
winds. *See* weather
Winter's Tale, The (Shakespeare), 238, 248–49
Wittenberg, 35, 45, 49
woods. *See* forests
world imagery, 156–64
wounds, 202–5
Wright, Edward, 177, 182